Progress in Inflammation Research

Series Editor

Prof. Dr. Michael J. Parnham
Senior Scientific Advisor
PLIVA dd
Prilaz baruna Filipovica 25
10000 Zagreb
Croatia

Advisory Board

Forthcoming titles:

Recent Advances in Pathophysiology of COPD, P.J. Barnes, T.T. Hansel (Editors), 2004
Cancer and Inflammation, D.W. Morgan, U. Forssmann, M. Nakada (Editors), 2004
Antirheumatic Therapy: Actions and Outcomes,
R.O. Day, D.E. Furst, P.L. van Riel (Editors), 2003
Cytokines and Joint Injury, P. Miossec, W.B. van den Berg (Editors), 2004
Antibiotics as Anti-inflammatory and Immunomodulatory Agents,
B. Rubin, J. Tamaoki (Editors), 2004

(Already published titles see last page.)

Arachidonate Remodeling and Inflammation

Alfred N. Fonteh
Robert L. Wykle

Editors

Springer Basel AG

Editors

Alfred N. Fonteh
Huntington Medical Research Institutes
Molecular Neurology Program
99 North El Molino Avenue
Pasadena, CA 91101-1830
USA

Robert L. Wykle
Wake Forest University
School of Medicine
Department of Biochemistry
Medical Center Boulevard
Winston-Salem, NC 27157-1016
USA

A CIP catalogue record for this book is available from the Library of Congress, Washington D.C., USA

Bibliographic information published by Die Deutsche Bibliothek
Die Deutsche Bibliothek lists this publication in the Deutsche Nationalbibliografie; detailed bibliographic data is available in the internet at http://dnb.ddb.de

ISBN 978-3-0348-9594-1 ISBN 978-3-0348-7848-7 (eBook)
DOI 10.1007/978-3-0348-7848-7

Originally published by Birkhäuser Verlag in 2004
Softcover reprint of the hardcover 1st Edition 2004
Printed on acid-free paper produced from chlorine-free pulp. TCF ∞
Cover design: Markus Etterich, Basel
Cover illustration: Arachidonic acid bound to P450BM-3, described in: Graham-Lorence SE et al (1997) *J Biol Chem* 272, 1127–1135. With the friendly permission of Julian A. Peterson, University of Texas, Southwestern Medical Center at Dallas, USA

ISBN 978-3-0348-9594-1

9 8 7 6 5 4 3 2 1 ww.birkhasuer-science.com

Contents

List of contributors

Salma Al-Darmaki, Department of Microbiology and Immunology, Virginia Commonwealth University, Box 980678, Richmond, VA 23298-0678, USA

María A. Balboa, Institute of Molecular Biology and Genetics (IBGM-CSIC), University of Valladolid School of Medicine, Avenida Ramón y Cajal 7, E-47005 Valladolid, Spain; e-mail: mbalboa@ibgm.uva.es

Jesús Balsinde, Institute of Molecular Biology and Genetics (IBGM-CSIC), University of Valladolid School of Medicine, Avenida Ramón y Cajal 7, E-47005 Valladolid, Spain; e-mail: jbalsinde@jbgm.uva.es

Suzanne E. Barbour, Department of Biochemistry, Virginia Commonwealth University, Box 980614, Richmond, VA 23298-0614, USA;
e-mail: sbarbour@hsc.vcu.edu

Nicolas G. Bazan, Neuroscience Center of Excellence, Louisiana State University Health Sciences Center, 2020 Gravier Street, Suite D, New Orleans, LA 70112, USA; e-mail: nbazan@lsuhsc.edu

Brooke J. Barham, Comprehensive Center for Inflammatory Diseases, Department of Dentistry, University of North Carolina, Chapel Hill, NC 27599, USA;
e-mail: jbarham@email.unc.edu

Nan Chiang, The Center for Experimental Therapeutics and Reperfusion Injury, Department of Anesthesiology, Perioperative and Pain Medicine, Brigham and Women's Hospital, Harvard Medical School, Boston, MA 02115, USA;
e-mail: nchiang@zeus.bwh.harvard.edu

Alfred N. Fonteh, Molecular Neurology Program, Huntington Medical Research Institutes, 99 North El Molino Avenue, Pasadena, CA 91101, USA;
e-mail: afonteh@hmri.org

Giorgio Giannattasio, Division of Clinical Immunology and Allergy, University of Naples Federico II, Via Pansini 5, 80131 Naples, Italy;
e-mail: giorgionap@yahoo.it

Maiko Gokoh, Faculty of Pharmaceutical Sciences, Teikyo University, Sagamiko, Tsukui-gun, Kanagawa 199-0195, Japan

Francescopaolo Granata, Division of Clinical Immunology and Allergy, University of Naples Federico II, Via Pansini 5, 80131 Naples, Italy;
e-mail: frapagra@hotmail.com

Michael G. Harrington, Molecular Neurology Program, Huntington Medical Research Institutes, Pasadena, CA 91101-1830, USA;
e-mail: mghworks@hmri.org

Margaret Johnson, Department of Medicine, Mayo Clinic, Jacksonville, Florida 32224, USA; e-mail: Johnson.Margaret2@mayo.edu

Seishi Kishimoto, Faculty of Pharmaceutical Sciences, Teikyo University, Sagamiko, Tsukui-gun, Kanagawa 199-0195, Japan; e-mail: skishimoto@pharm.teikyo-u.ac.jp

Stefania Loffredo, Division of Clinical Immunology and Allergy, University of Naples Federico II, Via Pansini 5, 80131 Naples, Italy;
e-mail: stefanialoffredo@hotmail.com

Alex D. Manguikian, Department of Biochemistry, Virginia Commonwealth University, Richmond, VA 23298-0614, USA

Chad R. Marion, Infectious Diseases, Department of Internal Medicine, Wake Forest University School of Medicine, Winston-Salem, North Carolina, 27157, USA;
e-mail: cmarion@wfubmc.edu

Gianni Marone, Division of Clinical Immunology and Allergy, University of Naples Federico II, Via Pansini 5, 80131 Naples, Italy: e-mail: marone@unina.it

Allen M. McAlexander, Department of Biology and Health Sciences, Meridith College, Raliegh, NC 27607, USA; e-mail: mcalexan@meredith.edu

Saori Oka, Faculty of Pharmaceutical Sciences, Teikyo University, Sagamiko, Tsukui-gun, Kanagawa 199-0195, Japan

Rebeca Pérez, Institute of Molecular Biology and Genetics (IBGM-CSIC), University of Valladolid School of Medicine, Avenida Ramón y Cajal 7, E-47005 Valladolid, Spain; e-mail: rperez@ibgm.uva.es

Francesca W. Rossi, Division of Clinical Immunology and Allergy, University of Naples Federico II, Via Pansini 5, 80131 Naples, Italy

Yolanda Sáez, Institute of Molecular Biology and Genetics (IBGM-CSIC), University of Valladolid School of Medicine, Avenida Ramón y Cajal 7, E-47005 Valladolid, Spain; e-mail: ysaez@ibgm.uva.es

Salvatore Salzano, Experimental Endocrinology and Oncology Center, Consiglio Nazionale delle Ricerche (CNR), Italy

Royal D. Saunders, Department of Biology, Xavier University of Louisiana, 1 Drexel Drive, Box 85B, New Orleans, LA 70125, USA; e-mail: rsaunder@xula.edu

Charles N. Serhan, The Center for Experimental Therapeutics and Reperfusion Injury, Department of Anesthesiology, Perioperative and Pain Medicine, Brigham and Women's Hospital, Harvard Medical School, Boston, MA 02115, USA; e-mail: cnserhan@zeus.bwh.harvard.edu

Takayuki Sugiura, Faculty of Pharmaceutical Sciences, Teikyo University, Sagamiko, Tsukui-gun, Kanagawa 199-0195, Japan; e-mail: sugiurat@pharm.teikyo-u.ac.jp

Massimo Triggiani, Division of Clinical Immunology and Allergy, University of Naples Federico II, Via Pansini 5, 80131 Naples, Italy, e-mail: triggian@unina.it

Keizo Waku, Faculty of Pharmaceutical Sciences, Teikyo University, Sagamiko, Tsukui-gun, Kanagawa 199-0195, Japan; e-mail: keiwaku@pharm.teikyo-u.ac.jp

Robert L. Wykle, Department of Biochemistry, Wake Forest University School of Medicine, Medical Center Boulevard, Winston-Salem, NC 27157, USA; e-mail: bwykle@wfubmc.edu [illegible] Bacteriology and Genetics (BGM) [illegible]

[illegible]

[illegible]

[illegible]

[illegible]

Preface

Arachidonic acid (AA) and other 20 or 22-carbon polyunsaturated fatty acids (PUFAs) are precursors of signaling molecules that are critical in disease processes and in regulating normal cell function. Remodeling is critical in homeostatic control and in dictating how PUFAs are converted to mediators. Thus, understanding remodeling will unravel better therapeutic targets for controlling inflammatory diseases.

The review chapters associate AA remodeling and the biosynthesis of mediators of inflammation or pain. By following the movement of AA rather than discussing a single enzyme and its product, the influence of upstream biosynthetic pathways on the formation of lipid mediators and interrelationship between all AA-derived mediators are examined in a comprehensive fashion. This approach innovates AA metabolism by describing new inhibitors, mode of action of these inhibitors and potentially efficacious targets not previously examined.

This volume is written by experts in the field to serve as a rich resource of knowledge for scientists and clinicians in academia, and researchers in the pharmaceutical industry involved in inflammation and pain research. Since AA is derived from essential fatty acids, this volume is also of interest to nutritionists.

July 2003

A.N. Fonteh
R.L. Wykle

An outline of arachidonate remodeling and its biological significance

Alfred N. Fonteh

HMRI – Molecular Neurology Program, Huntington Medical Research Institutes, 99 North El Molino Avenue, Pasadena, CA 91101, USA

Introduction

Arachidonic acid (AA) is a member of the n-6 family of essential fatty acids. Mammals obtain AA directly from vegetable or animal oils or from successive elongation and desaturation of linoleic acid (LA, 18:2, n-6) by the liver (Fig. 1) [1, 2]. Once presented to cells, AA is rapidly incorporated into glycerolipids. This incorporation is initiated by the ATP-dependent conversion of AA to arachidonyl CoA by arachidonyl CoA synthetases [3–6]. AA CoA can then be incorporated into lyso phospholipids by CoA-dependent acyl transferases [7–10]. This initial incorporation of AA into phospholipids is directed mainly into 1-acyl-linked subclasses. Once in these early AA pools, arachidonate is continuously and selectively transferred to 1-ether-linked phospholipids by CoA-independent transacylase (CoA-IT) and CoA-dependent transferase activities present in cells [11–14]. This selective transfer of arachidonate from 1-acyl-linked to 1-ether-linked phospholipids likely accounts for the asymmetrical distribution of arachidonate within phospholipid subclasses such that arachidonate is greater in ether-linked subclasses than in 1-acyl-linked subclasses in many cell types [11, 13–17]. While the most characterized remodeling processes have involved mammalian cells, many studies have shown that remodeling occurs in other cell types and may be a universal phenomenon in which the distribution of AA is maintained within glycerolipid classes. In most studies, remodeling has been characterized as the transfer of AA from 1-acyl-2-AA-GPC to ether-linked phospholipids [16]. However, the initial pool of AA that is labeled may be different in many cell types or in different species. For example, in breast cancer cells, AA is initially incorporated into PI and then remodeled into PE while in amoeba; AA is rapidly incorporated into PE and remodeled into PC. Regardless of the initial entry pool for AA into a cell type, the eventual outcome of the remodeling process is a shift of AA to predominantly ether-linked or triglyceride pools for storage or for utilization in mediator formation [14, 18, 19].

Arachidonate Remodeling and Inflammation, edited by Alfred N. Fonteh and Robert L. Wykle

Pro-inflammatory lipid mediators

Under resting conditions, the incorporation and remodeling process is well controlled such that very low levels of free AA are found within cells. However, cellular activation is accompanied by a rapid increase in AA remodeling concomitant with a drastic increase in free AA and AA-derived products [5, 12]. AA is recognized as the major fatty acid precursor of an important class of lipid mediators that are collectively known as eicosanoids. This class of lipids that include leukotrienes, prostaglandins and thromboxanes has been shown to be important as intra- and extra-cellular mediators of cell function. Eicosanoids are produced by a series of enzymatic reactions. Various phospholipase A_2s (PLA_2) may initiate this process by releasing AA from the *sn*-2 position of glycerophospholipids [20–25]. Once released, free AA can induce biological responses in neighboring cells or may be converted to leukotrienes or prostaglandins by 5-lipoxygenases and cyclo-oxygenases, respectively [26–28]. AA that is not converted to eicosanoids can undergo auto-oxidation [29–31] to form bioactive lipids, or it can be rapidly re-esterified into membrane glycerolipids. Under conditions where cells are exposed to high concentrations of AA, there is a shift in the incorporation/remodeling pattern such that

---

Figure 1
Overview of AA remodeling, role in lipid mediator formation and its potential biological significance
Free AA is obtained from vegetable or animal oils or from the elongation and/or desaturation of n-6 fatty acids by the liver (a). AA is converted to AA-CoA and incorporated into 1-acyl-linked phospholipid subclasses by AA-CoA synthetase (b) and AA-CoA-dependent acyl transferases (c), respectively. Under resting conditions, AA is gradually remodeled from 1-acyl-linked phospholipids to 1-ether-linked phospholipids by CoA-IT activity (d). During cell activation, the remodeling process is accelerated due to the formation of 1-alk-1enyl-2-lyso-GPE by PLA_2 *(e). Enhanced remodeling is also accompanied by an increase in the formation of 1-alkyl-2-lyso-PAF, which is converted, to PAF by acetyl transferase (f). A combination of fatty acid CoA synthetases (FA-CoA synthetases) and acyl trasferases acylate lyso phospholipid generated (g). Concomitantly, 5-LO utilizes free AA to form* LTB_4 *(h), or COX transforms AA to prostanoids and thromboxanes (i). P450 enzymes or autoxidation reactions also convert free AA to potent bioactive lipids (j). AA that is not utilized for mediator formation (AA**) is released from cells to act as a second messenger or to be reincorporated into the remodeling cycle. When there is an excess of free AA within cells, there is increase incorporation into monoglycerides (ligand of CB2 receptors), diglycerides and triglycerides (k). A CoA-, calcium- and ATP-independent transferase in combination with phosphodiesterase activity (l) is responsible for generating arachidonoylethanolamide (ligand of CB1 receptor). Changes in AA distribution within lipid classes and the formation of lipid mediators likely play important roles in inflammation, analgesia, cell maturation, apoptosis, cell migration and proliferation.*

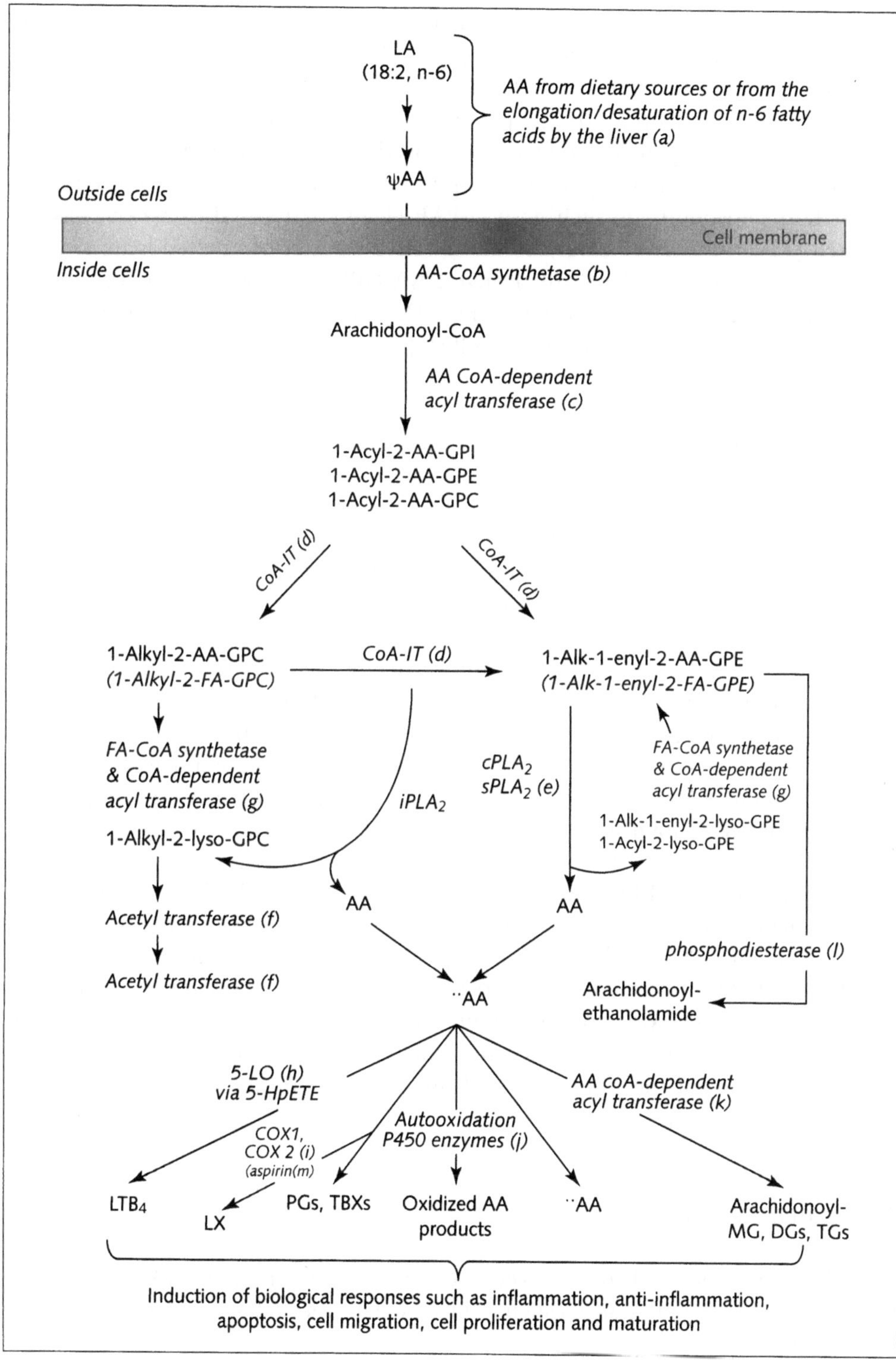
LA
(18:2, n-6)
AA from dietary sources or from the elongation/desaturation of n-6 fatty acids by the liver (a)
ψAA
Outside cells
Cell membrane
Inside cells
AA-CoA synthetase (b)
Arachidonoyl-CoA
AA CoA-dependent acyl transferase (c)
1-Acyl-2-AA-GPI
1-Acyl-2-AA-GPE
1-Acyl-2-AA-GPC
CoA-IT (d)
CoA-IT (d)
1-Alkyl-2-AA-GPC
(1-Alkyl-2-FA-GPC)
CoA-IT (d)
1-Alk-1-enyl-2-AA-GPE
(1-Alk-1-enyl-2-FA-GPE)
FA-CoA synthetase & CoA-dependent acyl transferase (g)
iPLA2
cPLA2
sPLA2 (e)
FA-CoA synthetase & CoA-dependent acyl transferase (g)
1-Alk-1-enyl-2-lyso-GPE
1-Acyl-2-lyso-GPE
1-Alkyl-2-lyso-GPC
AA
AA
Acetyl transferase (f)
Acetyl transferase (f)
phosphodiesterase (l)
··AA
Arachidonoyl-ethanolamide
5-LO (h) via 5-HpETE
COX1, COX 2 (i) (aspirin(m)
Autooxidation P450 enzymes (j)
AA coA-dependent acyl transferase (k)
LTB4
LX
PGs, TBXs
Oxidized AA products
··AA
Arachidonoyl-MG, DGs, TGs
Induction of biological responses such as inflammation, anti-inflammation, apoptosis, cell migration, cell proliferation and maturation

AA does not only accumulate in glycerophospholipids, but is rapidly expanded in neutral lipids such as diglycerides and triglycerides [5, 19].

In addition to AA that is converted to bioactive metabolites, PLA_2s also generates lyso phospholipids that are the precursor of another important class of bioactive lipid known as platelet activating factor (PAF). Several studies suggest that CoA-IT and reacylation/deacylation reactions may play a crucial role in the formation of lyso phospholipids such as lyso-PAF (1-alkyl-2-lyso-GPC) when cells are activated [32–36]. CoA-IT catalyzes the selective transfer of arachidonate from 1-alkyl-2-arachidonoyl-GPC to 1-alk-1-enyl-2-AA-GPE acceptors thereby forming the PAF intermediate 1-alkyl-2-lyso-GPC (Fig. 1). Acetylation orchestrated by acetyl transferase converts this intermediate to PAF [35].

Anti-inflammatory lipid mediators

Serhan and colleagues have extensively characterized another class of AA-derived mediators, known as lipoxins (LX) [37]. LX are trihydroxytetraene-containing eicosanoids formed by a combination of various LO activities. Interestingly, aspirin has been shown to trigger the biosynthesis of LX and novel receptors for these mediators have been cloned and characterized [38–40]. Various stable analogues of LX have been synthesized and shown to inhibit chemotaxis, adhesion and transmigration of inflammatory cells in various animal models of inflammation [41]. Together, LX and RX are distinguished from other eicosanoids by having anti-inflammatory properties that can be harnessed for the treatment of many diseases.

In addition to AA-derived anti-inflammatory mediators, other long chain PUFAs have recently been shown to be precursors of novel mediators [42]. For example, human COX-2 has been shown to generate stereo specific-oxygenated products of eicosapentaenoic acid (EPA) and docosahexaenoic acid (DHA). EPA and DHA are n-3 fatty acids that are implicated in anti-inflammation. DHA, a major PUFA in brain and retina, is released from mainly phosphatidylethanolamine after cell activation and is then converted to novel docosatrienes by glial cells *via* various hydroxyperoxy-DHA and epoxide-containing intermediates [43]. Transcellular metabolism of 17-HDHA results in its conversion to resolvins. Docosanoids and resolvins regulate the resolution of the acute phase of inflammation in many animal models of inflammation [42, 43].

Conjugated AA species

Recently, new classes of AA-derived metabolites consisting of AA attached to ethanolamine, or to amino acids such as glycine, have been described in mammalian brain tissues [44].

Arachidonoylethanolamide (also known as arachidonamide or anandamide) is the natural ligand of the cannabinoid CB1 receptor while lipoamino acids such as N-arachidonoyl glycine, N-arachidonoyl γ-aminobutyric acid and N-arachidonoylalanine lack affinity to CB1 receptors. Studies have shown that these widely distributed lipoamino acids have a physiological role in controlling analgesia [45, 46]. A novel calcium-, CoA- and ATP-independent pathway, which is highly selective for AA as the aliphatic constituent, is specific for ethanolamine as the polar group, catalyzes the synthesis of arachidonoylethanolamide. Enzymes responsible for the synthesis of arachidonoylethanolamide are located in brain microsomal and cytosolic fractions. Arachidonoylethanolamide is formed either by the direct conjugation of AA to ethanolamine or by the combined activities of transacylase and phosphodiesterase. Initially, AA located in diarachidonoyl GPC is transferred to ethanolamine head group by transacylase. Subsequently, the arachidonoylethanolamide is cleaved by phosphodiesterase activity to release it within brain tissues [44, 47]. Thus, the incorporation and remodeling of AA from specific phospholipid pools to PE or to specific membrane locations within the brain may be crucial in regulating the formation of arachidonoylethanolamide and these important lipoamino acids. Another AA-derived molecule, arachidonoylglycerol has recently been shown to be the natural ligand of the cannabinoid CB2 receptor and is likely formed by acyl transferase-dependent mechanisms [48].

Oxidized phospholipids and novel pathways

The classical pathway for eicosanoid biosynthesis assumes that AA must be released by lipases and made available for oxygenation. However, recent studies show that prostanoids and related oxygenated compounds can be found acylated to phospholipids [49–53]. Moreover, prostanoids and related compounds can also be formed independent of enzyme activity *via* auto-oxidative pathways [54]. The oxidized lipids are very potent mediators of biological function and are implicated in many disease conditions.

PUFA-derived products and lipidomics

While the above outline has concentrated on the remodeling of AA and the generation of AA products, it is worthwhile noting that other PUFAs, especially those of 20 or 22 carbon chain length, can also be utilized for generating lipid mediators as has been demonstrated by Serhan and colleagues [42, 43, 55]. One can envisage various COX or LO-derived products of these PUFAs. Interestingly, these products can also be conjugated to various amino acids leading to the formation of new classes of mediators. These new structures have different biological activities based on the affinities of the new compounds for novel receptors. The new field of structural elucidation of all lipids, recently termed lipidomics, will be advanced rapidly as new tools are being utilized for characterizing these new molecules. Of interest is the use of liquid chromatography coupled to electrospray ionization tandem mass spectrometry to identify over 500 molecular species of phospholipids [56], discover docosatrienes or to identify oxidized phospholipid species [52, 57]. A comprehensive determination of all lipid molecular species and oxidized products will be the backbone for a searchable database that will facilitate functional lipidomics. An understanding of the importance of lipids in diseases will be the basis for applied lipodomics where nutritional supplements can be used to control the signs and symptoms of diseases.

Summary

Overall, a critical look at AA metabolism suggests that it is a very complex process. First, AA levels are tightly controlled in resting cells by many enzyme activities. Secondly, AA is asymmetrically distributed between different phospholipid classes and subclasses. In most cells, the bulk of AA resides in ether-linked phospholipid subclasses. Thirdly, cellular activation or injury creates an imbalance in AA incorporation/remodeling leading to the release and buildup of free AA levels within cells. Under these conditions, the distribution of AA may be altered such that classes of glycerolipids that normally have low AA are substantially enriched with AA. Fourthly, cellular activation/injury is accompanied by rapid deacylation/reacylation of arachidonate within glycerolipid classes. These changes in arachidonate distribution are closely linked to pathological conditions. Fifthly, free AA is converted to bioactive lipid mediators. These mediators are implicated in inflammation, analgesia, apoptosis, cell migration, maturation and proliferation through processes linked to receptors, signaling pathways involving kinases and the activation of various transcription factors [14, 17, 45, 58–64]. Recent advances in technology are revealing that other PUFAs are remodeled in similar fashion as AA and can also be utilized in mediator formation. These important observations underscore the importance of the AA remodeling process in cell function. Details examining the critical roles of enzymes that regulate AA homeostasis (incorporation and remodeling) will be examined in detail in subsequent chapters of this book.

References

1 Parker-Barnes JM, Das T, Bobik E, Leonard AE, Thurmond JM, Chaung LT, Huang YS, Mukerji P (2000) Identification and characterization of an enzyme involved in the elon-

gation of n-6 and n-3 polyunsaturated fatty acids. *Proc Natl Acad Sci USA* 97: 8284–8289
2 Sprecher H (1986) The metabolism of (n-3) and (n-6) fatty acids and their oxygenation by platelet cyclo-oxygenase and lipoxygenase. *Prog Lipid Res* 25: 19–28
3 Neufeld EJ, Sprecher H, Evans RW, Majerus PW (1984) Fatty acid structural requirements for activity of arachidonoyl-CoA synthetase. *J Lipid Res* 25: 288–293
4 Taylor AS, Sprecher H, Russell JH (1985) Characterization of an arachidonic acid-selective acyl-CoA synthetase from murine T lymphocytes. *Biochim Biophys Acta* 833: 229–238
5 Chilton FH, Fonteh AN, Surette ME, Triggiani M, Winkler JD (1996) Control of arachidonate levels within inflammatory cells. *Biochim Biophys Acta* 1299: 1–15
6 Cao Y, Murphy KJ, McIntyre TM, Zimmerman GA, Prescott SM (2000) Expression of fatty acid-CoA ligase 4 during development and in brain. *FEBS Lett* 467: 263–267
7 Zimmerman WF, Keys S (1986) Acyl transferase and fatty acid coenzyme A synthetase activities within bovine rod outer segments. *Biochem Biophys Res Commun* 138: 988–994
8 Masuzawa Y, Sugiura T, Sprecher H, Waku K (1989) Selective acyl transfer in the reacylation of brain glycerophospholipids. Comparison of three acylation systems for 1-alk-1'-enylglycero-3-phosphoethanolamine, 1-acylglycero-3-phosphoethanolamine and 1-acylglycero-3-phosphocholine in rat brain microsomes. *Biochim Biophys Acta* 1005: 1–12
9 Sugiura T, Masuzawa Y, Nakagawa Y, Waku K (1987) Transacylation of lyso platelet-activating factor and other lysophospholipids by macrophage microsomes. Distinct donor and acceptor selectivities. *J Biol Chem* 262: 1199–1205
10 Sugiura T, Masuzawa Y, Waku K (1988) Coenzyme A-dependent transacylation system in rabbit liver microsomes. *J Biol Chem* 263: 17490–17498
11 Chilton FH, Murphy RC (1986) Remodeling of arachidonate-containing phosphoglycerides within the human neutrophil. *J Biol Chem* 261: 7771–7777
12 Fonteh AN, Chilton FH (1992) Rapid remodeling of arachidonate from phosphatidylcholine to phosphatidylethanolamine pools during mast cell activation. *J Immunol* 148: 1784–1791
13 MacDonald JI, Sprecher H (1991) Phospholipid fatty acid remodeling in mammalian cells. *Biochim Biophys Acta* 1084: 105–121
14 Yamashita A, Sugiura T, Waku K (1997) Acyltransferases and transacylases involved in fatty acid remodeling of phospholipids and metabolism of bioactive lipids in mammalian cells. *J Biochem (Tokyo)* 122: 1–16
15 Mueller HW, Purdon AD, Smith JB, Wykle RL (1983) 1-O-alkyl-linked phosphoglycerides of human platelets: distribution of arachidonate and other acyl residues in the ether-linked and diacyl species. *Lipids* 18: 814–819
16 Fonteh AN, Chilton FH (1993) Mobilization of different arachidonate pools and their roles in the generation of leukotrienes and free arachidonic acid during immunologic activation of mast cells. *J Immunol* 150: 563–570

17 Surette ME, Winkler JD, Fonteh AN, Chilton FH (1996) Relationship between arachidonate-phospholipid remodeling and apoptosis. *Biochemistry* 35: 9187–9196

18 Triggiani M, Oriente A, Seeds MC, Bass DA, Marone G, Chilton FH (1995) Migration of human inflammatory cells into the lung results in the remodeling of arachidonic acid into a triglyceride pool. *J Exp Med* 182: 181–1190

19 Johnson MM, Vaughn B, Triggiani M, Swan DD, Fonteh AN, Chilton FH (1999) Role of arachidonyl triglycerides within lipid bodies in eicosanoid formation by human polymorphonuclear cells. *Am J Respir Cell Mol Biol* 21: 253–258

20 Fonteh AN, Bass DA, Marshall LA, Seeds M, Samet JM, Chilton FH (1994) Evidence that secretory phospholipase A_2 plays a role in arachidonic acid release and eicosanoid biosynthesis by mast cells. *J Immunol* 152: 5438–5446

21 Balsinde J, Balboa MA, Dennis EA (1997) Antisense inhibition of group VI Ca^{2+}-independent phospholipase A_2 blocks phospholipid fatty acid remodeling in murine P388D1 macrophages. *J Biol Chem* 272: 29317–29321

22 Dennis EA (1997) The growing phospholipase A_2 superfamily of signal transduction enzymes. *Trends Biochem Sci* 22: 1–2

23 Murakami M, Nakatani Y, Atsumi G, Inoue K, Kudo I (1997) Regulatory functions of phospholipase A_2. *Crit Rev Immunol* 17: 225–283

24 Ramanadham S, Hsu FF, Bohrer A, Ma Z, Turk J (1999) Studies of the role of group VI phospholipase A_2 in fatty acid incorporation, phospholipid remodeling, lysophosphatidylcholine generation, and secretagogue-induced arachidonic acid release in pancreatic islets and insulinoma cells. *J Biol Chem* 274: 13915–13927

25 Ma Z, Turk J (2001) The molecular biology of the group VIA Ca^{2+}-independent phospholipase A_2. *Prog Nucleic Acid Res Mol Biol* 67: 1–33

26 Jakschik BA, Harper T, Murphy RC (1982) The 5-lipoxygenase and leukotriene forming enzymes. *Methods Enzymol* 86: 30–37

27 Peters-Golden M (1998) Cell biology of the 5-lipoxygenase pathway. *Am J Respir Crit Care Med* 157: S227–S231

28 Smith WL, DeWitt DL, Garavito RM (2000) Cyclooxygenases: Structural, cellular, and molecular biology. *Annu Rev Biochem* 69: 145–182

29 Fitzpatrick FA, Murphy RC (1988) Cytochrome P-450 metabolism of arachidonic acid: formation and biological actions of 'epoxygenase'-derived eicosanoids. *Pharmacol Rev* 40: 229–241

30 O'Flaherty JT, Kuroki M, Nixon AB, Wijkander J, Yee E, Lee SL, Smitherman PK, Wykle RL, Daniel LW (1996) 5-Oxo-eicosanoids and hematopoietic cytokines cooperate in stimulating neutrophil function and the mitogen-activated protein kinase pathway. *J Biol Chem* 271: 17821–17828

31 Pratico D, MY LV, Trojanowski JQ, Rokach J, Fitzgerald GA (1998) Increased F_2-isoprostanes in Alzheimer's disease: evidence for enhanced lipid peroxidation *in vivo*. *FASEB J* 12: 1777–1783

32 Kramer RM, Jakubowski JA, Deykin D (1988) Hydrolysis of 1-alkyl-2-arachidonoyl-sn-

glycero-3-phosphocholine, a common precursor of platelet-activating factor and eicosanoids, by human platelet phospholipase A_2. *Biochim Biophys Acta* 959: 269–279
33 Reinhold SL, Zimmerman GA, Prescott SM, McIntyre TM (1989) Phospholipid remodeling in human neutrophils. Parallel activation of a deacylation/reacylation cycle and platelet-activating factor synthesis. *J Biol Chem* 264: 21652–21659
34 Uemura Y, Lee TC, Snyder F (1991) A coenzyme A-independent transacylase is linked to the formation of platelet-activating factor (PAF) by generating the lyso-PAF intermediate in the remodeling pathway. *J Biol Chem* 266: 8268–8272
35 Nieto ML, Venable ME, Bauldry SA, Greene DG, Kennedy M, Bass DA, Wykle RL (1991) Evidence that hydrolysis of ethanolamine plasmalogens triggers synthesis of platelet-activating factor *via* a transacylation reaction. *J Biol Chem* 266: 18699–18706
36 Blank ML, Smith ZL, Fitzgerald V, Snyder F (1995) The CoA-independent transacylase in PAF biosynthesis: Tissue distribution and molecular species selectivity. *Biochim Biophys Acta* 1254: 295–301
37 Serhan CN, Wong PY, Samuelsson B (1987) Nomenclature of lipoxins and related compounds derived from arachidonic acid and eicosapentaenoic acid. *Prostaglandins* 34: 201–204
38 Serhan CN, Takano T, Clish CB, Gronert K, Petasis N (1999) Aspirin-triggered 15-epi-lipoxin A_4 and novel lipoxin B_4 stable analogs inhibit neutrophil-mediated changes in vascular permeability. *Adv Exp Med Biol* 469: 287–293
39 Jozsef L, Zouki C, Petasis NA, Serhan CN, Filep JG (2002) Lipoxin A4 and aspirin-triggered 15-epi-lipoxin A_4 inhibit peroxynitrite formation, NF-kappa B and AP-1 activation, and IL-8 gene expression in human leukocytes. *Proc Natl Acad Sci USA* 99: 13266–13271
40 Serhan CN, Levy BD, Clish CB, Gronert K, Chiang N (2000) Lipoxins, aspirin-triggered 15-epi-lipoxin stable analogs and their receptors in anti-inflammation: A window for therapeutic opportunity. *Ernst Schering Res Found Workshop* 143–185
41 Serhan CN, Takano T, Maddox JF (1999) Aspirin-triggered 15-epi-lipoxin A_4 and stable analogs on lipoxin A_4 are potent inhibitors of acute inflammation. Receptors and pathways. *Adv Exp Med Biol* 447: 133–149
42 Serhan CN, Hong S, Gronert K, Colgan SP, Devchand PR, Mirick G, Moussignac RL (2002) Resolvins: A family of bioactive products of omega-3 fatty acid transformation circuits initiated by aspirin treatment that counter pro-inflammation signals. *J Exp Med* 196: 1025–1037
43 Hong S, Gronert K, Devchand PR, Moussignac RL, Serhan CN (2003) Novel docosatrienes and 17S-resolvins generated from docosahexaenoic acid in murine brain, human blood, and glial cells. Autacoids in anti-inflammation. *J Biol Chem* 278: 14677–14687
44 Kruszka KK, Gross RW (1994) The ATP- and CoA-independent synthesis of arachidonoylethanolamide. A novel mechanism underlying the synthesis of the endogenous ligand of the cannabinoid receptor. *J Biol Chem* 269: 14345–14348
45 Huang SM, Bisogno T, Petros TJ, Chang SY, Zavitsanos PA, Zipkin RE, Sivakumar R,

Coop A, Maeda DY, De Petrocellis L et al (2001) Identification of a new class of molecules, the arachidonyl amino acids, and characterization of one member that inhibits pain. *J Biol Chem* 276: 42639–42644

46 Walker JM, Huang SM, Strangman NM, Tsou K, Sanudo-Pena MC (1999) Pain modulation by release of the endogenous cannabinoid anandamide. *Proc Natl Acad Sci USA* 96: 12198–12203

47 Sugiura T, Kondo S, Sukagawa A, Tonegawa T, Nakane S, Yamashita A, Ishima Y, Waku K (1996) Transacylase-mediated and phosphodiesterase-mediated synthesis of N-arachidonoylethanolamine, an endogenous cannabinoid-receptor ligand, in rat brain microsomes. Comparison with synthesis from free arachidonic acid and ethanolamine. *Eur J Biochem* 240: 53–62

48 Sugiura T, Kondo S, Kishimoto S, Miyashita T, Nakane S, Kodaka T, Suhara Y, Takayama H, Waku K (2000) Evidence that 2-arachidonoylglycerol but not N-palmitoylethanolamine or anandamide is the physiological ligand for the cannabinoid CB2 receptor. Comparison of the agonistic activities of various cannabinoid receptor ligands in HL-60 cells. *J Biol Chem* 275: 605–612

49 Morrow JD, Awad JA, Boss HJ, Blair IA, Roberts LJ (1992) Non-cyclooxygenase-derived prostanoids (F_2-isoprostanes) are formed *in situ* on phospholipids. *Proc Natl Acad Sci USA* 89: 10721–10725

50 Davies SS, Pontsler AV, Marathe GK, Harrison KA, Murphy RC, Hinshaw JC, Prestwich GD, Hilaire AS, Prescott SM, Zimmerman GA et al (2001) Oxidized alkyl phospholipids are specific, high affinity peroxisome proliferator-activated receptor gamma ligands and agonists. *J Biol Chem* 276:16015–16023

51 Kayganich-Harrison KA, Rose DM, Murphy RC, Morrow JD, Roberts LJ (1993) Collision-induced dissociation of F_2-isoprostane-containing phospholipids. *J Lipid Res* 34: 1229–1235

52 Khaselev N, Murphy RC (2000) Structural characterization of oxidized phospholipid products derived from arachidonate-containing plasmenyl glycerophosphocholine. *J Lipid Res* 41: 564–572

53 Marathe GK, Harrison KA, Murphy RC, Prescott SM, Zimmerman GA, McIntyre TM (2000) Bioactive phospholipid oxidation products. *Free Radic Biol Med* 28: 1762–1770

54 Gao L, Zackert WE, Hasford JJ, Danekis ME, Milne GL, Remmert C, Reese J, Yin H, Tai HH, Dey SK et al (2003) Formation of prostaglandin E_2 and prostaglandin D_2 *via* the isoprostane pathway: A mechanism for the generation of bioactive prostaglandins independent of the cyclo-oxygenase. *J Biol Chem* 278: 28479–28489

55 Serhan CN, Clish CB, Brannon J, Colgan SP, Gronert K, Chiang N (2000) Anti-microinflammatory lipid signals generated from dietary N-3 fatty acids *via* cyclo-oxygenase-2 and transcellular processing: A novel mechanism for NSAID and N-3 PUFA therapeutic actions. *J Physiol Pharmacol* 51: 643–654

56 Taguchi R, Hayakawa J, Takeuchi Y, Ishida M (2000) Two-dimensional analysis of phospholipids by capillary liquid chromatography/electrospray ionization mass spectrometry. *J Mass Spectrom* 35: 953–966

57 Murphy RC, Khaselev N, Nakamura T, Hall LM (1999) Oxidation of glycerophospholipids from biological membranes by reactive oxygen species: Liquid chromatographic-mass spectrometric analysis of eicosanoid products. *J Chromatogr B Biomed Sci Appl* 731: 59–71

58 Winkler JD, Fonteh AN, Sung CM, Heravi JD, Nixon AB, Chabot-Fletcher M, Griswold D, Marshall LA, Chilton FH (1995) Effects of CoA-independent transacylase inhibitors on the production of lipid inflammatory mediators. *J Pharmacol Exp Ther* 274: 1338–1347

59 Triggiani M, De MV, de Crescenzo G, Marone G (1997) Arachidonic acid remodeling in human inflammatory cells migrating to the lung *in vivo*. *Int Arch Allergy Immunol* 113: 190–192

60 Chilton FH, Surette ME, Winkler JD (1996) Arachidonate-phospholipid remodeling and cell proliferation. *Adv Exp Med Biol* 416: 169–172

61 Fonteh AN, LaPorte T, Swan D, McAlexander MA (2001) A decrease in remodeling accounts for the accumulation of arachidonic acid in murine mast cells undergoing apoptosis. *J Biol Chem* 276: 1439–1449

62 Samet JM, Fonteh AN, Galli SJ, Tsai M, Fasano MB, Chilton FH (1996) Alterations in arachidonic acid metabolism in mouse mast cells induced to undergo maturation *in vitro* in response to stem cell factor. *J Allergy Clin Immunol* 97:1329–1341

63 Trimboli AJ, Waite BM, Atsumi G, Fonteh AN, Namen AM, Clay CE, Kute TE, High KP, Willingham MC, Chilton CH (1999) Influence of coenzyme A-independent transacylase and cyclo-oxygenase inhibitors on the proliferation of breast cancer cells. *Cancer Res* 59: 6171–6177

64 Boilard E, Surette ME (2001) Anti-CD3 and concanavalin A-induced human T cell proliferation is associated with an increased rate of arachidonate-phospholipid remodeling. Lack of involvement of group IV and group VI phospholipase A_2 in remodeling and increased susceptibility of proliferating T cells to CoA-independent transacyclase inhibitor-induced apoptosis. *J Biol Chem* 276: 17568–17575

Phospholipase A_2 and remodeling in inflammatory cells

Suzanne E. Barbour[2], *Salma Al-Darmaki*[1], *and Alex D. Manguikian*[2]

[1]Department of Microbiology and Immunology and [2]Department of Biochemistry, Virginia Commonwealth University, Box 980614, Richmond, VA 23298-0614, USA

Acyl chain composition of glycerophospholipids

For many years, lipids were long thought to be merely the bricks and mortar that composed cell membranes. However, in the past 20 years it has become evident that these molecules are not merely passive building blocks but contain information that can be unleashed when they are metabolized upon cell activation. The signaling properties of specific glycerophospholipids and their metabolites highlight the importance of lipid composition in cell homeostasis. For example, phospholipid to sterol ratios are tightly maintained and the loss of these ratios is thought to contribute to foam cell death and the progression of atherosclerosis [1, 2]. Different cell types not only vary in the relative abundance of each phospholipid class, but also in the fatty acyl composition within each class of phospholipids [3, 4].

Glycerophospholipids consist of a glycerol backbone to which substituents are esterified (Fig. 1). The *sn*-3 carbon is occupied by a phosphate ester that in turn is associated with a polar alcohol (e.g., choline, ethanolamine, phosphorylated inositol). The glycerophospholipid class is determined by the polar alcohol at the *sn*-3 position (choline, ethanolamine, etc.). Long chain hydrocarbons (fatty acids) are esterified at the *sn*-1 and *sn*-2 carbons. Typically, the *sn*-1 carbon is occupied by a saturated fatty acid, while some phospholipid subclasses contain ether- or vinyl ether-linked hydrocarbons at the *sn*-1 position [3, 4]. In mammalian glycerophospholipids, the *sn*-1 acyl chain length is typically C-16 or C-18 [3, 4]. The acyl chain in the *sn*-2 position can vary tremendously, ranging from the minimal *sn*-2 acetyl moiety of platelet-activating factor (PAF) to acyl chain lengths that exceed 20 carbons. Although saturated fatty acids can occupy the *sn*-2 position of glycerophospholipids, this position is often enriched in long chain unsaturated species [3, 4]. Inflammatory cells (macrophages, neutrophils) in particular are noted for the prevalence of polyunsaturated fatty acids in the *sn*-2 position [3, 5]. The fatty acyl composition of the glycerophospholipids of inflammatory cells has important implications for the functioning of these cells, as some of these substituents (arachidonic acid, in particular) are precursors of bioactive lipids that regulate the inflammatory response [5, 6].

Arachidonate Remodeling and Inflammation, edited by Alfred N. Fonteh and Robert L. Wykle

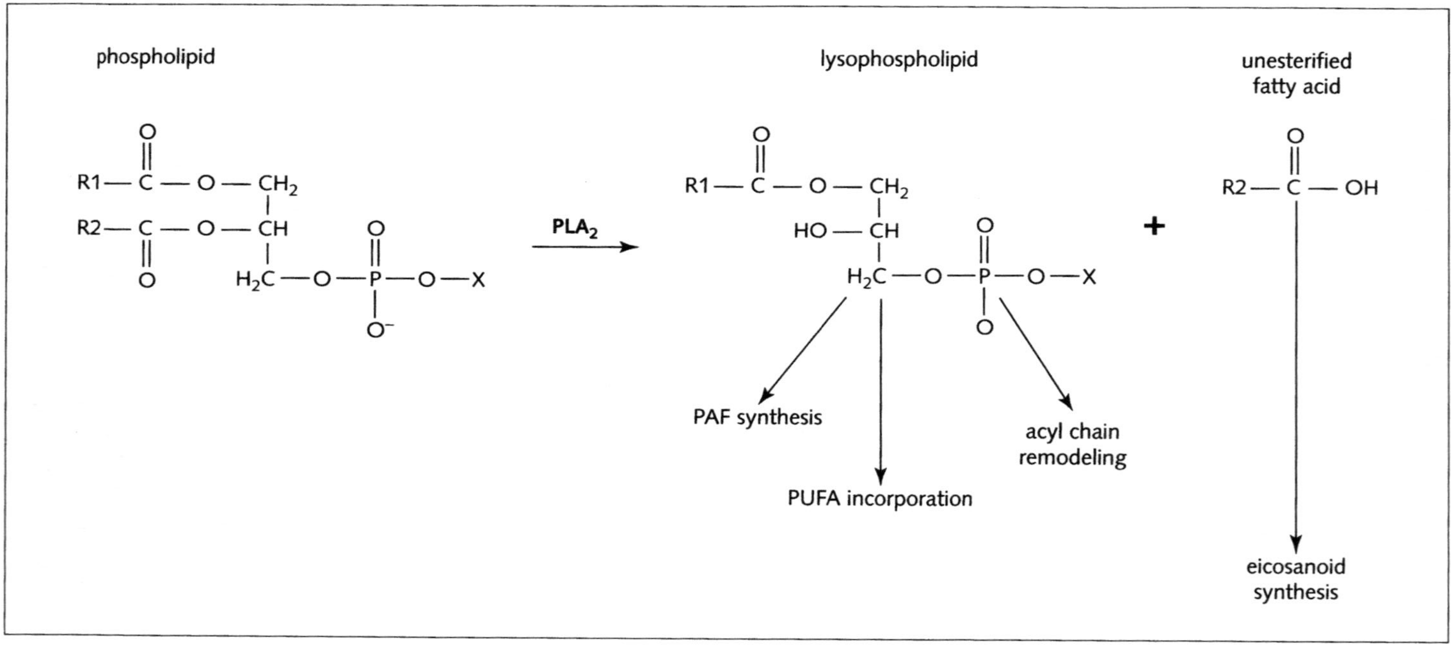

Figure 1
The phospholipase A_2 reaction and its effects on glycerophospholipid metabolism
The phospholipases A_2 (PLA_2) catalyze the hydrolysis of sn-2 *acyl chains from glycerophospholipids, producing lysophospholipid and unesterified fatty acid as products. Oxygenated derivatives of one* sn-2 *fatty acid, arachidonic acid, promote the inflammatory response through their interactions with G protein coupled receptors. In addition, nuclear receptors for arachidonic acid metabolites have been associated with the regulation of cell cycle progression. The lysophospholipid product is associated with regulation of the acyl chain composition of glycerophospholipids as it is essential for both the initial incorporation of polyunsaturated fatty acids (PUFA) into glycerophospholipid and for acyl chain remodeling. In addition, lysophospholipid is involved in the synthesis of platelet-activating factor (PAF), another bioactive lipid.*

The acyl chain composition of glycerophospholipids is determined both by the endogenous fatty acids that are produced by a cell and by exogenous fatty acids that are derived from the diet. Acyl chain composition is also determined by the specificity of the biosynthetic enzymes that generate glycerophospholipids. In mammalian cells in general, saturated and monounsaturated fatty acids are incorporated during the *de novo* synthesis of glycerophospholipids [3, 4]. However, the incorporation of long chain polyunsaturated fatty acids such as arachidonic acid occurs much later and is largely dependent on the Lands pathway [3, 4, 7, 8]. In the Lands pathway, the *sn*-2 acyl chain is hydrolyzed from intact glycerophospholipid generating a lysophospholipid acceptor for polyunsaturated fatty acid. Typically, the source of the polyunsaturated fatty acid is a coenzyme A (CoA) adduct in a reaction catalyzed by a CoA-dependent acyltransferase. Although none of these enzymes has yet been cloned or purified to homogeneity, it is clear that they exhibit acyl chain specificity [3–5, 9–11]. This acyl chain specificity is of particular importance in inflammatory cells in which acyltransferases with specificity for arachidonic acid (C20:4) and other long chain polyunsaturated fatty acids have been identified [12–14]. The substrate specificity of the acyltransferases in inflammatory cells also impinges on the observation that *sn*-2 acyl chains are not static, but are remodeled during the lifetime of the glycerol backbone. For example, arachidonic acid is initially incorporated into diacylphosphatidylcholine (PC) but over time is transferred into ether-linked species of PC and phosphatidylethanolamine (PE) [5, 14–17]. This process of fatty acid remodeling typically involves CoA-independent transacylases that transfer polyunsaturated acyl chains directly from intact PC to lysophosphatidylethanolamine (lysoPE), most likely through an enzyme-fatty acid intermediate [3, 4]. As a result of these reactions, the ether-linked subclasses of PE and PC in macrophages and neutrophils are enriched in polyunsaturated fatty acids [18–21]. As noted above, this enrichment has functional consequences for inflammatory cells, as it generates a pool of precursor for such bioactive lipids as the eicosanoids and platelet-activating factor (PAF).

The phospholipases A_2

Both the initial incorporation of polyunsaturated fatty acids into glycerophospholipids and their remodeling among phospholipid subclasses are dependent on sources of lysophospholipid [3–5]. The generation of these lysophospholipids is most closely associated with the phospholipase A_2 (PLA_2) family of enzymes [22, 23]. These enzymes are defined by their hydrolytic actions on the *sn*-2 carbonyl of glycerophospholipids. Although the PLA_2s vary in their substrate specificities, all of these enzymes generate lysophospholipids and unesterified fatty acids as the products of their reactions (see Fig. 1). As noted below, some of the PLA_2 enzymes also possess lysophospholipase activity and therefore can limit the accumulation of

Table 1: Families of phospholipase A_2 enzymes

Enzyme	M.W. (kDa)	Ca^{2+}?	Active site	Expression	Location	Regulation
$sPLA_2$	14–18	mM	His-Asp	Inducible	Extracellular	Expression
$cPLA_2$	85 (α)	μM	GLSGS	Constitutive	Cytosol	Ca^{2+}, PO_4 (α,β)
$iPLA_2$	80	None	GxSxG	Constitutive	Cytosol	Oligomerization?

detergent-like lysophospholipids. In addition, as lysophospholipids serve as the initial acceptors for polyunsaturated fatty acids, the PLA_2s play important roles in determining the acyl chain composition of glycerophospholipids.

The PLA_2s vary tremendously in their primary, secondary, and tertiary structures, catalytic mechanisms, substrate specificities, subcellular localization, and regulatory mechanisms [22, 23]. To date, at least twelve PLA_2 family members have been identified and these enzymes are found in a diverse array of species ranging from bacteria and invertebrates to reptiles and mammals. The PLA_2s are typically grouped into three broad families of enzymes: the secreted or $sPLA_2$s, the cytosolic or $cPLA_2$s, and the calcium-independent or $iPLA_2$s (Tab. 1). As will be noted below, some of these monikers are rather unfortunate as they describe members of several families of PLA_2.

The $sPLA_2$s were the first family to be identified and studied intensely, in large part due to the accessibility of these enzymes [24]. As their name implies, the $sPLA_2$s are all secreted enzymes and are found in high concentrations in snake and bee venoms [23, 24]. In the late 1980s, it became apparent that mammalian cells also express sPLA2s and that the expression and secretion of these enzymes are induced in inflammation [25–31]. The $sPLA_2$s are relatively small enzymes, ranging from ~12–18 kDa in molecular mass. A common structural characteristic shared by all of the $sPLA_2$s is the presence of multiple (5–8) disulfide bonds [23, 32, 33]. Thus, these enzymes have a rather rigid structure and are only active when sequestered away from the reducing environment of the cytosol. The catalytic activity of the $sPLA_2$s is strictly dependent on calcium at mM concentrations and this cation plays an essential role in the catalytic mechanisms of these enzymes. The association with calcium is mediated through the "calcium binding loop", a glycine-rich stretch of amino acids [23, 24, 32]. The calcium ion has been shown to co-ordinate with the *sn*-3 phosphate and *sn*-2 carbonyl and thereby to stabilize the transition state of substrate in the active site of the $sPLA_2$ [34, 35]. The active site itself consists of a histidine-aspartic acid pair that participates in a charge relay system resulting in the polarization of a water molecule, the nucleophile that actually attacks the *sn*-2 carbonyl [34, 35]. The $sPLA_2$s show little specificity for the acyl chain in the *sn*-2 position, although some $sPLA_2$s appear to preferentially hydrolyze anionic phospholipids [36–38]. Importantly, the expression of these enzymes is typically very low in resting cells but is induced upon activation with such pro-inflammatory stimuli as interleukin 1 (IL-1), interleukin-6 (IL-6), and bacterial lipopolysaccharides (LPS) [26, 39–43]. This regulatory mechanism impinges on the functioning of the $sPLA_2$s in the lipid metabolism of activated mammalian cells, in particular on their roles in the production of eicosanoids and other pro-inflammatory lipids.

The $cPLA_2$ family was initially named based on its localization in the cytosol of mammalian cells. This turns out to be a rather unfortunate choice, as it is now known that other PLA_2s (the $iPLA_2$s, in particular) are also localized to the cytosol [22, 23]. As a result, it has also been proposed that this family be called the calcium-dependent cytosolic PLA_2s. This choice is also inappropriate, as at least one $cPLA_2$ family member is known to be calcium-independent [44, 45]. Thus, $cPLA_2$ is a rather diverse class of enzymes that are grouped together based on rather loose structural homologies [22, 23]. The first family member to be described (and the best studied and understood) is the group IVA $cPLA_2$-α. Although the catalytic activity of this enzyme is calcium dependent, requiring μM concentrations like the $sPLA_2$s, it has no structural homology and has a distinct catalytic mechanism. In fact, the catalytic mechanism of $cPLA_2$-α is more closely akin to that of general lipases, as an active site serine serves as the nucleophile that attacks the *sn*-2 carbonyl [46–48]. The calcium-dependence of $cPLA_2$-α is related to substrate binding. The enzyme has an N-terminal C2 domain and has been shown to translocate from cytosol to membrane (substrate) in a calcium-dependent manner [49–51]. Several lines of evidence indicate that $cPLA_2$-α activity is augmented when the enzyme is phosphorylated on serine 505, although there is some discrepancy regarding the kinases that mediate this phosphorylation, the relationship of phosphorylation to calcium dependence/translocation, and the physiological relevance of phosphorylation [52–57]. Together, the regulation of $cPLA_2$-α by calcium and phosphorylation suggest that this enzyme plays a role in the lipid metabolism of activated cells. This hypothesis is borne out by mounting evidence that the enzyme is activated in response to receptor-dependent stimuli and that it plays an essential role in the production of bioactive lipids by activated cells [58–64]. To this end, it is of interest that $cPLA_2$-α shows a preference for hydrolyzing arachidonic acid in the *sn*-2 position of glycerophospholipid and therefore contributes to eicosanoid production by inflammatory cells [65–67].

Two other $cPLA_2$ family members have been cloned, $cPLA_2$-β and $cPLA_2$-γ [44, 45]. Little is known about the regulatory mechanisms of these enzymes and they do not appear to be as widely expressed as is $cPLA_2$-α. However, we have demonstrated that $cPLA_2$-γ is most likely the isoform expressed by *Naegleria fowleri* amoebae, suggesting that this calcium-independent "paralog" may have important roles in the lipid metabolism of lower organisms [68].

Like $cPLA_2$, the $iPLA_2$ family consists of a diverse array of enzymes with the group VIA enzyme being the most intensively studied and understood [22, 23, 69].

As its name implies, the $iPLA_2$ family shows absolutely no requirement for calcium in its catalytic mechanism. Like $cPLA_2$, this family uses an active site serine as the nucleophile that attacks the *sn*-2 carbonyl and also acts as a lysophospholipase [70–75]. Unlike $cPLA_2$-α and $cPLA_2$-γ, the group VIA $iPLA_2$ shows little or no specificity for either the *sn*-2 fatty acid or head group of its substrate, although it has been suggested that some $iPLA_2$s preferentially hydrolyze plasmalogens [70, 71, 76]. A key feature that distinguishes $iPLA_2$ from the other two families is its regulatory mechanism, as this does not appear to be dependent on cell activation. Indeed, although ATP has been shown to "stabilize" $iPLA_2$ activity in cell homogenates, there is no evidence that this involves a phosphorylation event [70, 77]. Interestingly, the $iPLA_2$ pre-mRNA has been shown to undergo alternative splicing events that may impinge on the regulation of enzyme activity [78, 79]. In this regard, it is useful to consider the structure of $iPLA_2$ by dividing the protein roughly in half. The C-terminal portion of the protein contains the active site serine. Much of the N-terminal portion consists of a series of ankyrin repeats that are thought to mediate the oligomerization of $iPLA_2$ monomers into a ~320 kDa active homotetramer [70, 79]. Two $iPLA_2$ splice variants have been described that contain the N-terminal ankyrin repeats but lack the active site serine in the C-terminus [78,79] (see Fig. 2). When over expressed with the full length $iPLA_2$ (containing both ankryin repeats and catalytic domain), the splice variant proteins suppress $iPLA_2$ activity, presumably by acting in a dominant negative fashion to inhibit the association of full length monomers into active tetramers [79]. These observations suggest that the catalytic activity of $iPLA_2$ may be regulated through alternative splicing. Indeed, our preliminary experiments indicate that the catalytic activity of $iPLA_2$ is cell cycle dependent and may correlate with the accumulation of splice variant proteins (Manguikian and Barbour, unpublished).

Regulation of glycerophospholipid composition by PLA_2s

As noted earlier, the glycerophospholipid composition of mammalian cell membranes is tightly controlled, both at the level of the phospholipid classes and the acyl chains esterified to the glycerol backbones of each class of glycerophospholipid. Recent studies indicate that the PLA_2s, in particular $iPLA_2$, may play major roles in the regulation of glycerophospholipid composition of mammalian cells. The involvement of $iPLA_2$ in these processes is consistent with the constitutive expression and regulatory mechanisms of this enzyme. Unlike the $sPLA_2$s or $cPLA_2$, $iPLA_2$ activity is not controlled by events associated with cell activation and hence this appears to be a homeostatic enzyme [22, 69]. In fact, $iPLA_2$ appears to play several roles in glycerophospholipid homeostasis, ranging from the regulation of phospholipid mass to the regulation of *sn*-2 acyl chain composition, to the accumulation of lipids for daughter cell membranes.

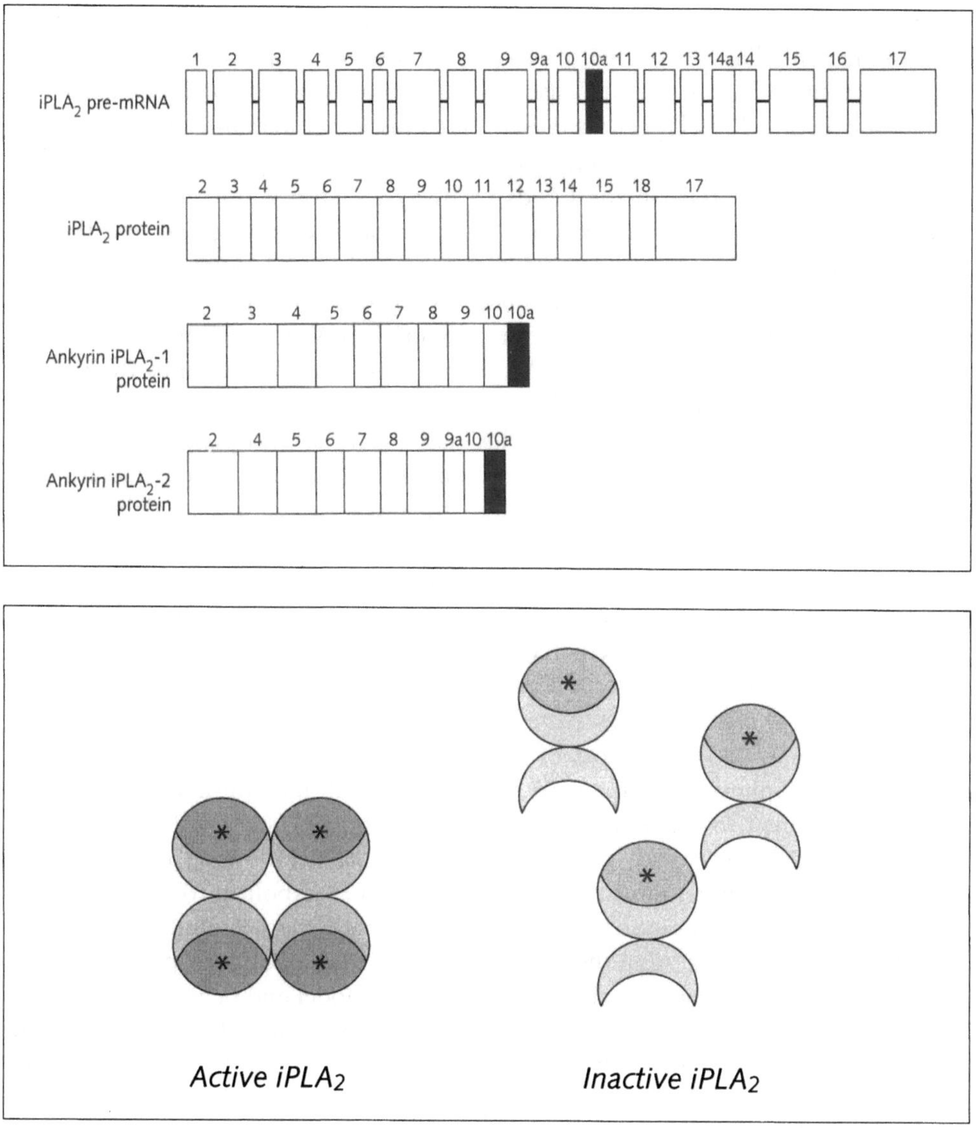

Figure 2

Regulation of $iPLA_2$ activity

The structure of the $iPLA_2$ gene is shown along with exon usage for catalytically active iPLA2 and two splice variants (ankryin-$iPLA_2$-1 and ankyrin-$iPLA_2$-2) that do not contain the active site and may inhibit $iPLA_2$ activity through a dominant negative mechanism.

Proposed models for catalytically active and inactive $iPLA_2$ are shown. The active enzyme is thought to be a homotetramer. The truncated splice variant proteins are proposed to interact with full length proteins through their N-terminal ankryin repeats and block the formation of active tetramers.

The role of $iPLA_2$ in acyl chain composition has been most extensively studied by Balsinde et al. These investigators have used a selective inhibitor of $iPLA_2$, bromoenol lactone (BEL) and an antisense oligonucleotide strategy to study the enzyme in macrophage-like $P388D_1$ cells [80–82]. Like most other mammalian cells, macrophages do not incorporate arachidonic acid into phospholipids through *de novo* pathways, but rather rely on the Lands pathway to esterify arachidonic acid into existing glycerophospholipid backbones [3, 4]. Balsinde et al. showed that this initial incorporation of arachidonic acid was inhibited in cells with reduced $iPLA_2$ activity and/or expression, presumably because the enzyme was necessary to generate the lysophospholipid acceptor for arachidonyl-CoA. In macrophages, arachidonic acid is initially esterified into PC, but is rapidly remodeled into the various subclasses of PE and phosphatidylinositol (PI). Interestingly, neither the $iPLA_2$ inhibitor nor the antisense oligonucleotide had effects on this process. These data suggest that while $iPLA_2$ is essential for the initial incorporation of polyunsaturated fatty acids into macrophage phospholipids, it is not involved in remodeling *per se*. To date, the identity of the PLA_2 that generates the lysoPE acceptor for remodeled arachidonic acid remains unknown. Although it remains possible that remodeling is mediated by $sPLA_2$, $cPLA_2$, or one of the other known forms of $iPLA_2$, it is likely that this important process is mediated by an as yet undescribed form of PLA_2 [80].

In addition to its role in determining the acyl chain composition of glycerophospholipids, several recent studies have also implicated $iPLA_2$ in the regulation of phospholipid mass [83–85]. Most of these studies have focused on PC, the most abundant glycerophospholipid in mammalian cell membranes [86]. The initial studies were performed by Walkey and Cornell who showed that over expression of CTP – phosphocholine cytidylyltransferase (CT), the rate limiting enzyme in PC synthesis – resulted in a compensatory increase in PC catabolism to maintain mass [87]. The increase in catabolism was associated with the accumulation of glycerophosphocholine, suggesting the involvement of a PLA_2. Baburina and Jackowski [83] subsequently over-expressed CT in a HeLa cell model and demonstrated that a suicide inhibitor of $iPLA_2$, bromoenol lactone (BEL), prevented the accumulation of glycerophosphocholine. Similarly, BEL treatment both induced PC accumulation in, and allowed the survival of, mutant Chinese hamster ovary (CHO) cells that could not synthesize PC [85]. Our laboratory extended these studies by demonstrating the induction of $iPLA_2$ protein in CHO cells that over expressed CT [84]. Importantly, this induction was specific for $iPLA_2$, as neither $cPLA_2$ or lysophospholipase activity was induced in the over-expressing cells. Conversely, $iPLA_2$ activity was suppressed in mutant MT58 cells that exhibited little CT activity, although the expression of the enzyme remained unchanged (Barbour, unpublished observation). Based on these studies, we propose that $iPLA_2$ is co-ordinately regulated with the activity of CT and other synthetic enzymes to maintain lipid homeostasis as is shown for PC in Figure 3. This model proposes that changes in glycerophospholipid synthesis are balanced by changes in $iPLA_2$ activity such that the mass of a given glycerophos-

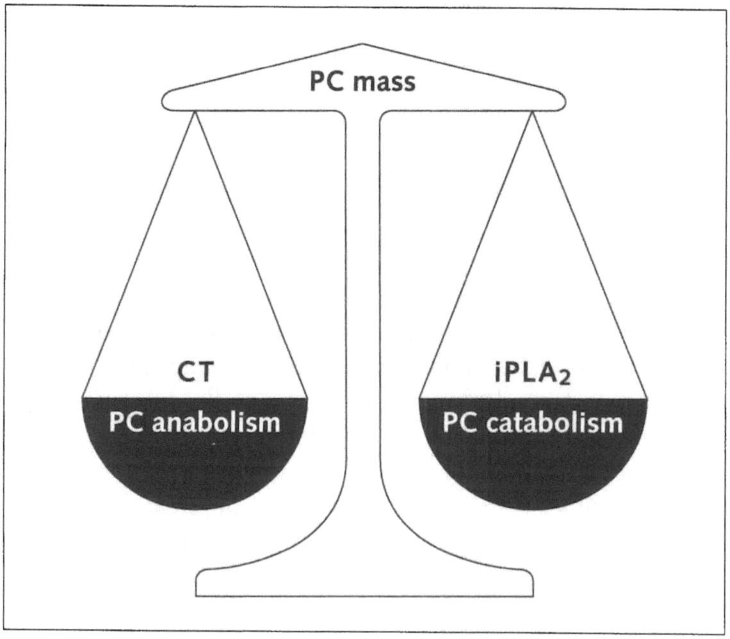

Figure 3
Proposed model for the regulation of glycerophospholipid mass
The model shown is based on phosphatidylcholine metabolism, but is thought to be applicable to other glycerophospholipids. The activities of enzymes involved in glycerophospholipid synthesis (for example, CT, CTP : phosphocholine cytidylyltransferase) are co-ordinately regulated with iPLA$_2$ activity so that glycerophospholipid mass remains balanced.

pholipid remains relatively constant. This hypothesis is in accord with the observation that the ratios of glycerophospholipid species in mammalian cell membranes are tightly maintained [4, 12]. It should be noted that a recent report refutes this model and suggests that changes in iPLA$_2$ expression do not necessarily induce compensatory changes in PC synthesis [88]. While the experiments in this report were performed in transiently transfected COS cells, the proposed model is based on studies of phospholipid metabolism in stably transfected cells. Thus, it is possible that this discrepancy can be attributed to chronic *versus* acute alterations in phospholipid metabolism.

At least one circumstance exists in which homeostasis is not maintained: when lipids are accumulated during the cell cycle for daughter cell membranes. Several lines of evidence indicate that this accumulation occurs during S phase and is not the result of increased glycerophospholipid synthesis but rather a decrease in catabolism [89–94]. Preliminary experiments in our laboratory (Manguikian and Barbour, unpublished) and others [95] indicate that the accumulation of PC during S phase may be mediated through a decline in $iPLA_2$ activity. Hence, $iPLA_2$ not only maintains lipid mass in resting cells, but also controls the accumulation of glycerophospholipids in cells that are actively dividing.

Relevance of lipid composition to the functioning of inflammatory cells

The glycerophospholipids of inflammatory cells such as macrophages and neutrophils are relatively enriched in arachidonic acid (C20:4) and other long chain polyunsaturated fatty acids [3, 5]. This fatty acyl composition almost certainly has important implications for the physicochemical properties of membranes in these cells, as polyunsaturated fatty acids tend to increase membrane fluidity [96–99]. It is also possible that the degree of unsaturation may contribute to the formation of lipid rafts, discrete areas of the plasma membrane that are enriched in sphingolipids, cholesterol, and signaling proteins [100]. Apart from these physical effects, the enrichment of polyunsaturated fatty acids also provides macrophages and neutrophils with raw materials necessary for producing bioactive lipids associated with inflammation. Among these bioactive lipids are the eicosanoids (oxygenated derivatives of arachidonic acid) and platelet-activating factor (PAF), a derivative of PC. Inflammatory cells not only produce these bioactive lipids, but they respond to them as well, typically through G protein coupled receptors.

Two pathways for the synthesis of PAF have been described, a *de novo* pathway and a pathway that is dependent on acyl chain remodeling. The remodeling pathway is thought to be the most relevant in inflammatory cells [3, 5, 15, 17 101–104]. The goal of the remodeling pathway is to generate the PAF precursor 1-O-alkyl-2-lyso-*sn*-glycero-3-phosphocholine (lyso-PAF), a substrate for the CoA-dependent acetyltransferase that generates PAF [101]. A prerequisite for this goal is the removal of the acyl chain at the *sn*-2 position of 1-O-alkyl-PC. This is accomplished, at least in part, through transfer of the acyl chain in a reaction catalyzed by coenzyme A independent transacylase (CoAIT). The CoAIT reaction requires a lyso-PE acceptor and it is thought that this substrate is generated through the actions of a PLA_2.

This mechanism was implicated in early studies in human mesanglial cells and HL60 granulocytes (a human neutrophil model) in which inhibitors of $sPLA_2$ (p-bromophenacyl bromide, mepacrine) were shown to inhibit the synthesis of PAF [64, 105]. The role of $sPLA_2$ in PAF synthesis was later supported in a variety of primary cell types (human monocytes, PMN, and endothelial cells, for example) treated with scalaradial or SB203347, specific inhibitors of $sPLA_2$ [106–108]. In contrast, arachidonyltrifluoromethyl ketone (AACOCF3) and methyl arachidonyl fluorophosphonate (MAFP), inhibitors of $cPLA_2$ and $iPLA_2$ have no effect on PAF synthesis by human monocytes, alveolar macrophages, or vascular endothelial growth factor- (VEGF-) stimulated endothelial cells [107–110]. Furthermore, antisense oligonucleotides against $cPLA_2$ have been shown to suppress expression of the enzyme and the production of prostanoids but have no effect on the synthesis of PAF [107, 109]. Conversely, PAF synthesis is induced when U937 monocytes are treated with recombinant human $sPLA_2$ while $cPLA_2$ has no such effect [107]. Together, these data indicate that although $sPLA_2$ and $cPLA_2$ are both able to produce lyso-PE, the $sPLA_2$ reaction is more relevant for PAF synthesis. These observations are somewhat surprising given that PE is enriched in arachidonic acid, that arachidonic acid release is typically observed in concert with PAF production, and that $cPLA_2$ has been shown to selectively hydrolyze arachidonate containing glycerophospholipids [65–67]. It is possible that the pool of lyso-PE generated by $cPLA_2$ is rapidly reacylated with long chain fatty acids and therefore is not available as a substrate for CoAIT. Alternatively, the lyso-PE generated by $cPLA_2$ may be rapidly converted to glycerophosphocholine by the inherent lysophospholipase activity of the enzyme [74, 75]. There is at least one cell type in which $cPLA_2$ activity may contribute to the synthesis of PAF. We have recently reported that monocyte-derived dendritic cells synthesize large amounts of PAF [111] and our preliminary experiments indicate that this may be related to the high levels of $cPLA_2$ activity that are exhibited by these cells (Al-Darmaki and Barbour, unpublished).

As the $sPLA_2$ inhibitors do not target specific $sPLA_2$ isoforms, it is not known if the production of lyso-PE for PAF synthesis is mediated by specific $sPLA_2$s. At least one isoform of $sPLA_2$, the group IIa enzyme preferentially hydrolyzes PE and this may be relevant to the production of lyso-PE for PAF synthesis [36–38]. However monocytes express primarily the group V $sPLA_2$ [112], suggesting that this may be the relevant isoform for PAF synthesis. Another cautionary note regarding the role of PLA_2 in PAF synthesis regards endothelial cells. A recent report [113] suggests that thrombin-stimulated endothelial cells produce PAF in an $iPLA_2$-dependent manner. However, the production of PAF by VEGF stimulated endothelial cells is not inhibited by bromoenol lactone (BEL), a specific inhibitor of $iPLA_2$ [108]. Taken together, existing data are consistent with a role for PLA_2 in the acyl chain remodeling that precedes PAF synthesis, although the specific PLA_2s involved may vary with different cell types and stimuli.

The production of PAF by inflammatory cells is typically accompanied by the production of eicosanoids, oxygenated derivates of arachidonic acid [3, 5, 101]. The two major families of eicosanoids, prostaglandins and leukotrienes, are produced through the addition of molecular oxygen to arachidonic acid (C20:4) in reactions catalyzed by the cyclooxygenase and lipoxygenase enzymes, respectively. Receptor-dependent stimulation of inflammatory cells induces eicosanoids that then modulate cell behavior and biological responses through their binding to G protein coupled receptors (GPCR) [114, 115]. For example, activated macrophages produce PGE_2, a bioactive lipid that induces such diverse responses as the production of Th2 cytokines, interferon-γ production, bone resorption, and the production of IgG2 antibodies [116–119]. It has been proposed that the rate-limiting step in the production of both families of eicosanoids is the release of arachidonic acid from intact phospholipids by a PLA_2 [120,121]. However, it can also be argued that the limiting step is the initial incorporation of arachidonic acid into glycerophospholipid pools, as this is clearly a prerequisite for arachidonate release. As noted above, this incorporation is mediated through the Lands pathway and involves the reacylation of lysophospholipid [3–5]. In macrophages, $iPLA_2$ activity is essential for generation of the lysophospholipid acceptor for arachidonyl-CoA [80–82]. Although most studies indicate that the release of arachidonic acid from glycerophospholipids of activated cells is mediated by either $sPLA_2$ or $cPLA_2$, $iPLA_2$ has been implicated in these reactions as well [58, 69, 122–124]. Thus, $iPLA_2$ may play two distinct roles in eicosanoid production by inflammatory cells: the initial incorporation of arachidonate into glycerophospholipid and its release upon cell activation. Interestingly, although $iPLA_2$ is involved in the release of arachidonic acid by activated pancreatic islet cells, it is not required for the initial incorporation into phospholipids [125–127]. This again suggests that the roles of the various PLA_2s in acyl chain remodeling may be cell type specific.

For many years, it has been known that the eicosanoids can regulate cell behavior by binding to GPCR in the plasma membrane [114, 115]. More recently, intracellular receptors for the eicosanoids have been identified. These receptors, the peroxisome proliferator-activated receptors, PPAR, are DNA binding proteins that, when bound to ligand, translocate to the nucleus and regulate gene expression [128–130]. Three families of PPAR have been described and each family of receptors binds a different spectrum of bioactive lipids or lipophilic drugs. For example, PPAR-α is the target of the fibrate class of hypolipidemic drugs and has been shown to regulate fatty acid metabolism [131, 132]. Endogenous ligands of PPAR-α include linolenic acid and 8S-HETE [128–130]. The biological function of the PPAR-β/δ subtype remains somewhat debatable, although this receptor has also been shown to bind linolenic acid [128–130]. A variety of functions have been ascribed to PPAR-γ, including the regulation of adipogenesis and antagonism of the inflammatory response [133]. Most recently, the ligation of PPAR-γ has been linked to cell cycle arrest through a novel mechanism that explains a large body of data [134].

For many years it has been known that exogenous prostanoids induce growth arrest in a variety of tumor cell types. Specifically, prostaglandin A_2 (PGA_2), prostaglandin D_2 (PDG_2), and 15-deoxy-$\Delta^{12,14}$ prostaglandin J_2 (PGJ_2) have been shown to arrest cells at the border of G1 and S phases [134–140]. This growth arrest has been linked to reduced expression of cyclin D and cdk4 proteins and the induction of p21, a cdk inhibitor [134, 135]. However the molecular mechanism of prostanoid-induced growth arrest has remained elusive until recently. Using MCF 7 breast carcinoma cells as a model, Wang et al. recently demonstrated that PGD_2 and PGJ_2 suppress cyclin D1 expression through their association with PPAR-γ [134]. In these experiments, ligated PPAR-γ was shown to recruit p300 away from a critical c-Fos binding site in the cyclin D1 promoter and thereby suppress transcription of the

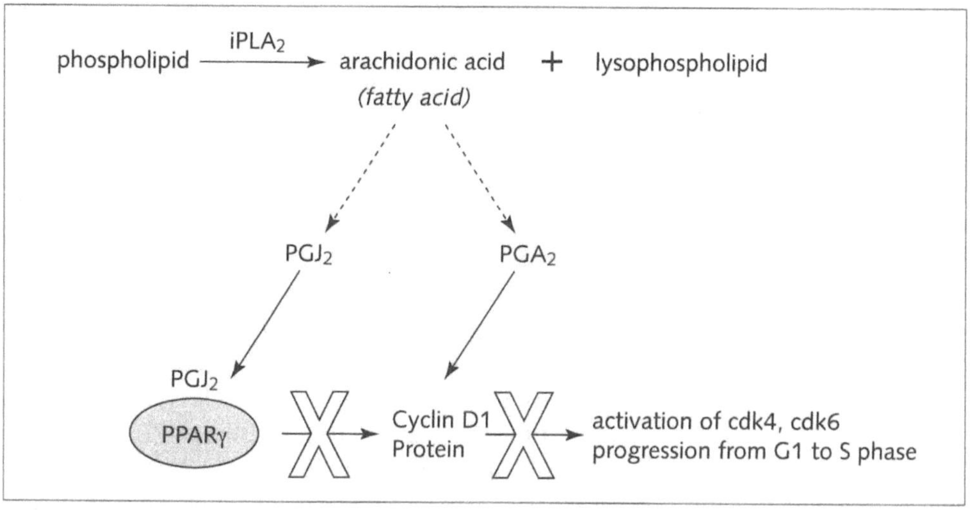

Figure 4
Regulation of cell cycle progression by derivatives of arachidonic acid
Once arachidonic acid has been released from glycerophospholipids by iPLA$_2$, it can be converted into oxygenated derivatives such as prostaglandin A_2 (PGA$_2$) and 15-deoxy-$\Delta^{12,14}$ prostaglandin J_2 (15-deoxy PGJ$_2$). These bioactive lipids are known to down-regulate the expression of cyclin D1 and other proteins that are essential for progression from G1 to S phase and thereby to limit cell cycle progression.

gene. Together, these data suggest that a cell's ability to generate prostanoids can have a major impact on its ability to proliferate. As prostanoids are derived from arachidonic acid, this in turn suggests that cell growth and viability are linked to the acyl chain composition of glycerophospholipid. We have recently demonstrated that the catalytic activity of iPLA$_2$ (the PLA$_2$ that provides the lysophospholipid acceptor for the initial incorporation of arachidonic acid into glycerophospholipid) exhibits cell cycle dependence and is lowest at the border of G1 and S phase (Manguikian and Barbour, unpublished). Our work is consistent with a recently published study in human T-cells [141]. In contrast, other phospholipases (e.g., cPLA$_2$) do not exhibit cell cycle dependent activity. We predict that loss of iPLA$_2$ activity during G1 results in reduced levels of endogenous prostanoids and that this in turn facilitates entry into S phase (see Fig. 4). There are at least two potential roles for iPLA$_2$ in this regard, as it can both modulate the amount of arachidonic acid in glycerophospholipids and release incorporated arachidonic acid from glycerophospholipids, thereby regulating the amount of substrate available for the synthesis of PGJ$_2$, PGA$_2$, and other prostanoids. Thus, iPLA$_2$ regulation may be essential for cell cycle progression.

Conclusion and perspectives for the future

For many years, it has been known that the distribution of fatty acids among glycerophospholipid species is tightly controlled and is maintained in a cell type specific manner. Recently, it has become clear that this regulation is essential not only for maintaining membrane structure and fluidity, but also to control the way that cells interact with their environments. This is of special significance for inflammatory cells such as macrophages and neutrophils as these cells produce a battery of bioactive lipids and are also capable of responding to these molecules in ways that dramatically affect the host immune response. Among these bioactive lipids are oxygenated derivatives of arachidonic acid and platelet-activating factor (PAF). Acyl chain remodeling is essential for production of both of these classes of bioactive lipids. This process is dependent on the activities of various forms of PLA_2, the family of hydrolases that generate lysophospholipid acceptors for acyl chain remodeling. Hence, identification of the specific forms of PLA_2 involved in remodeling and a clearer understanding of their catalytic mechanisms and regulation may facilitate the development of novel strategies to control inflammatory cells and the bioactive lipids that they produce. In addition, as some derivatives of arachidonic acid have been implicated in the regulation of cell cycle progression, such studies may also provide insights into novel mechanisms to control the proliferation of transformed cells.

References

1 Ridgway ND, Byers DM, Cook HW, and Storey MK (1999) Integration of phospholipid and sterol metabolism in mammalian cells. *Prog Lipid Res* 38: 337–360
2 Tabas I (2000) Cholesterol and phospholipid metabolism in macrophages. *Biochim Biophys Acta* 1529: 164–174
3 MacDonald JI, Sprecher H (1991) Phospholipid fatty acid remodeling in mammalian cells. *Biochim Biophys Acta* 1084: 105–121
4 Yamashita A, Sugiura T, Waku K (1997) Acyltransferases and transacylases involved in fatty acid remodeling of phospholipids and metabolism of bioactive lipids in mammalian cells. *J Biochem (Tokyo)* 122: 1–16
5 Chilton FH, Fonteh AN, Surette ME, Triggiani M,Winkler JD (1996) Control of arachidonate levels within inflammatory cells. *Biochim Biophys Acta Lipids Lipid Metab* 1299: 1–15
6 Seeds MC, Bass DA (1999) Regulation and metabolism of arachidonic acid. Clinical Reviews. *Allergy Immunol* 17: 5–26
7 Lands WE (2000) Stories about acyl chains. *Biochim Biophys Acta* 1483: 1–14
8 Hill EE, Lands WE (1968) Incorporation of long-chain and polyunsaturated acids into phosphatidate and phosphatidylcholine. *Biochim Biophys Acta* 152: 645–648

9 McKean ML, Smith JB, Silver MJ (1982) Phospholipid biosynthesis in human platelets. Formation of phosphatidylcholine from 1-acyl lysophosphatidylcholine by acyl-CoA:1-acyl-*sn*-glycero-3-phosphocholine acyltransferase. *J Biol Chem* 257: 11278–11283

10 Iritani N, Ikeda Y, Kajitani H (1984) Selectivities of 1-acylglycerophosphorylcholine acyltransferase and acyl-CoA synthetase for n-3 polyunsaturated fatty acids in platelets and liver microsomes. *Biochim Biophys Acta* 793: 416–422

11 Inoue M, Murase S, and Okuyama H (1984) Acyl coenzyme A: Phospholipid acyltransferases in porcine platelets discriminate between omega-3 and omega-6 unsaturated fatty acids. *Arch Biochem Biophys* 231: 29–37

12 MacDonald JI, Sprecher H (1989) Distribution of arachidonic acid in choline- and ethanolamine-containing phosphoglycerides in subfractionated human neutrophils. *J Biol Chem* 264: 17718–17726

13 Lapetina EG, Billah MM, Cuatrecasas P (1980) Rapid acylation and deacylation of arachidonic acid into phosphatidic acid of horse neutrophils. *J Biol Chem* 255: 10966–10970

14 MacDonald JI, Sprecher H (1989) Studies on the incorporation and transacylation of various fatty acids in choline and ethanolamine-containing phosphoacylglycerol subclasses in human neutrophils. *Biochim Biophys Acta* 1004: 151–157

15 Chilton FH, Murphy RC (1986) Remodeling of arachidonate-containing phosphoglycerides within the human neutrophil. *J Biol Chem* 261: 7771–7777

16 Sugiura T, Katayama O, Fukui J, Nakagawa Y, Waku K (1984) Mobilization of arachidonic acid between diacyl and ether phospholipids in rabbit alveolar macrophages. *FEBS Lett* 165: 273–276

17 Robinson M, Blank ML, Snyder F (1985) Acylation of lysophospholipids by rabbit alveolar macrophages. Specificities of CoA-dependent and CoA-independent reactions. *J Biol Chem* 260: 7889–7895

18 Mueller HW, O'Flaherty JT, Greene DG, Samuel MP, Wykle RL (1984) 1-O-Alkyl-linked glycerophospholipids of human neutrophils: Distribution of arachidonate and other acyl residues in the ether-linked and diacyl species. *J Lipid Res* 25: 383–388

19 Mueller HW, O'Flaherty JT, Wykle RL (1982) Ether lipid content and fatty acid distribution in rabbit polymorphonuclear neutrophil phospholipids. Lipids 17: 72–77

20 Sugiura T, Onumam Y, Sekiguchi N, Waku K (1982) Ether phospholipids in guinea pig polymorphonuclear leukocytes and macrophages. Occurence of high levels of 1-O-alkyl-2-acyl-*sn*-glycero-3-phosphocholine. *Biochim Biophys Acta* 712: 515–522

21 Nakagawa Y, Sugiura T, Waku K (1985) The molecular species composition of diacyl-, alkylacyl- and alkenylacylglycerophospholipids in rabbit alveolar macrophages. High amounts of 1-O-hexadecyl-2-arachidonyl molecular species in alkyl-acyl-glycerophosphocholine. *Biochim Biophys Acta* 833: 323–329

22 Kudo I, Murakami M (2002) Phospholipase A_2 enzymes. *Prostaglandins Other Lipid Mediat* 68–69: 3–58

23 Six DA, Dennis EA (2000) The expanding superfamily of phospholipase $A(_2)$ enzymes: Classification and characterization. *Biochim Biophys Acta* 1488: 1–19

24 Valentin E, Lambeau G (2000) Increasing molecular diversity of secreted phospholipases A(2) and their receptors and binding proteins. *Biochim Biophys Acta* 1488: 59–70
25 Andreani M, Olivier JL, Berenbaum F, Raymondjean M, Bereziat G (2000) Transcriptional regulation of inflammatory secreted phospholipases A(2). *Biochimica et Biophysica Acta – Molecular and Cell Biology Biochimica et Biophysi* 1488: 149–158
26 Crowl RM, Stoller TJ, Conroy RR, Stoner CR (1991) Induction of phospholipase A_2 gene expression in human hepatoma cells by mediators of the acute phase response. *J Biol Chem* 266: 2647–2651
27 Nakano T, Ohara O, Teraoka H, Arita H (1990) Group II phospholipase A_2 mRNA synthesis is stimulated by two distinct mechanisms in rat vascular smooth muscle cells. *FEBS Lett* 261: 171–174
28 Seilhamer JJ, Pruzanski W, Vadas P, Plant S, Miller JA, Kloss J, Johnson LK (1989) Cloning and recombinant expression of phospholipase A_2 present in rheumatoid arthritic synovial fluid. *J Biol Chem* 264: 5335–5338
29 Ulevitch RJ, Watanabe Y, Sano M, Lister MD, Deems RA, Dennis EA (1988) Solubilization, purification, and characterization of a membrane-bound phospholipase A_2 from the P388D1 macrophage-like cell line. *J Biol Chem* 263: 3079–3085
30 Kramer RM, Hession C, Johansen B, Hayes G, McGray P, Chow EP, Tizard R, Pepinsky RB (1989) Structure and properties of a human non-pancreatic phospholipase A_2. *J Biol Chem* 264: 5768–5775
31 Kudo I, Chang HW, Hara S, Murakami M, Inoue K (1989) Characteristic and pathophysiological roles of extracellular phospholipase A_2 in inflamed sites. *Dermatologica* 179 (Suppl 1): 72–76
32 Tischfield JA (1997) A thematic series on phospholipases 5. A reassessment of the low molecular weight phospholipase A_2 gene family in mammals. *J Biol Chem* 272: 17247–17250
33 Schalkwijk CG, Marki F, Wiesenberg I, van den BH (1991) The detection of multimeric forms of phospholipase A_2 upon sodium dodecylsulfate-polyacrylamide electrophoresis. *J Lipid Mediat* 4: 83–95
34 Scott DL, White SP, Otwinowski Z, Yuan W, Gelb MH, Sigler PB (1990) Interfacial catalysis: The mechanism of phospholipase A_2. *Science* 250: 1541–1546
35 Wery JP, Schevitz RW, Clawson DK, Bobbitt JL, Dow ER, Gamboa G, Goodson TJR, Hermann RB, Kramer RM, McClure DB (1991) Structure of recombinant human rheumatoid arthritic synovial fluid phospholipase A_2 at 2.2 A resolution. *Nature* 352: 79–82
36 Bayburt T, Yu BZ, Lin HK, Browning J, Jain MK, Gelb MH (1993) Human nonpancreatic secreted phospholipase A_2: Interfacial parameters, substrate specificities, and competitive inhibitors. *Biochemistry* 32: 573–582
37 Mounier C, Vargaftig BB, Franken PA, Verheij HM, Bon C, Touqui L (1994) Platelet secretory phospholipase A_2 fails to induce rabbit platelet activation and to release arachidonic acid in contrast with venom phospholipases A2. *Biochim Biophys Acta* 1214: 88–96

38 Baker SF, Othman R, Wilton DC (1998) Tryptophan-containing mutant of human (group IIa) secreted phospholipase A_2 has a dramatically increased ability to hydrolyze phosphatidylcholine vesicles and cell membranes. *Biochemistry* 37: 13203–13211

39 Shinohara H, Balboa MA, Johnson CA, Balsinde J, Dennis EA (1999) Regulation of delayed prostaglandin production in activated P388D1 macrophages by Group IV cytosolic and Group V secretory phospholipase A_2s. *J Biol Chem* 274: 12263–12268

40 Vial D, Senorale-Pose M, Havet N, Molio L, Vargaftig BB, Touqui L (1995) Expression of the type-II phospholipase A_2 in alveolar macrophages. Down-regulation by an inflammatory signal. *J Biol Chem* 270: 17327–17332

41 Tan XD, Wang H, Gonzalez-Crussi FX, Chang H, Gonzalez-Crussi F, Hsueh W (1996) Platelet-activating factor and endotoxin increase the enzyme activity and gene expression of type II phospholipase A_2 in the rat intestine – Role of polymorphonuclear leukocytes. *J Immunol* 156: 2985–2990

42 Lyons-Giordano B, Pratta MA, Galbraith W, Davis GL, Arner EC (1993) Interleukin-1 differentially modulates chondrocyte expression of cyclooxygenase-2 and phospholipase A_2. *Exp Cell Res* 206: 58–62

43 Schalkwijk C, Pfeilschifter J, Marki F, van den Bosch H (1992) Interleukin-1 beta- and forskolin-induced synthesis and secretion of group II phospholipase A_2 and prostaglandin E_2 in rat mesangial cells is prevented by transforming growth factor-beta 2. *J Biol Chem* 267: 8846–8851

44 Pickard RT, Strifler BA, Kramer RM, Sharp JD (1999) Molecular cloning of two new human paralogs of 85-kDa cytosolic phospholipase A_2. *J Biol Chem* 274: 8823–8831

45 Underwood KW, Song CZ, Kriz RW, Chang XJ, Knopf JL, Lin LL (1998) A novel calcium-independent phospholipase A_2, cPLA2-gamma, that is prenylated and contains homology to $cPLA_2$. *J Biol Chem* 273: 21926–21932

46 Dessen A, Tang J, Schmidt H, Stahl M, Clark JD, Seehra J, Somers WS (1999) Crystal structure of human cytosolic phospholipase A_2 reveals a novel topology and catalytic mechanism. *Cell* 97: 349–360

47 Dessen A (2000) Structure and mechanism of human cytosolic phospholipase $A(_2)$. *Biochim Biophys Acta* 1488: 40–47

48 Sharp JD, Pickard RT, Chiou XG, Manetta JV, Kovacevic S, Miller JR, Varshavsky AD, Roberts EF, Strifler BA, Brems DN et al (1994) Serine 228 is essential for catalytic activities of 85-kDa cytosolic phospholipase A_2. *J Biol Chem* 269: 23250–23254

49 Xu GY, McDonagh T, Yu HA, Nalefski EA, Clark JD, Cumming DA (1998) Solution structure and membrane interactions of the C2 domain of cytosolic phospholipase A_2. *J Mol Biol* 280: 485–500

50 Clark JD, Lin LL, Kriz RW, Ramesha CS, Sultzman LA, Lin AY, Milona N, Knopf JL (1991) A novel arachidonic acid-selective cytosolic PLA_2 contains a Ca^{2+}-dependent translocation domain with homology to PKC and GAP. *Cell* 65: 1043–1051

51 Davletov B, Perisic O, Williams RL (1998) Calcium-dependent membrane penetration is a hallmark of the C2 domain of cytosolic phospholipase A_2 whereas the C2A domain of synaptotagmin binds membranes electrostatically. *J Biol Chem* 273: 19093–19096

52 Lin LL, Wartmann M, Lin AY, Knopf JL, Seth A, Davis RJ (1993) cPLA2 is phosphorylated and activated by MAP kinase. *Cell* 72: 269–278

53 De Carvalho MGS, McCormack AL, Olson E, Ghomashchi F, Gelb MH, Yates JR, III, Leslie CC (1996) Identification of phosphorylation sites of human 85-kDa cytosolic phospholipase A_2 expressed in insect cells and present in human monocytes. *J Biol Chem* 271: 6987–6997

54 Qiu Z-H., De Carvalho M, Leslie CC (1993) Regulation of phospholipase A_2 activation by phosphorylation in mouse peritoneal macrophages. *J Biol Chem* 268: 24506–24513

55 Qiu ZH, Gijón MA, De Carvalho M, Spencer DM, Leslie CC (1998) The role of calcium and phosphorylation of cytosolic phospholipase A_2 in regulating arachidonic acid release in macrophages. *J Biol Chem* 273: 8203–8211

56 Kramer RM, Roberts EF, Um SL, Borsch-Haubold AG, Watson SP, Fisher MJ, Jakubowski JA (1996) p38 mitogen activated protein kinase phosphorylates cytosolic phospholipase A_2 ($cPLA_2$) in thrombin-stimulated platelets. *J Biol Chem* 271: 27723–27729

57 Kramer RM, Roberts EF, Manetta JV, Hyslop PA, Jakubowski JA (1993) Thrombin-induced phosphorylation and activation of Ca^{2+}-sensitive cytosolic phospholipase A_2 in human platelets. *J Biol Chem* 268: 26796–26804

58 Murakami M, Shimbara S, Kambe T, Kuwata H, Winstead MV, Tischfield JA, Kudo I (1998) The functions of five distinct mammalian phospholipase A_2s in regulating arachidonic acid release. *J Biol Chem* 273: 14411–14423

59 Balsinde J, Dennis EA (1996) Distinct roles in signal transduction for each of the phospholipase A_2 enzymes present in P388D1 macrophages. *J Biol Chem* 271: 6758–6765

60 Bartoli F, Lin HK, Ghomashchi F, Gelb MH, Jain MK, Apitz-Castro R (1994) Tight binding inhibitors of 85-kDa phospholipase A_2 but not 14-kDa phospholipase A_2 inhibit release of free arachidonate in thrombin-stimulated human platelets. *J Biol Chem* 269: 15625–15630

61 Xing MB, Insel PA (1996) Protein kinase C-dependent activation of cytosolic phospholipase A_2 and mitogen-activated protein kinase by alpha1-adrenergic receptors in Madin-Darby canine kidney cells. *J Clin Invest* 97: 1302–1310

62 Gilbert JJ, Stewart A, Courtney CA, Fleming MC, Reid P, Jackson CG, Wise A, Wakelam MJO, Harnett MM (1996) Antigen receptors on immature, but not mature, B and T cells are coupled to cytosolic phospholipase A_2 activation – Expression and activation of cytosolic phospholipase A_2 correlate with lymphocyte maturation. *J Immunol* 156: 2054–2061

63 Lin LL, Lin AY, Knopf JL (1992) Cytosolic phospholipase A_2 is coupled to hormonally regulated release of arachidonic acid. *Proc Natl Acad Sci USA* 89: 6147–6151

64 Suga K, Kawasaki T, Blank ML, Snyder F (1990) An arachidonoyl (polyenoic)-specific phospholipase A_2 activity regulates the synthesis of platelet-activating factor in granulocytic HL-60 cells. *J Biol Chem* 265: 12363–12371

65 Leslie CC, Voelker DR, Channon JY, Wall MM, Zelarney PT (1988) Properties and

purification of an arachidonoyl-hydrolyzing phospholipase A_2 from a macrophage cell line, RAW 264.7. *Biochim Biophys Acta* 963: 476–492

66 Clark JD, Milona N, Knopf JL (1990) Purification of a 110-kilodalton cytosolic phospholipase A_2 from the human monocytic cell line U937. *Proc Natl Acad Sci USA* 87: 7708–7712

67 Sundler R, Winstedt D, Wijkander J (1994) Acyl-chain selectivity of the 85 kDa phospholipase A_2 and of the release process in intact macrophages. *Biochem J* 301: 455–458

68 Barbour SE, Marciano-Cabral F (2001) *Naegleria fowleri* amoebae express a membrane-associated calcium-independent phospholipase A(2). *Biochim Biophys Acta* 1530: 123–133

69 Winstead MV, Balsinde J, Dennis EA (2000) Calcium-independent phospholipase A(2): Structure and function. *Biochim Biophys Acta* 1488: 28–39

70 Ackermann EJ, Kempner ES, and Dennis EA (1994) Ca^{2+}-independent cytosolic phospholipase A_2 from macrophage-like P388D1 cells. Isolation and characterization. *J Biol Chem* 269: 9227–9233

71 Tang J, Kriz RW, Wolfman N, Shaffer M, Seehra J, Jones SS (1997) A novel cytosolic calcium-independent phospholipase A_2 contains eight ankyrin motifs. *J Biol Chem* 272: 8567–8575

72 Wolf MJ, Gross RW (1996) Expression, purification, and kinetic characterization of a recombinant 80-kDa intracellular calcium-independent phospholipase A_2. *J Biol Chem* 271: 30879–30885

73 Lio YC, Dennis EA (1998) Interfacial activation, lysophospholipase and transacylase activity of group VI Ca^{2+}-independent phospholipase A_2. *Biochim Biophys Acta* Lipids Lipid Metab 1392: 320–332

74 de Carvalho MG, Garritano J, Leslie CC (1995) Regulation of lysophospholipase activity of the 85-kDa phospholipase A_2 and activation in mouse peritoneal macrophages. *J Biol Chem* 270: 20439–20446

75 Fujimori Y, Kudo I, Fujita KI, Inoue K (1994) Characteristics of lysophospholipase activity expressed by cytosolic phospholipase A_2. *Eur J Biochem* 218: 629–635

76 Hazen SL, Stuppy RJ, Gross RW (1990) Purification and characterization of canine myocardial cytosolic phospholipase A_2. A calcium-independent phospholipase with absolute f1-2 regiospecificity for diradyl glycerophospholipids. *J Biol Chem* 265: 10622–10630

77 Hazen SL, Gross RW (1991) ATP-dependent regulation of rabbit myocardial cytosolic calcium-independent phospholipase A_2. *J Biol Chem* 266: 14526–14534

78 Forsell PK, Kennedy BP, Claesson HE (1999) The human calcium-independent phospholipase A_2 gene – Multiple enzymes with distinct properties from a single gene. *Eur J Biochem* 262: 575–585

79 Larsson PKA, Claesson HE, Kennedy BP (1998) Multiple splice variants of the human calcium-independent phospholipase A_2 and their effect on enzyme activity. *J Biol Chem* 273: 207–214

80 Balsinde J (2002) Roles of various phospholipases A(2) in providing lysophospholipid

acceptors for fatty acid phospholipid incorporation and remodelling. *Biochem J* 364: 695–702

81 Balsinde J, Balboa MA, Dennis EA (1997) Antisense inhibition of group VI Ca^{2+}-independent phospholipase A_2 blocks phospholipid fatty acid remodeling in murine P388D1 macrophages. *J Biol Chem* 272: 29317–29321

82 Balsinde J, Bianco ID, Ackermann EJ, Conde-Frieboes K, Dennis EA (1995) Inhibition of calcium-independent phospholipase A_2 prevents arachidonic acid incorporation and phospholipid remodeling in P388D1 macrophages. *Proc Natl Acad Sci USA* 92: 8527–8531

83 Baburina I, Jackowski S (1999) Cellular responses to excess phospholipid. *J Biol Chem* 274: 9400–9408

84 Barbour SE, Kapur A, Deal CL (1999) Regulation of phosphatidylcholine homeostasis by calcium-independent phospholipase A_2. *Biochim Biophys Acta* 1439: 77–88

85 Waite KA, Vance DE (2000) Why expression of phosphatidylethanolamine N-methyltransferase does not rescue chinese hamster ovary cells that have an impaired CDP-choline pathway. *J Biol Chem* 275: 21197–21202

86 Vance DE (1990) Boehringer Mannheim Award lecture. Phosphatidylcholine metabolism: masochistic enzymology, metabolic regulation, and lipoprotein assembly. *Biochem Cell Biol* 68: 1151–1165

87 Walkey CJ, Kalmar GB, Cornell RB (1994) Over-expression of rat liver CTP:phosphocholine cytidylyltransferase accelerates phosphatidylcholine synthesis and degradation. *J Biol Chem* 269: 5742–5749

88 Chiu CH, Jackowski S (2001) Role of calcium-independent phospholipases (iPLA($_2$)) in phosphatidylcholine metabolism. *Biochem Biophys Res Commun* 287: 600–606

89 Gross G, Danzl M, Fischer W, Brand K (1988) Alterations of cellular lipids in rat thymocytes during cell cycle progression. *Biochim Biophys Acta* 962: 220–226

90 Jackowski S (1996) Cell cycle regulation of membrane phospholipid metabolism. *J Biol Chem* 271: 20219–20222

91 Jackowski S (1994) Coordination of membrane phospholipid synthesis with the cell cycle. *J Biol Chem* 269: 3858–3867

92 Terce F, Brun H, Vance DE (1994) Requirement of phosphatidylcholine for normal progression through the cell cycle in C3H/10T1/2 fibroblasts. *J Lipid Res* 35: 2130–2142

93 Bergeron JJ, Warmsley AM, Pasternak CA (1970) Phospholipid synthesis and degradation during the life-cycle of P815Y mast cells synchronized with excess of thymidine. *Biochem J* 119: 489–492

94 Northwood IC, Tong AHY, Crawford B, Drobnies AE, Cornell BB (1999) Shuttling of CTP: Phosphocholine cytidylyltransferase between the nucleus and endoplasmic reticulum accompanies the wave of phosphatidylcholine synthesis during the G(0) → G(1) transition. *J Biol Chem* 274: 26240–26248

95 Roshak AK, Capper EA, Stevenson C, Eichman C, Marshall LA (2000) Human calcium-independent phospholipase A_2 mediates lymphocyte proliferation. *J Biol Chem* 275: 35692–35698

96 Mitchell DC, Litman BJ (1998) Effect of cholesterol on molecular order and dynamics in highly polyunsaturated phospholipid bilayers. *Biophys J* 75: 896–908

97 Caiveau O, Fortune D, Cantrel C, Zachowski A, Moreau F (2001) Consequences of omega-6-oleate desaturase deficiency on lipid dynamics and functional properties of mitochondrial membranes of Arabidopsis thaliana. *J Biol Chem* 276: 5788–5794

98 Schaeffer BE, Zadunaisky JA (1979) Stimulation of chloride transport by fatty acids in corneal epithelium and relation to changes in membrane fluidity. *Biochim Biophys Acta* 556: 131–143

99 Galla HJ, Hartmann W, Theilen U, Sackmann E (1979) On two-dimensional passive random walk in lipid bilayers and fluid pathways in biomembranes. *J Membr Biol* 48: 215–236

100 Brown DA, London E (1998) Functions of lipid rafts in biological membranes. *Annu Rev Cell Dev Biol* 14: 111–136

101 Venable ME, Zimmerman GA, McIntyre TM, Prescott SM (1993) Platelet-activating factor: A phospholipid autacoid with diverse actions. *J Lipid Res* 34: 691–702

102 Naraba H, Imai Y, Kudo I, Nakagawa Y, Oh-ishi S (1995) Activation of phospholipase A_2 and acylation of lysophospholipids: The major regulators for platelet activating factor production in rat neutrophils. *J Biochem (Tokyo)* 118: 442–447

103 Reinhold SL, Zimmerman GA, Prescott SM, McIntyre TM (1989) Phospholipid remodeling in human neutrophils. Parallel activation of a deacylation/reacylation cycle and platelet-activating factor synthesis. *J Biol Chem* 264: 21652–21659

104 Nieto ML, Venable ME, Bauldry SA, Greene DG, Kennedy M, Bass DA, Wykle RL (1991) Evidence that hydrolysis of ethanolamine plasmalogens triggers synthesis of platelet-activating factor *via* a transacylation reaction. *J Biol Chem* 266: 18699–18706

105 Camussi G, Biancone L, Iorio EL, Silvestro L, Da Col R, Capasso C, Rossano F, Servillo L, Balestrieri C, Tufano MA (1992) Porins and lipopolysaccharide stimulate platelet activating factor synthesis by human mesangial cells. *Kidney Int* 42: 1309–1318

106 Marshall LA, Hall RH, Winkler JD, Badger A, Bolognese B, Roshak A, Flamberg PL, Sung CM, Chabot-Fletcher M, Adams JL et al (1995) SB 203347, an inhibitor of 14 kDa phospholipase A_2, alters human neutrophil arachidonic acid release and metabolism and prolongs survival in murine endotoxin shock. *J Pharmacol Exp Ther* 274: 1254–1262

107 Winkler JD, Bolognese BJ, Roshak AK, Sung CM, Marshall LA (1997) Evidence that 85 kDa phospholipase A_2 is not linked to CoA-independent transacylase-mediated production of platelet-activating factor in human monocytes. *Biochim Biophys Acta Lipids Lipid Metab* 1346: 173–184

108 Bernatchez PN, Winstead MV, Dennis EA, Sirois MG (2001) VEGF stimulation of endothelial cell PAF synthesis is mediated by group V 14 kDa secretory phospholipase A(2). *Br J Pharmacol* 134: 197–205

109 Marshall LA, Bolognese B, Roshak A (1999) Respective roles of the 14 kDa and 85 kDa phospholipase A_2 enzymes in human monocyte eicosanoid formation. *Adv Exp Med Biol* 469: 215–219

110 Shamsuddin M, Chen E, Anderson J, Smith LJ (1997) Regulation of leukotriene and

platelet-activating factor synthesis in human alveolar macrophages. *J Lab Clin Med* 130: 615–626

111 Al Darmaki S, Schenkein HA, Tew JG, Barbour SE (2003) Differential expression of platelet-activating factor acetylhydrolase in macrophages and monocyte-derived dendritic cells. *J Immunol* 170 (1): 167–173

112 Balboa MA, Balsinde J, Winstead MV, Tischfield JA, Dennis EA (1996) Novel group V phospholipase A_2 involved in arachidonic acid mobilization in murine P388D1 macrophages. *J Biol Chem* 271: 32381–32384

113 McHowat J, Kell PJ, O'Neill HB, Creer MH (2001) Endothelial cell PAF synthesis following thrombin stimulation utilizes Ca^{2+}-independent phospholipase $A(_2)$. *Biochemistry* 40: 14921–14931

114 Di Marzo V (1995) Arachidonic acid and eicosanoids as targets and effectors in second messenger interactions. *Prostaglandins Leukot Essent Fatty Acids* 53: 239–254

115 Versteeg HH, Henegouwen PM, Van Deventer SJ, Peppelenbosch MP (1999) Cyclo-oxygenase-dependent signalling: Molecular events and consequences. *FEBS Lett* 445: 1–5

116 Harris SG, Padilla J, Koumas L, Ray D, Phipps RP (2002) Prostaglandins as modulators of immunity. *Trends Immunol* 23: 144–150

117 Cavanaugh PF Jr, Meredith MP, Buchanan W, Doyle MJ, Reddy MS, Jeffcoat MK (1998) Coordinate production of PGE_2 and IL-1α in the gingival crevicular fluid of adults with periodontitis: Its relationship to alveolar bone loss and disruption by twice daily treatment with ketorolac tromethamine oral rinse. *J Periodont Res* 33: 75–82

118 Ishihara Y, Zhang JB, Quinn SM, Schenkein HA, Best AM, Barbour SE, Tew JG (2000) Regulation of immunoglobulin G2 production by prostaglandin E-2 and platelet-activating factor. *Infect Immun* 68: 1563–1568

119 Bloom D, Jabrane-Ferrat N, Zeng L, Wu A, Li L, Lo D, Turck CW, An S, Goetzl EJ (1999) Prostaglandin E_2 enhancement of interferon-gamma production by antigen-stimulated type 1 helper T cells. *Cell Immunol* 194: 21–27

120 Chaminade B, Le Balle F, Fourcade O, Nauze M, Delagebeaudeuf C, Gassama-Diagne A, Simon MF, Fauvel J, Chap H (1999) New developments in phospholipase $A(_2)$. *Lipids* 34: S49–S55

121 Dennis EA, Rhee SG, Billah MM, Hannun YA (1991) Role of phospholipase in generating lipid second messengers in signal transduction. *FASEB J* 5: 2068–2077

122 Atsumi G, Murakami M, Kojima K, Hadano A, Tajima M, Kudo I (2000) Distinct roles of two intracellular phospholipase $A_{(2)s}$ in fatty acid release in the cell death pathway – Proteolytic fragment of type IVA cytosolic phospholipase $A_{(2)}$alpha inhibits stimulus-induced arachidonate release, whereas that of type VICa^{2+}-independent phospholipase $A(_2)$ augments spontaneous fatty acid release. *J Biol Chem* 275: 18248–18258

123 Forsell PK, Runarsson G, Ibrahim M, Bjrkholm M, Claesson HE (1998) On the expression of cytosolic calcium-independent phospholipase A_2 (88 kDa) in immature and mature myeloid cells and its role in leukotriene synthesis in human granulocytes. *FEBS Lett* 434: 295–299

124 Akiba S, Mizunaga S, Kume K, Hayama M, Sato T (1999) Involvement of group

VICa^{2+}-independent phospholipase A($_2$) in protein kinase C-dependent arachidonic acid liberation in zymosan-stimulated macrophage-like P388D(1) cells. *J Biol Chem* 274: 19906–19912
125 Ramanadham S, Hsu FF, Bohrer A, Ma ZM, Turk J (1999) Studies of the role of group VI phospholipase A_2 in fatty acid incorporation, phospholipid remodeling, lysophosphatidylcholine generation, and secretagogue-inducecd arachidonic acid release in pancreatic islets and insulinoma cells. *J Biol Chem* 274: 13915–13927
126 Ma ZM, Ramanadham S, Wohltmann M, Bohrer A, Hsu FF, Turk J (2001) Studies of insulin secretory responses and of arachidonic acid incorporation into phospholipids of stably transfected insulinoma cells that over-express group VIA phospholipase A($_2$) (iPLA($_2$)beta) indicate a signaling rather than a housekeeping role for iPLA($_2$)beta. *J Biol Chem* 276: 13198–13208
127 Ma ZM, Bohrer A, Wohltmann M, Ramanadham S, Hsu FF, Turk J (2001) Studies of phospholipid metabolism, proliferation, and secretion of stably transfected insulinoma cells that overexpress group VIA phospholipase A($_2$). *Lipids* 36: 689–700
128 Dussault I and Forman BM (2000) Prostaglandins and fatty acids regulate transcriptional signaling *via* the peroxisome proliferator activated receptor nuclear receptors. *Prostaglandins Other Lipid Mediat* 62: 1–13
129 Escher P, Wahli W (2000) Peroxisome proliferator-activated receptors: Insight into multiple cellular functions. *Mutat Res* 448: 121–138
130 Corton JC, anderson SP, stauber A (2002) Central role of peroxisome proliferator-activated receptors in the actions of peroxisome proliferators. *Annu Rev Pharmacol Toxicol* 40: 491–518
131 Peters JM, Hennuyer N, Staels B, Fruchart JC, Fievet C, Gonzalez FJ, Auwerx J (1997) Alterations in lipoprotein metabolism in peroxisome proliferator-activated receptor alpha-deficient mice. *J Biol Chem* 272: 27307–27312
132 Aoyama T, Peters JM, Iritani N, Nakajima T, Furihata K, Hashimoto T, Gonzalez FJ (1998) Altered constitutive expression of fatty acid-metabolizing enzymes in mice lacking the peroxisome proliferator-activated receptor alpha (PPARalpha). *J Biol Chem* 273: 5678–5684
133 Lowell BB (1999) PPARgamma: an essential regulator of adipogenesis and modulator of fat cell function. *Cell* 99: 239–242
134 Wang CG, Fu MF, D'Amico M, Albanese C, Zhou JN, Brownlee M, Lisanti MP, Chatterjee VKK, Lazar MA, Pestell RG (2001) Inhibition of cellular proliferation through I kappa B kinase-independent and peroxisome proliferator-activated receptor gamma-dependent repression of cyclin D1. *Mol Cell Biol* 21: 3057–3070
135 Gorospe M, Liu Y, Xu Q, Chrest FJ, Holbrook NJ (1996) Inhibition of G1 cyclin-dependent kinase activity during growth arrest of human breast carcinoma cells by prostaglandin A_2. *Mol Cell Biol* 16: 762–770
136 Clay CE, Namen AM, Atsumi G, Willingham MC, High KP, Kute TE, Trimboli AJ, Fonteh AN, Dawson PA, Chilton FH (1999) Influence of J series prostaglandins on apoptosis and tumorigenesis of breast cancer cells. *Carcinogenesis* 20: 1905–1911

137 Bhuyan BK, Badiner GJ, Adams EG, Chase R (1986) Cytotoxicity of combinations of prostaglandin D_2 (PGD_2) and antitumor drugs for B16 melanoma cells in culture. *Invest New Drugs* 4: 315–323

138 Miwa Y, Sasaguri T, Inoue H, Taba Y, Ishida A, Abumiya T (2000) 15-Deoxy-delta 12,14-prostaglandin J_2 induces G1 arrest and differentiation marker expression in vascular smooth muscle cells. *Mol Pharmacol* 58: 837–844

139 Narumiya S, Fukushima M (1985) Delta 12-prostaglandin J_2, an ultimate metabolite of prostaglandin D2 exerting cell growth inhibition. *Biochem Biophys Res Commun* 127: 739–745

140 Todo S, Hashida T, Shimizu Y, Imashuku S, Takamatsu T, Fujita S (1986) Cell kinetic studies of PGD_2 cytotoxicity on the *in vitro* growth of human neuroblastoma. *Prostaglandins Leukot Med* 23: 55–65

141 Roshak AK, Capper EA, Stevenson C, Eichman C, Marshall LA (2000) Human calcium-independent phospholipase $A(_2)$ mediates lymphocyte proliferation. *J Biol Chem* 275: 35692–35698

Enzymatic and receptor mediated effects of secretory phospholipase A_2 on the pathophysiology of inflammatory diseases

Chad R. Marion[1] *and Alfred N. Fonteh*[2]

[1]Infectious Diseases, Department of Internal Medicine, Wake Forest University School of Medicine, Winston-Salem, North Carolina, USA; [2]Molecular Neurology Program, Huntington Medical Research Institutes, 99 N. El Molino Avenue, Pasadena, CA 91101, USA

Overview of phospholipases

Phospholipases A_2 are enzymes that share the common attribute to hydrolyze fatty acids from the *sn*-2 position of glycerol phospholipids [1–3]. Groups I, II, V and X PLA_2 are four sets of enzymes in a highly conserved family of secreted PLA_2 found in mammals [4–14]. Other non-secreted PLA_2 enzymes include group IV, cytosolic PLA_2 ($cPLA_2$) [15–17], group VI, calcium-independent PLA_2 ($iPLA_2$) [18–22] and groups VII and VIII, selective acetyl hydrolases [23–28]. The secretory family of enzymes has a number of features that distinguish them from other PLA_2 families including a relatively low molecular weight (~ 14 kDa), high disulfide bond content and a requirement for relatively high concentrations of calcium for maximal activation [29, 30]. In contrast, cytosolic enzymes are generally higher molecular weight proteins and require no calcium or very low calcium concentrations for optimal activation [18, 22]. Many $sPLA_2$ isotypes are synthesized as proenzymes that contain a signal peptide sequence that facilitates its release from cells. $sPLA_2$ isotypes have been studied extensively in mammals and in snake venoms, yet there is no clear understanding of their physiological and pathophysiological roles. Inspection of numerous publications dealing with $sPLA_2$s reveals that they have potential to mediate a wide range of biological activities including:

1) Producers of AA that contributes to eicosanoid formation [31–36];
2) Generation of lysophospholipids that contribute to electrophysiologic alteration that lead to arrythmogenesis in the heart or altered airway permeability and surfactant properties in the lung [37–48];
3) Potent antibacterial effects and implications in viral infections [49–54];
4) Key components in glycerophospholipid digestion [55];
5) Serum markers and potential regulators of severe illnesses such as sepsis, shock, organ injury and pancreatitis, all of which are linked to the development of adult respiratory distress syndrome or multiple organ failure [56–75];

Arachidonate Remodeling and Inflammation, edited by Alfred N. Fonteh and Robert L. Wykle

6) Regulators of platelet aggregation in hemorrhagic diseases [76–78];
7) Prevention of apoptosis of inflammatory cells and initiators of cell proliferation in several cancer cell lines [79–82];
8) A potent modifying locus in intestinal tumorigenesis in mice that is absent in human [83];
9) Pro-inflammatory components in diseases such as rheumatoid arthritis and asthma [84-93].

This overwhelming list of biological activities and diseases raises deep-seated questions as to whether $sPLA_2$ cause or is merely associated with the aforesaid effects. It also raises questions about molecular mechanisms that this family of enzymes could influence in order to control such a wide range of biological activities.

Role of $sPLA_2$ on cell function and animal physiology

With the milieu of so many potential biological activities, inhibitors, antibodies, antisense oligonucleotides and genetic models have been used to better define the essential processes induced by $sPLA_2$ secretion into sites of inflammation [94–105]. While some inhibitors, antibodies and antisense oligonucleotides have been developed that block $sPLA_2$ activity and inflammatory processes, the lack of selectivity among these reagents against different $sPLA_2$ isotypes make data interpretation ambiguous. Various genetic models have been discovered or developed in order to address the complex issue of the role $sPLA_2$ isotypes may play in diseases [106–111]. For example, peritoneal macrophages from mice with targeted gene disruption of group IV $cPLA_2$ show a marked reduction in their capacity to synthesize leukotriene B_4 (LTB_4), leukotriene C_4 (LTC_4), prostaglandin E_2 (PGE_2) and platelet-activating factor (PAF) [111]. These animals have attenuation in pulmonary responses and hyper-responsiveness after allergen challenge. In terms of $sPLA_2$, Nevalainen and colleagues designed experiments where transgenic mice expressed more than eighty fold more group II PLA_2 in most tissues including liver, lung, kidney and skin than non-transgenic littermates. Histopathological analysis of these animals revealed a disorder in skin consisting of hyperkeratosis, epidermal and adrenectral hypoplasia [112]. Chronic hypoplasia and hyperkeratosis observed in these animals is similar to that seen in a variety of skin disorders including human psoriasis. Certain mouse strains (C57BL/6, 129, A/J, C58 and P/J) have been shown to have a natural disruption of group IIA PLA_2 gene [106]. Thus, these strains are deficient in functional group IIA PLA_2 and have been used to determine the need for this enzyme in cell function. Interestingly, mast cells obtained from $PLA_2g2a^{+/+}$ and $PLA_2g2a^{-/-}$ mice both contained $sPLA_2$ activity and release similar quantities of AA upon antigen stimulation. Studies using these animals and antisense oligonucleotides reveal that group V $sPLA_2$, and not IIA, is likely an important $sPLA_2$ isotype in mast cell immune activation [113]. Similar studies using antisense oligonucleotide specific for group V PLA_2 in macrophages also demonstrated that group V PLA_2 has an important role in extracellular AA release after endotoxin and PAF stimulation [114]. Fonteh and colleagues have also shown that cells over-expressing group IIA PLA_2 and group V PLA_2 release more AA that mock-transfected cells [115]. Together, these studies show that various $sPLA_2$ isotypes can induce AA release from a variety of inflammatory cells.

Cytokine-like effects of $sPLA_2$ in inflammatory diseases

$sPLA_2$ may contribute to the pathogenesis of an inflammatory disease such as asthma in one of the following ways. First, $sPLA_2$ may induce lipid mediator formation. Second, $sPLA_2$ may induce degranulation of inflammatory cells leading to the release of preformed mediators such as histamine. Third, $sPLA_2$ may induce the synthesis of inflammatory cytokines. Our data show that $sPLA_2$ can also induce the formation of cytokines that prevent mast cells from undergoing apoptosis, thus preventing the resolution of inflammation. These potential effects of $sPLA_2$ summarized in Figure 1 and discussed in detail below can have significant ramifications in the management of inflammatory diseases.

Eicosanoid biosynthesis

Although the existence of $sPLA_2$ receptors has been recognized for several years, few studies have focused on the significance of receptor occupancy or the signaling mechanisms associated with receptor occupancy and how these events manifests themselves in inflammatory disease processes. Early work on $sPLA_2$ receptors focused on the neurotoxic effects of snake venom $sPLA_2$ acting through high affinity for $sPLA_2$ receptors. It has been speculated that somewhere along the evolution of $sPLA_2$, the mannose receptor may have been duplicated to accommodate other forms of and other functions of $sPLA_2$. These $sPLA_2$ receptors and their agonists ($sPLA_2$) regulate events in inflammatory cells that are critical to the pathogenesis of diseases such as asthma. Thus, $sPLA_2$ isotypes can act through their receptors in both autocrine and cytokine-like fashion. For example, mast cells contain and release group V PLA_2 during antigen activation and our data reveal that mast cells contain plasma membrane receptors and respond to receptor occupancy by $sPLA_2$ [116, 117]. A $sPLA_2$ receptor pathway could clarify several studies in the literature. For example, Arm and colleagues have shown that there are two phases of prostaglandin production in cultured mast cells that are primed with *c-kit* ligand and stimulated with antigen [118, 119]. AA is supplied for the first phase of prostanoids synthesis by $cPLA_2$ activation and AA for the second phase is provided by $sPLA_2$. It has

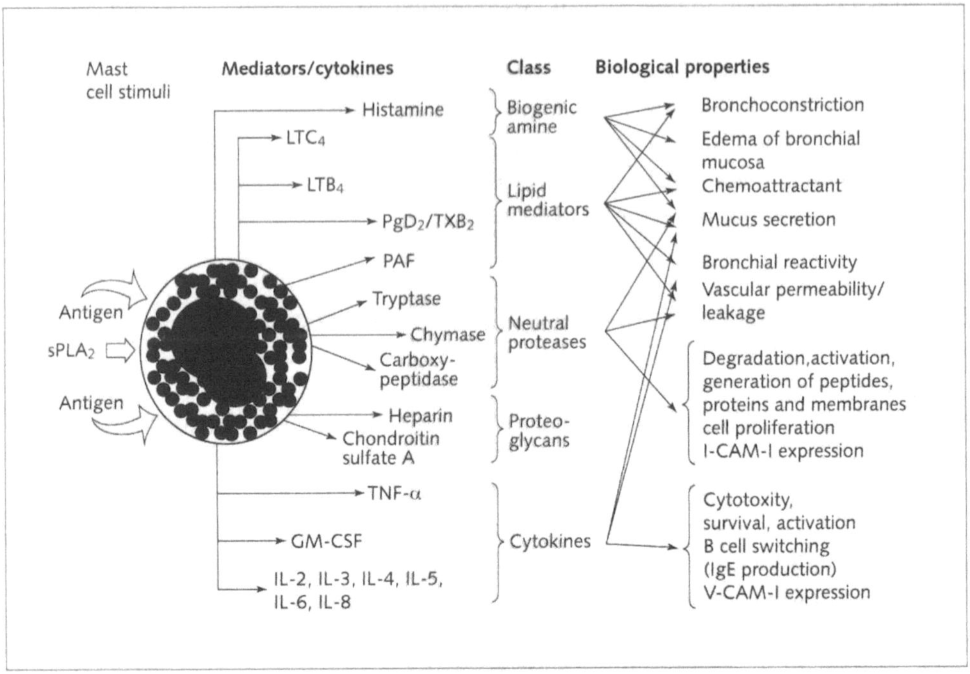

Figure 1
Stimulation of mast cells by sPLA$_2$, formation of mediators and their biological properties sPLA$_2$ receptor occupancy results in the activation of mast cells. Mediators released include preformed biogenic amines (histamine), newly formed lipid mediators of inflammation (leukotrienes (LTC$_4$, LTB$_4$), prostaglandins (PGD$_2$), thromboxane (TXB$_2$) and platelet activating factor (PAF)), neutral proteases, proteoglycans and cytokines. These mediators induce various biological functions that are linked to disease processes.

been proposed that sPLA$_2$ hydrolyzes AA from membrane phospholipids to supply the substrate for PGD$_2$ biosynthesis. Fonteh and colleagues have provided evidence that sPLA$_2$ binds to cell surface receptors on mast cells that may initiate subsequent activation of cPLA$_2$, cyclo-oxygenase and lipoxygenase enzymes needed for AA release, prostaglandin formation and leukotriene biosynthesis, respectively [115, 116]. sPLA$_2$ also could have cytokine-like roles in that it could be released from one cell type and subsequently act on a variety of other cells types expressing sPLA$_2$ receptors. For example, it is well documented that high levels of sPLA$_2$ are found in serum of patients with sepsis, shock, organ injury or pancreatitis. Tools for studying these diseases suggest that group IIA sPLA$_2$ is involved. However, many of these studies used antibodies that are non-specific recognizing group IIA, V and X PLA$_2$. Thus, it is not clear which of the various sPLA$_2$ isotypes play a role in these diseases.

We have postulated that $sPLA_2$ isotypes released from mast cells after antigen challenge are found in sites of inflammation and these $sPLA_2$ isotypes induce lipid mediator formation or influence the recruitment and function of cells that participate in airway diseases. A study by Reddy and colleagues showing that mast cells can provide $sPLA_2$ to fibroblasts for prostaglandin production supports this postulation [120]. Cells other than mast cells can also provide $sPLA_2$ in airways. For example, using an enzyme-linked immunoassay specific for groups IIA and V PLA_2, has shown that human eosinophils have approximately 14-fold more of the enzymes than human neutrophils and this activity is released very rapidly upon cell activation [121]. In a related study, Hundley and colleagues have shown that $sPLA_2$ is released from human basophils and likely participates in leukotriene generation [122]. Thus, there is also potential for $sPLA_2$ from eosinophils or basophils to have both autocrine and cytokine effects in airway disease by inducing the formation of eicosanoids.

Degranulation of inflammatory cells

In addition to eicosanoid production, $sPLA_2$ has been shown to induce degranulation of several cells. Fonteh and colleagues showed that incubation of mast cells with different $sPLA_2$ isotypes resulted in the release of histamine. Likewise Triggiani and colleagues have duplicated these mast cell studies using macrophages, monocytes and eosinophils [123–125]. Their studies show that $sPLA_2$ induce the release of β-glucuronidase and the production of IL-6, IL-8, IL-12 and TNF-α by these cells. They conclude that this process is mediated *via* the mannose receptor and another receptor based on experiments showing that $sPLA_2$ isotypes are not cytotoxic to macrophages that were used in their studies. Together, these studies suggest that $sPLA_2$ may induce immune and inflammatory responses in cells by inducing exocytosis resulting in the release of mediators such as histamine and cytokines. Mast cells and macrophages also contain other proteases, which are released by activated inflammatory cells [126–129]. Induction of the release of these can result in bronchoconstriction, edema of the bronchial mucosa, chemoattraction, mucus secretion, vascular permeability and leakage, cell proliferation and expression of adhesion molecules (Fig. 1). Together, these biological properties constitute events related to inflammatory diseases of the airway.

Induction of cytokine formation

An important observation in mast cells, macrophages, monocytes and eosinophils incubated with $sPLA_2$ is that these cells while releasing AA still remain viable, indicating that these enzymes are not cytotoxic to these cells. This paradoxical effect can be explained by the fact that $sPLA_2$ acts on receptors to induce processes that prevent apoptosis. For example, $sPLA_2$ acting on $sPLA_2$ receptors may activate mitogen-activated protein kinase (MAP kinase) and this process may result in the biosynthesis of cytokines, which prevent apoptosis. We confirmed this process by showing that incubation of bone marrow-derived mast cells with low concentrations of $sPLA_2$ resulted in the induction of IL-3 release into supernatant fluids. Only $sPLA_2$ isotypes that induced IL-3 production prevent apoptosis of mast cells [130]. A similar study showed that some $sPLA_2$ isotypes induce IL-6, IL-8, IL-12 and TNF-α formation by other inflammatory cells [123–125]. In addition to preventing apoptosis, cytokines induced by $sPLA_2$ also induce several biological effects including bronchial reactivity, chemoattraction, and activation of other inflammatory cells and the induction of adhesion molecules. Prevention of apoptosis of inflammatory cells such as mast cells and macrophages keep these cells longer in the site of inflammation and thus prevent quick resolution of the inflammatory process.

Mechanisms that account for the biologic functions of $sPLA_2$

$sPLA_2$ receptors

To date, most of the biological activities of $sPLA_2$ have been attributed to its capacity to hydrolyze membrane phospholipids. However, several of the biological functions described above cannot be easily reconciled with enzymatic activity alone. For example, intradermal injection of inactivated $sPLA_2$ causes similar phenotypic changes in skin to those observed when the fully active enzyme was injected [131]. Similarly, others have shown that the physiologic action of $sPLA_2$ is not due to hydrolytic activity [125, 132]. We have demonstrated that very low concentrations of $sPLA_2$ (low nanomolar levels) of certain $sPLA_2$ isotypes induce AA release, histamine release, and proliferation of some cells and enhance the survival of other cells in a receptor-mediated fashion [82]. Our studies also show that $sPLA_2$'s cause the selective release of AA and not other more abundant fatty acids from cells that express $sPLA_2$ receptors [115]. In contrast, cells that do not express $sPLA_2$ receptors do not selectively release AA when incubated with low amounts of $sPLA_2$.

Recently, different subtypes of membrane receptors for $sPLA_2$ have been identified in a variety of cells by determining their affinities for various types of $sPLA_2$. Arita and colleagues described the existence of a specific receptor family termed PLA_2-I receptor that is abundant in brain and several other tissues and has high affinity for the binding of pancreatic-type PLA_2 [133–136]. More recently, receptors have been divided into two classes termed N-type receptors (neuronal) or M-type receptor (muscle). Lambeau and colleagues report that a major difference between N-type and M-type receptors is their capacity to bind group III PLA_2 from bee venom [137, 138]. N-type receptor associates very tightly with both pancreatic

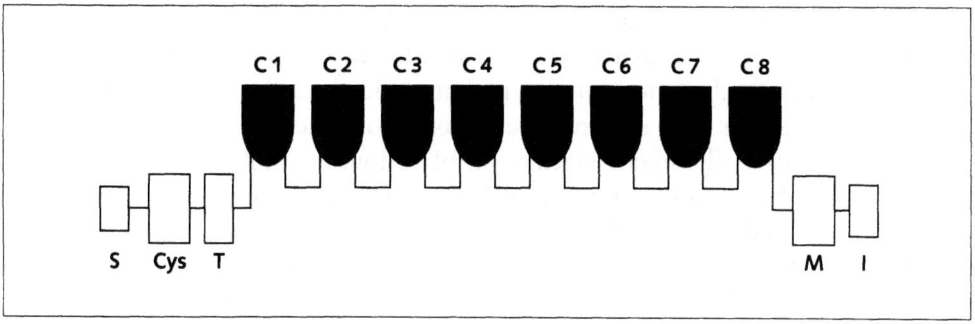

Figure 2
Domain organization of the $sPLA_2$ receptor
S, signal sequence; Cys, cysteine-rich domain; T, fibronectin type repeat; C1–C8, carbohydrate-like domain; M, membrane spanning domain; I, intracellular region

$sPLA_2$ and bee venom group III PLA_2 while rabbit muscle M-type receptor tightly binds human synovial fluid group II PLA_2 but does not bind bee venom $sPLA_2$. Our studies show that bee venom $sPLA_2$ at extremely low concentrations induced the selective release of AA from mast cells suggesting that this mast cell line expresses a protein that is similar if not identical to N-type receptors. Both membrane and plasma bound $sPLA_2$ receptors have been described [139]. The plasma bound $sPLA_2$ receptors seem to prevent the LPS-induced inflammatory process by binding $sPLA_2$ and thus preventing septic shock during bacterial infections [140].

Despite the fact that these receptor subtypes show somewhat different $sPLA_2$ binding profiles, their amino acid sequences are strikingly similar with as much as 82% homology. Additionally, the sequences of the cloned $sPLA_2$ receptor is homologous to that of the macrophage mannose receptor and DEC-205, suggesting that these proteins constitute a new family of membrane proteins [137]. The $sPLA_2$ receptor is composed of an N-terminal cysteine rich domain (Cys) a fibronectin-like type II domain (T), eight carbohydrate recognition domains (C1–C8), a membrane spanning domain (M) and an intracellular tail (I) (Fig. 2). $sPLA_2$ is thought to bind to the M-type receptor *via* the carbohydrate domains (particular C5) [137]. Although occupancy of the $sPLA_2$ receptor has been suggested to enhance cell survival, proliferation, cell migration, much remains to be learned about the molecular events and physiological ramifications of $sPLA_2$ receptor activation.

Hydrolytic activity of $sPLA_2$

In addition to receptor binding, the hydrolytic activity of PLA_2 may also play an important role in releasing fatty acids from cells. There are several distinct features that distinguish receptor-mediated effects of $sPLA_2$ from enzymatic activity. First, whereas $sPLA_2$ receptor-mediated release is specific for AA [35, 115, 116], hydrolytic activity releases other more abundant fatty acids and degrades phospholipids [141]. Various reports have shown more release of oleic acid than AA in cells where hydrolytic activity is the major mechanism of action. Secondly, very low levels of $sPLA_2$ (nanomolar amounts) are required for receptor-mediated release of AA. In contrast, 1,000 fold more $sPLA_2$ is needed to release fatty acids by hydrolytic activity. Thirdly, disruption of cell membrane is not required for receptor-mediated AA release, while perturbation of cellular membranes is needed for hydrolytic activity [79, 130, 142]. Disruption of cell membranes or loss of membranes phospholipid asymmetry is usually accomplished using cell-activating agents such as ionophore and thrombin in the case of platelets or antigen in the case of mast cells [143, 144]. Additionally, there is alteration in membrane asymmetry when cells are undergoing apoptosis. A combination of disruptive agents and $sPLA_2$ treatment, or treatment of apoptotic cells, usually results in enhanced fatty acid mobilization. Fourthly, $sPLA_2$ receptor-mediated AA release is predominantly from the phosphatidylethanolamine pool whereas hydrolytic release favors phosphatidylcholine [35, 145, 146]. As shown in Figure 3, very low amounts of $sPLA_2$ (0.1 nM) induce the selective formation of lysophosphatidylethanolamine from [^{3}H]-ethanolamine-labeled mast cells while higher concentrations (100 nM) are required to significantly form lysophosphatidylcholine from [^{3}H]-choline labeled mast cells. It is important to note that the receptor-mediated release from a pool of phospholipid that is usually found within the inner bilayer of cell membranes will only be possible if there is recruitment of another lipase activity within cells. Importantly, it is worth noting that the profile of AA release in mast cells incubated with $sPLA_2$ (receptor-mediated) is similar to that of IgE-receptor mediated release of AA. Finally, $sPLA_2$-receptor mediated processes may lead to enhanced cell survival or cell proliferation while hydrolysis inevitably results in cell death as a result of lysis of cell membranes. Similarities between receptor-mediated and hydrolytic activity revolve around the fact that both processes release AA from phospholipid pools and also form lysophospholipids. These lysophospholipids can be acetylated to form PAF or can act as mediators of several processes in cells [147–149]. Thus, AA and lysophospholipids released by hydrolytic action of $sPLA_2$ can be converted to eicosanoid or PAF, respectively and these lipid mediators can then induce several biological effects at sites of inflammation.

$sPLA_2$ receptor-mediated signaling pathways

Although several $sPLA_2$ binding proteins have been described in various mammalian cells, little is currently known about signaling events that are initiated once $sPLA_2$ isotypes or mannose receptors are occupied by ligands. Although little is known about binding of group I PLA_2 to N-type receptors, it is clear that calcium is not

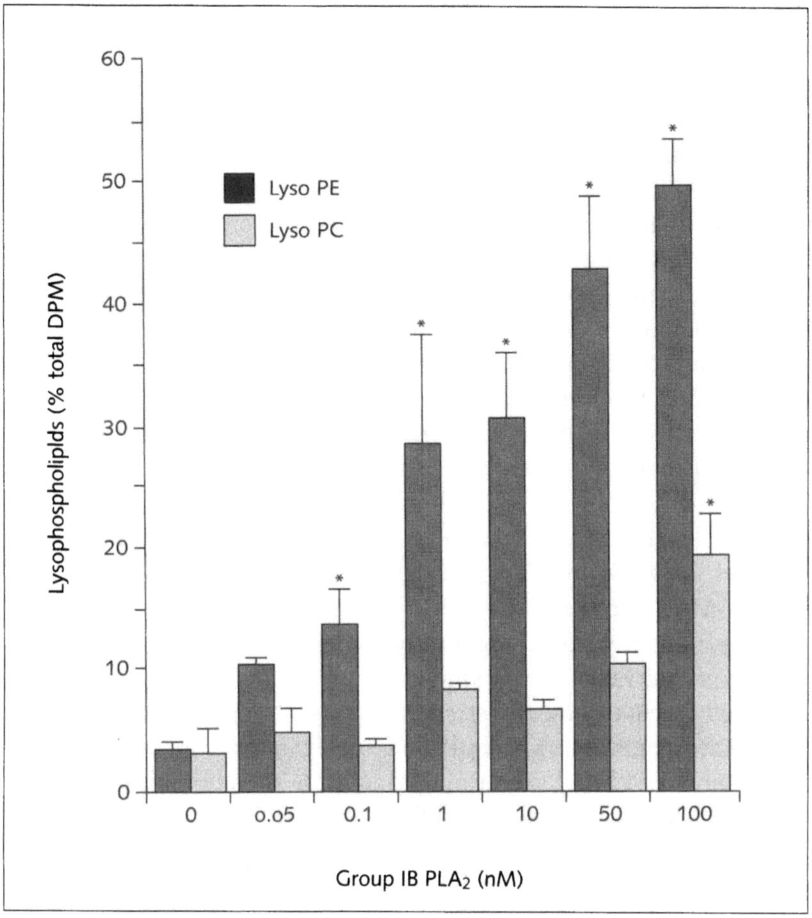

Figure 3
Mobilization of lysophospholipids by group IB PLA_2
Mast cells labeled with [^{3}H]-ethanolamine or [^{3}H]-choline are incubated with increasing concentrations of group IB $sPLA_2$. Lysophosphatidylethanolamine (Lyso PE) and lysophosphatidylcholine (Lyso PC) were isolated using thin layer chromatography and radioactivity in these lipid species determined using lipid scintillation counting ($p < 0.05$).*

required, while calcium is required for the mannose receptor. Mannose receptor occupancy is linked with tyrosine phosphorylation while $sPLA_2$ receptors have been recently shown to mediate cell proliferation and AA release by MAP kinase activation [116, 150]. Demonstration of $sPLA_2$ receptor function as opposed to catalytic action is based on studies using catalytically inactive $sPLA_2$ or ligands of $sPLA_2$ receptors that are devoid of hydrolytic activity. In these studies, catalytically inactive $sPLA_2$ induced the selective release of AA from mast cells and prevented apoptosis of these same cells when they were cultured in cytokine-depleted cell culture medium. Likewise, a ligand of the mannose receptor, *p*-amino-phenyl-D-mannopyranoside BSA (APDM-BSA) is shown to induce AA release from mast cells and to compete with $sPLA_2$ in this process. Binding studies also show that only $sPLA_2$ isotypes that selectively induce AA release can compete with each other for specific binding. Interestingly, APDM-BSA does not prevent mast cells from undergoing apoptosis. This suggests that there are at least two $sPLA_2$ receptor subtypes in mast cells, one that is linked to AA release (binds APDM-BSA) and another that prevents apoptosis. Alternatively, there are multiple signaling pathways in mast cells activated differentially likely due to difference in receptor affinity of the different ligands. These signaling events are reviewed below.

In receptor-mediated AA·release, another lipase activity must be recruited if $sPLA_2$ activity is not required. As described above, AA is mobilized from phospholipid pools (mainly PE) that are normally found within cells. Therefore, an ideal PLA_2 that can release this AA pool is the hormonally regulated cytosolic PLA_2 ($cPLA_2$). $cPLA_2$ is translocated to a membrane location in response to an increase in cytosolic calcium and activated by phosphorylation of serine 505 or other phosphorylation sites by MAP kinases [151–154]. The extracellular signal-regulated kinase (ERKs, p42/p44) initially was thought to be the major kinases responsible for $cPLA_2$ phosphorylation. However, recent studies suggest that p38 kinase pathway, which can be activated by environmental stresses and inflammatory cytokines, may also phosphorylate/activate $cPLA_2$. In stimulated platelets, inhibitors of p38 kinase have been shown to prevent $cPLA_2$ activation while these same inhibitors indicate that ERKs and not p38 kinase may activate $cPLA_2$ in other cell types [155, 156]. Wykle and colleagues have shown in human neutrophils that both ERKs and p38 kinases are important in $cPLA_2$ activation depending on the stimuli used. Fonteh and colleagues have also shown that tyrosine kinase inhibitors attenuate $sPLA_2$-induced $cPLA_2$ and Ras activation [116]. Since Ras activation is upstream of MAP kinase activation, we have proposed the signaling pathway depicted in Figure 4A for $cPLA_2$ recruitment and AA release after $sPLA_2$ receptor occupancy. We have proposed that the $sPLA_2$ receptor is similar to other protein tyrosine kinase (PTK) receptors that may have an intrinsic kinase activity or may be able to recruit kinases from cytosol upon ligand binding. Once tyrosine kinases are activated, a sequence of events including Ras, ERKs or p38 activation lead to the phosphorylation and translocation of $cPLA_2$ from cytosol to membranes [116]. This results in the mobilization of AA that is utilized for eicosanoid formation (Fig. 4A). It is likely that the c-Jun pathway may also be linked to $cPLA_2$ activation. As the tools become available for studying and discriminating between the various signaling pathways, it will become clearer whether c-Jun kinases phosphorylate $cPLA_2$ and induce AA release from cells.

The high affinity IgE receptor is a well-characterized membrane bound protein that belongs to the multi-chain system of receptors involved in hypersensitivity reac-

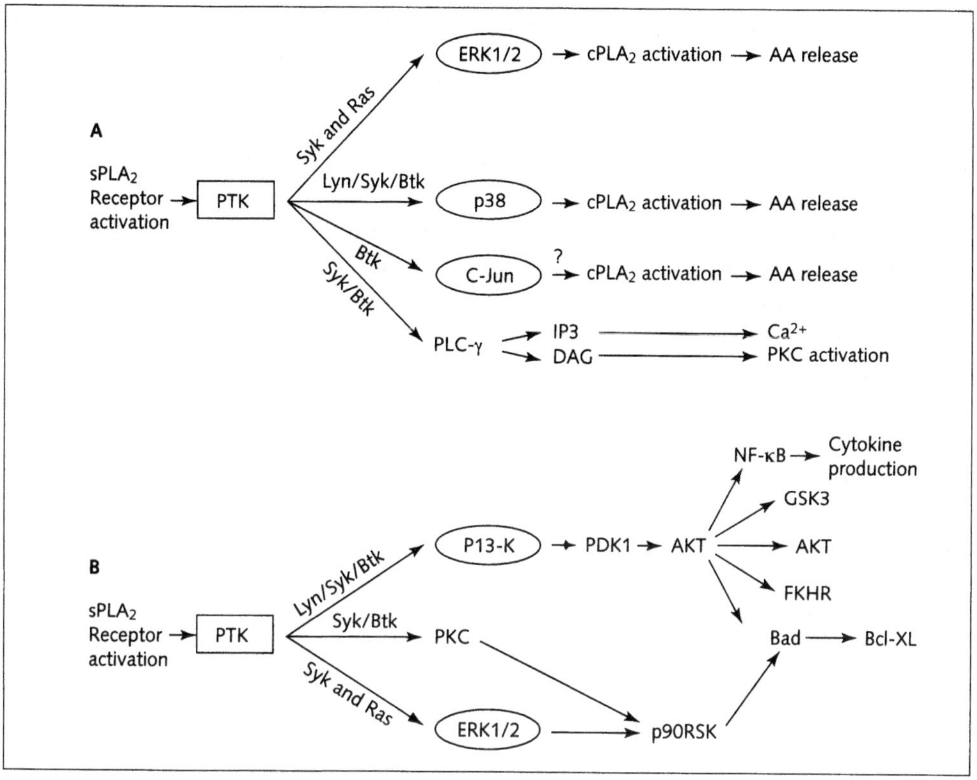

Figure 4
$sPLA_2$-receptor mediated signaling pathways
A) Cytosolic PLA_2 activation. Binding of $sPLA_2$ to its receptors results in the activation of kinase pathways (protein tyrosine kinase (PTK), tyrosine kinases (Syk, Syk and Btk), MAP kinases (ERK1/2, p38, c-Jun)). These kinases activate cytosolic phospholipase A_2 ($cPLA_2$) via phosphorylation of various amino acid residues. Activated $cPLA_2$ translocates to membrane fraction of cells and releases AA that is used for eicosanoid biosynthesis. Other second messengers (DAG and Ca^{2+}) are also initiated via PLCγ.
B) Anti-apoptotic signaling pathways. $sPLA_2$ binds to its receptors and activates kinase pathways (phosphoinositide 3-kinase (PI3-K), phosphatidylinositol dependent kinase (PDK, glycogen synthase kinase (GSK)), transcription factors (nuclear factor kappa B (NF-κB) and anti-apoptotic proteins such as Bad. Activation of these pathways prevents apoptosis of cells and/or induces cell growth and differentiation.

tions [157]. These receptors lack intrinsic tyrosine kinase activity and so have to recruit cytoplasmic tyrosine kinase that phosphorylates the receptor at sites known as immunoreceptor tyrosine-based activation motifs (ITAMs). Phosphorylation of ITAMs results in protein tyrosine kinase activation. Three major PTKs have been described in mast cells including Lyn, Syk and Burton's tyrosine kinase (Btk) that are upstream of the three subfamilies of MAK kinases (ERKs, p38 and c-Jun NH_2 terminal kinases) [158–160]. Activation of PTKs and their respective downstream MAP kinases result in $cPLA_2$ activation and the formation of pro-inflammatory mediators described above. Another group of mast cell receptors characterized by *kit*, have intrinsic tyrosine kinase activity and are involved in mast cell survival and proliferation [160, 161]. Similar to *kit*, the high affinity nerve growth factor (NGF, 165 kDa) receptor autophosphorylates tyrosine residues to activate multiple downstream effectors including PLCγ, MAP kinases and phosphoinositide-3-kinase (PI3-K/Akt). Of the many signaling pathways influenced by *kit* and NGF, the PI3-K/Akt pathway has been implicated in mast cell survival and growth. PI3-K is a dual kinase consisting of an 85 kDa regulatory unit and a 110 kDa catalytic unit. PI3-K adds a phosphate molecule specifically to the 3 position of the inositol ring of phosphatidylinositols resulting in the formation of products that have been implicated in survival, proliferation or cell migration. There are striking similarities between *kit*, NGF and the $sPLA_2$ receptor when one examines mast cell survival. First, the cloned $sPLA_2$ receptor (180 kDa) has one membrane-spanning domain, as does the NGF receptor. Secondly, NGF prevents apoptosis of mast cells, as does $sPLA_2$ isotypes that bind specifically to $sPLA_2$ receptors. Thirdly, NGF activates PI3-K/Akt pathway. We have shown that PI3-K specific inhibitors reverse the anti-apoptotic effects observed in mast cells incubated with very low levels of $sPLA_2$. Moreover, $sPLA_2$ also induce Akt phosphorylation in mast cells while inhibitors of nuclear factor kappa B (NF-κB) are shown to prevent $sPLA_2$ effects on mast cells. Both active and catalytically inactive $sPLA_2$ induce IL-3 production from mast cells and NF-κB inhibitors reverse this property of $sPLA_2$ [130]. Taken together, these data show that $sPLA_2$ produce IL-3 by activating the PTK/PI3-K/Akt/NF-κB pathway. Similarly antigen-stimulated mast cells have been shown to produce cytokines *via* PTK/PI3-K/Akt/NF-κB activation.

Conclusions

The first step in designing new pharmaceutical agents is the identification of a candidate target. Elucidation of all biological properties of the identified target plays a crucial role in conceptualizing new strategies for selectively blocking disease-related events.

The above review examines $sPLA_2$ from mast cells as a candidate ligand and $sPLA_2$ receptor as a potential target responsible for important biological functions linking $sPLA_2$ to inflammatory diseases. To our knowledge, the concept that $sPLA_2$ receptors play an important role in inflammatory diseases is novel and potentially interesting to pursue in the development of agents to avert allergic and inflammatory reactions associated with these diseases.

Acknowledgements
We are grateful for expert technical assistance by Tiffany Mendahall, Michelle Edens, Brooke Barham and Dennis Swan.

References

1 Dennis EA (1997) The growing phospholipase A_2 superfamily of signal transduction enzymes. *Trends Biochem Sci* 22: 1–2
2 Murakami M, Nakatani Y, Atsumi G, Inoue K, Kudo I (1997) Regulatory functions of phospholipase A_2. *Crit Rev Immunol* 17: 225–283
3 Kramer RM (1993) Structure, function and regulation of mammalian phospholipases A_2. *Adv Second Messenger Phosphoprotein Res* 28: 81–89
4 Kramer RM, Hession C, Johansen B, Hayes G, McGray P, Chow EP, Tizard R, Pepinsky RB (1989) Structure and properties of a human non-pancreatic phospholipase A_2. *J Biol Chem* 264: 5768–5775
5 Cupillard L, Koumanov K, Mattei MG, Lazdunski M, Lambeau G (1997) Cloning, chromosomal mapping, and expression of a novel human secretory phospholipase A_2. *J Biol Chem* 272: 15745–15752
6 Balboa MA, Balsinde J, Winstead MV, Tischfield JA, Dennis EA (1996) Novel group V phospholipase A_2 involved in arachidonic acid mobilization in murine P388D1 macrophages. *J Biol Chem* 271: 32381–32384
7 Cho W, Han SK, Lee BI, Snitko Y, Dua R (1999) Purification and assay of mammalian group I and group IIa secretory phospholipase A_2. *Methods Mol Biol* 109: 31–38
8 Han SK, Lee BI, Cho W (1997) Bacterial expression and characterization of human pancreatic phospholipase A_2. *Biochim Biophys Acta* 1346: 185–192
9 Chen J, Shao C, Lazar V, Srivastava CH, Lee WH, Tischfield JA (1997) Localization of group IIc low molecular weight phospholipase A_2 mRNA to meiotic cells in the mouse. *J Cell Biochem* 64: 369–375
10 Tischfield JA (1997) A reassessment of the low molecular weight phospholipase A_2 gene family in mammals. *J Biol Chem* 272: 17247–17250
11 Seeds MC, Jones DF, Chilton FH, Bass DA (1998) Secretory and cytosolic phospholipases A_2 are activated during TNF priming of human neutrophils. *Biochim Biophys Acta* 1389: 273–284
12 Shimbara S, Murakami M, Kambe T, Kudo I (1999) Comparison of recombinant types IIA, V and IIC phospholipase A_2S, the three related mammalian secretory phospholipase A_2 isozymes [In Process Citation]. *Adv Exp Med Biol* 469: 209–214
13 Gelb MH, Valentin E, Ghomashchi F, Lazdunski M, Lambeau G (2000) Cloning and recombinant expression of a structurally novel human secreted phospholipase A_2. *J Biol Chem* 275: 39823–39826
14 Tischfield JA (1997) A reassessment of the low molecular weight phospholipase A_2 gene family in mammals. *J Biol Chem* 272: 17247–17250

15 Kramer RM, Roberts EF, Hyslop PA, Utterback BG, Hui KY, Jakubowski JA (1995) Differential activation of cytosolic phospholipase A_2 ($cPLA_2$) by thrombin and thrombin receptor agonist peptide in human platelets. Evidence for activation of $cPLA_2$ independent of the mitogen-activated protein kinases ERK1/2. *J Biol Chem* 270: 14816–14823
16 Sharp JD, White DL, Chiou XG, Goodson T, Gamboa GC, McClure D, Burgett S, Hoskins J, Skatrud PL, Sportsman JR (1991) Molecular cloning and expression of human Ca(2+)-sensitive cytosolic phospholipase A_2. *J Biol Chem* 266: 14850–14853
17 Pickard RT, Strifler BA, Kramer RM, Sharp JD (1999) Molecular cloning of two new human paralogs of 85-kDa cytosolic phospholipase A_2. *J Biol Chem* 274: 8823–8831
18 Ackermann EJ, Dennis EA (1995) Mammalian calcium-independent phospholipase A_2. *Biochim Biophys Acta* 1259: 125–136
19 Gross RW (1998) Activation of calcium-independent phospholipase A_2 by depletion of internal calcium stores. *Biochem Soc Trans* 26: 345–349
20 Hazen SL, Stuppy RJ, Gross RW (1990) Purification and characterization of canine myocardial cytosolic phospholipase A_2. A calcium-independent phospholipase with absolute f1-2 regiospecificity for diradyl glycerophospholipids. *J Biol Chem* 265: 10622–10630
21 Portilla D, Crew MD, Grant D, Serrero G, Bates LM, Dai G, Sasner M, Cheng J, Buonanno A (1998) cDNA cloning and expression of a novel family of enzymes with calcium-independent phospholipase A_2 and lysophospholipase activities. *J Am Soc Nephrol* 9: 1178–1186
22 Underwood KW, Song C, Kriz RW, Chang XJ, Knopf JL, Lin LL (1998) A novel calcium-independent phospholipase A_2, $cPLA_2$-gamma, that is prenylated and contains homology to $cPLA_2$. *J Biol Chem* 273: 21926–21932
23 Endo S, Inada K, Yamashita H, Takakuwa T, Nakae H, Kasai T, Kikuchi M, Ogawa M, Uchida K, Yoshida M (1994) Platelet-activating factor (PAF) acetylhydrolase activity, type II phospholipase A_2, and cytokine levels in patients with sepsis. *Res Commun Chem Pathol Pharmacol* 83: 289–295
24 MacPhee CH, Moores KE, Boyd HF, Dhanak D, Ife RJ, Leach CA, Leake DS, Milliner KJ, Patterson RA, Suckling KE et al (1999) Lipoprotein-associated phospholipase A_2, platelet-activating factor acetylhydrolase, generates two bioactive products during the oxidation of low-density lipoprotein: Use of a novel inhibitor. *Biochem J* 338 (Pt 2): 479–487
25 Nijssen JG, Roosenboom CF, van den? BH (1986) Identification of a calcium-independent phospholipase A_2 in rat lung cytosol and differentiation from acetylhydrolase for 1-alkyl-2-acetyl-*sn*-glycero-3-phosphocholine (PAF-acether). *Biochim Biophys Acta* 876: 611–618
26 Soubeyrand S, Lazure C, Manjunath P (1998) Phospholipase A_2 from bovine seminal plasma is a platelet-activating factor acetylhydrolase. *Biochem J* 329 (Pt 1): 41–47
27 Tjoelker LW, Eberhardt C, Unger J, Trong HL, Zimmerman GA, McIntyre TM, Stafforini DM, Prescott SM, Gray PW (1995) Plasma platelet-activating factor acetylhydrolase is a secreted phospholipase A_2 with a catalytic triad. *J Biol Chem* 270: 25481–25487

28 Touqui L, Herpin-Richard N, Gene RM, Jullian E, Aljabi D, Hamberger C, Vargaftig BB, Dessange JF (1994) Excretion of platelet activating factor-acetylhydrolase and phospholipase A_2 into nasal fluids after allergenic challenge: Possible role in the regulation of platelet activating factor release. *J Allergy Clin Immunol* 94: 109–119
29 Fonteh AN, Bass DA, Marshall LA, Seeds M, Samet JM, Chilton FH (1994) Evidence that secretory phospholipase A_2 plays a role in arachidonic acid release and eicosanoid biosynthesis by mast cells. *J Immunol* 152: 5438–5446
30 Scott DL, Sigler PB (1994) Structure and catalytic mechanism of secretory phospholipases A_2. *Adv Protein Chem* 45: 53–88
31 Bomalaski JS, Baker DG, Brophy L, Resurreccion NV, Spilberg I, Muniain M, Clark MA (1989) A phospholipase A_2-activating protein (PLAP) stimulates human neutrophil aggregation and release of lysosomal enzymes, superoxide, and eicosanoids. *J Immunol* 142: 3957–3962
32 Hsueh W, Tan XD, Qu XW, Sun XM, Gonzalez-Crussi F (1997) Injurious and protective mechanisms in the gut. Interaction of PAF, phospholipase A_2, eicosanoids, and nitric oxide synthase. *Adv Exp Med Biol* 407: 365–369
33 Kramer RM, Jakubowski JA, Deykin D (1988) Hydrolysis of 1-alkyl-2-arachidonoyl-*sn*-glycero-3-phosphocholine, a common precursor of platelet-activating factor and eicosanoids, by human platelet phospholipase A_2. *Biochim Biophys Acta* 959: 269–279
34 Nakae H, Endo S, Inada K, Yamashita H, Yamada Y, Takakuwa T, Kasai T, Ogawa M, Uchida K (1995) Plasma concentrations of type II phospholipase A_2, cytokines and eicosanoids in patients with burns. *Burns* 21: 422–426
35 Fonteh AN, Bass DA, Marshall LA, Seeds M, Samet JM, Chilton FH (1994) Evidence that secretory phospholipase A_2 plays a role in arachidonic acid release and eicosanoid biosynthesis by mast cells. *J Immunol* 152: 5438–5446
36 Fonteh AN, Samet JM, Chilton FH (1995) Regulation of arachidonic acid, eicosanoid, and phospholipase A_2 levels in murine mast cells by recombinant stem cell factor. *J Clin Invest* 96: 1432–1439
37 Balsinde J (2002) Roles of various phospholipases A_2 in providing lysophospholipid acceptors for fatty acid phospholipid incorporation and remodelling. *Biochem J* 364: 695–702
38 Metz SA (1986) Lysophosphatidylinositol, but not lysophosphatidic acid, stimulates insulin release. A possible role for phospholipase A_2 but not *de novo* synthesis of lysophospholipid in pancreatic islet function. *Biochem Biophys Res Commun* 138: 720–727
39 Arbibe L, Koumanov K, Vial D, Rougeot C, Faure G, Havet N, Longacre S, Vargaftig BB, Bereziat G, Voelker DR et al (1998) Generation of lyso-phospholipids from surfactant in acute lung injury is mediated by type-II phospholipase A_2 and inhibited by a direct surfactant protein A-phospholipase A_2 protein interaction. *J Clin Invest* 102: 1152–1160
40 Edelson JD, Vadas P, Villar J, Mullen JB, Pruzanski W (1991) Acute lung injury induced

by phospholipase A_2. Structural and functional changes. *Am Rev Respir Dis* 143: 1102–1109
41 Fisher AB, Dodia C, Chander A, Jain M (1992) A competitive inhibitor of phospholipase A_2 decreases surfactant phosphatidylcholine degradation by the rat lung. *Biochem J* 288 (Pt 2): 407–411
42 Furue S, Mikawa K, Nishina K, Shiga M, Ueno M, Tomita Y, Kuwabara K, Teshirogi I, Ono T, Hori Y et al (2001) Therapeutic time-window of a group IIA phospholipase A_2 inhibitor in rabbit acute lung injury: Correlation with lung surfactant protection. *Crit Care Med* 29: 719–727
43 Koike K, Yamamoto Y, Hori Y, Ono T (2000) Group IIA phospholipase A_2 mediates lung injury in intestinal ischemia-reperfusion. *Ann Surg* 232: 90–97
44 Liu L (1999) Regulation of lung surfactant secretion by phospholipase A_2. *J Cell Biochem* 72: 103–110
45 Chilton FH, Averill FJ, Hubbard WC, Fonteh AN, Triggiani M, Liu MC (1996) Antigen-induced generation of lyso-phospholipids in human airways. *J Exp Med* 183: 2235–2245
46 De Windt LJ, Willems J, Roemen TH, Coumans WA, Reneman RS, Van Der Vusse GJ, Van Bilsen M (2001) Ischemic-reperfused isolated working mouse hearts: Membrane damage and type IIA phospholipase A_2. *Am J Physiol Heart Circ Physiol* 280: H2572–H2580
47 Fletcher JE, Yang CC, Rosenberg P (1982) Basic phospholipase A_2 from Naja nigricollis snake venom: Phospholipid hydrolysis and effects on electrical and contractile activity of the rat heart. *Toxicol Appl Pharmacol* 66: 39–54
48 Van Bilsen M, Van Der Vusse GJ (1995) Phospholipase-A_2-dependent signalling in the heart. *Cardiovasc Res* 30: 518–529
49 Gronroos JO, Laine VJ, Nevalainen TJ (2002) Bactericidal group IIA phospholipase A_2 in serum of patients with bacterial infections. *J Infect Dis* 185: 1767–1772
50 Juffrie M, Meer GM, Hack CE, Haasnoot K, Sutaryo Veerman AJ, Thijs LG (2001) Inflammatory mediators in dengue virus infection in children: Interleukin-6 and its relation to C-reactive protein and secretory phospholipase A_2. *Am J Trop Med Hyg* 65: 70–75
51 Laine VJ, Grass DS, Nevalainen TJ (2000) Resistance of transgenic mice expressing human group II phospholipase A_2 to *Escherichia coli* infection. *Infect Immun* 68: 87–92
52 Rintala EM, Nevalainen TJ (1993) Group II phospholipase A_2 in sera of febrile patients with microbiologically or clinically documented infections. *Clin Infect Dis* 17: 864–870
53 Tunaz H, Park Y, Buyukguzel K, Bedick JC, Nor Aliza AR, Stanley DW (2003) Eicosanoids in insect immunity: Bacterial infection stimulates hemocytic phospholipase A_2 activity in tobacco hornworms. *Arch Insect Biochem Physiol* 52: 1–6
54 Degousee N, Ghomashchi F, Stefanski E, Singer A, Smart BP, Borregaard N, Reithmeier R, Lindsay TF, Lichtenberger C, Reinisch W et al (2002) Groups IV, V, and X phospholipases A_2s in human neutrophils: Role in eicosanoid production and gram-negative bacterial phospholipid hydrolysis. *J Biol Chem* 277: 5061–5073

55 Borgstrom B (1980) Importance of phospholipids, pancreatic phospholipase A_2, and fatty acid for the digestion of dietary fat: *In vitro* experiments with the porcine enzymes. *Gastroenterology* 78: 954–962
56 Gijon MA, Perez C, Mendez E, Sanchez CM (1995) Phospholipase A_2 from plasma of patients with septic shock is associated with high-density lipoproteins and C3 anaphylatoxin: Some implications for its functional role. *Biochem J* 306 (Pt 1): 167–175
57 Gonzalez RJ, Moore EE, Ciesla DJ, Meng X, Biffl WL, Silliman CC (2001) Post-hemorrhagic shock mesenteric lymph lipids prime neutrophils for enhanced cytotoxicity via phospholipase A_2. *Shock* 16: 218–222
58 Green JA, Smith GM, Buchta R, Lee R, Ho KY, Rajkovic IA, Scott KF (1991) Circulating phospholipase A_2 activity associated with sepsis and septic shock is indistinguishable from that associated with rheumatoid arthritis. *Inflammation* 15: 355–367
59 Marshall LA, Hall RH, Winkler JD, Badger A, Bolognese B, Roshak A, Flamberg PL, Sung CM, Chabot-Fletcher M, Adams JL et al (1995) SB 203347, an inhibitor of 14 kDa phospholipase A_2, alters human neutrophil arachidonic acid release and metabolism and prolongs survival in murine endotoxin shock. *J Pharmacol Exp Ther* 274: 1254–1262
60 Sorensen J, Kald B, Tagesson C, Lindahl M (1994) Platelet-activating factor and phospholipase A_2 in patients with septic shock and trauma. *Intensive Care Med* 20: 555–561
61 Vadas P, Pruzanski W, Stefanski E (1988) Extracellular phospholipase A_2: Causative agent in circulatory collapse of septic shock? *Agents Actions* 24: 320–325
62 Vadas P, Pruzanski W (1991) Serum phospholipase A_2 values and septic shock. *Crit Care Med* 19: 988–990
63 Xu D, Lu Q, Deitch EA (1995) Calcium and phospholipase A_2 appear to be involved in the pathogenesis of hemorrhagic shock-induced mucosal injury and bacterial translocation. *Crit Care Med* 23: 125–131
64 Aufenanger J, Samman M, Quintel M, Fassbender K, Zimmer W, Bertsch T (2002) Pancreatic phospholipase A_2 activity in acute pancreatitis: A prognostic marker for early identification of patients at risk. *Clin Chem Lab Med* 40: 293–297
65 Bird NC, Goodman AJ, Johnson AG (1989) Serum phospholipase A_2 activity in acute pancreatitis: An early guide to severity. *Br J Surg* 76: 731–732
66 Buchler M, Malfertheiner P, Schadlich H, Nevalainen TJ, Friess H, Beger HG (1989) role of phospholipase A_2 in human acute pancreatitis. *Gastroenterology* 97: 1521–1526
67 Eskola JU, Nevalainen TJ (1986) Pancreatic phospholipase A_2 in human acute pancreatitis. *Mater Med Pol* 18: 132–135
68 Friess H, Shrikhande S, Riesle E, Kashiwagi M, Baczako K, Zimmermann A, Uhl W, Buchler MW (2001) Phospholipase A_2 isoforms in acute pancreatitis. *Ann Surg* 233: 204–212
69 Gronroos JM, Nevalainen TJ (1992) Increased concentrations of synovial-type phospholipase A_2 in serum and pulmonary and renal complications in acute pancreatitis. *Digestion* 52: 232–236
70 Kemppainen E, Hietaranta A, Puolakkainen P, Sainio V, Halttunen J, Haapiainen R, Kivilaakso E, Nevalainen T (1999) Bactericidal/permeability-increasing protein and

group I and II phospholipase A_2 during the induction phase of human acute pancreatitis. *Pancreas* 18: 21–27

71 Matsuda Y, Ogawa M, Nishijima J, Miyauchi K, Mori T (1986) Usefulness of determination of serum immunoreactive pancreatic phospholipase A_2 content for early identification of severe acute pancreatitis. *Hepatogastroenterology* 33: 214–216

72 Miura M, Endo S, Kaku LL, Inoue Y, Sato N, Wakabayshi G, Baba E, Katsuya H, Inada K, Sato S (2001) Plasma type II phospholipase A_2 levels in patients with acute pancreatitis. *Res Commun Mol Pathol Pharmacol* 109: 159–164

73 Nevalainen TJ (1989) The role of phospholipase A_2 in human acute pancreatitis. *Klin Wochenschr* 67: 180–182

74 Schuppisser JP, Grotzinger U, Reichlin B, Tondelli P (1985) The role of phospholipase A2 in respiratory failure of acute pancreatitis. *Helv Chir Acta* 51: 665–667

75 Touqui L, Arbibe L (1999) A role for phospholipase A_2 in ARDS pathogenesis. *Mol Med Today* 5: 244–249

76 Blackwell GJ (1978) Phospholipase A_2 and platelet aggregation. *Adv Prostaglandin Thromboxane Res* 3: 137–142

77 Hazlett TL, Deems RA, Dennis EA (1990) Activation, aggregation, inhibition and the mechanism of phospholipase A_2. *Adv Exp Med Biol* 279: 49–64

78 Nakano T, Hanasaki K, Matsumoto S, Arita H (1988) Retinol induces platelet aggregation *via* activation of phospholipase A_2. *Biochem Biophys Res Commun* 154: 1075–1080

79 Atsumi G, Murakami M, Tajima M, Shimbara S, Hara N, Kudo I (1997) The perturbed membrane of cells undergoing apoptosis is susceptible to type II secretory phospholipase A_2 to liberate arachidonic acid. *Biochim Biophys Acta* 1349: 43–54

80 Yagami T, Ueda K, Asakura K, Hayasaki-Kajiwara Y, Nakazato H, Sakaeda T, Hata S, Kuroda T, Takasu N, Hori Y (2002) Group IB secretory phospholipase A_2 induces neuronal cell death via apoptosis. *J Neurochem* 81: 449–461

81 Yagami T, Ueda K, Asakura K, Hata S, Kuroda T, Sakaeda T, Takasu N, Tanaka K, Gemba T, Hori Y (2002) Human group IIA secretory phospholipase A_2 induces neuronal cell death *via* apoptosis. *Mol Pharmacol* 61: 114–126

82 Fonteh AN, Marion CR, Barham BJ, Edens MB, Atsumi G, Samet JM, High KP, Chilton FH (2001) Enhancement of mast cell survival: A novel function of some secretory phospholipase A_2 isotypes. *J Immunol* 167: 4161–4171

83 Riggins GJ, Markowitz S, Wilson JK, Vogelstein B, Kinzler KW (1995) Absence of secretory phospholipase A2 gene alterations in human colorectal cancer. *Cancer Res* 55: 5184–5186

84 Bomalaski JS, Clark MA (1990) Activation of phospholipase A_2 in rheumatoid arthritis. *Adv Exp Med Biol* 279: 231–238

85 Bowton DL, Seeds MC, Fasano MB, Goldsmith B, Bass DA (1997) Phospholipase A_2 and arachidonate increase in bronchoalveolar lavage fluid after inhaled antigen challenge in asthmatics. *Am J Respir Crit Care Med* 155: 421–425

86 Busse W (1998) The role and contribution of leukotrienes in asthma. *Ann Allergy Asthma Immunol* 81: 17–26
87 Calabrese C, Triggiani M, Marone G, Mazzarella G (2000) Arachidonic acid metabolism in inflammatory cells of patients with bronchial asthma. *Allergy* 55 (Suppl 61): 27–30
88 Hurt-Camejo E, Paredes S, Masana L, Camejo G, Sartipy P, Rosengren B, Pedreno J, Vallve JC, Benito P, Wiklund O (2001) Elevated levels of small, low-density lipoprotein with high affinity for arterial matrix components in patients with rheumatoid arthritis: Possible contribution of phospholipase A_2 to this atherogenic profile. *Arthritis Rheum* 44: 2761–2767
89 Komatsubara T, Tojo H, Ying Z, Tomita T, Ochi T, Okamoto M (1995) Serum phospholipase A_2 activity and immunoreactive group II phospholipase A_2 in rheumatoid arthritis. *Clin Chim Acta* 236: 109–112
90 Kortekangas P, Aro HT, Nevalainen TJ (1994) Group II phospholipase A_2 in synovial fluid and serum in acute arthritis. *Scand J Rheumatol* 23: 68–72
91 Kramer RM, Pepinsky RB (1991) Assay and purification of phospholipase A_2 from human synovial fluid in rheumatoid arthritis. *Methods Enzymol* 197: 373–381
92 Lin MK, Farewell V, Vadas P, Bookman AA, Keystone EC, Pruzanski W (1996) Secretory phospholipase A_2 as an index of disease activity in rheumatoid arthritis. Prospective double blind study of 212 patients. J *Rheumatol* 23: 1162–1166
93 Pruzanski W, Vadas P (1988) Secretory synovial fluid phospholipase A_2 and its role in the pathogenesis of inflammation in arthritis. *J Rheumatol* 15: 1601–1603
94 Barbour SE, Dennis EA (1993) Antisense inhibition of group II phospholipase A_2 expression blocks the production of prostaglandin E_2 by P388D1 cells. *J Biol Chem* 268: 21875–21882
95 Bayburt T, Yu BZ, Lin HK, Browning J, Jain MK, Gelb MH (1993) Human nonpancreatic secreted phospholipase A_2: Interfacial parameters, substrate specificities, and competitive inhibitors. *Biochemistry* 32: 573–582
96 Bernard P, Pintore M, Berthon JY, Chretien JR (2001) A molecular modeling and 3D QSAR study of a large series of indole inhibitors of human non-pancreatic secretory phospholipase A_2. *Eur J Med Chem* 36: 1–19
97 Blanchard SG, Andrews RC, Brown PJ, Gan LS, Lee FW, Sinhababu AK, Wheeler TN (1998) Discovery of bioavailable inhibitors of secretory phospholipase A_2. *Pharm Biotechnol* 11: 445–463
98 Flower R (1978) Steroidal anti-inflammatory drugs as inhibitors of phospholipase A_2. *Adv Prostaglandin Thromboxane Res* 3: 105–112
99 Glaser KB (1995) Regulation of phospholipase A_2 enzymes: Selective inhibitors and their pharmacological potential. *Adv Pharmacol* 32: 31–66
100 Hansford KA, Reid RC, Clark CI, Tyndall JD, Whitehouse MW, Guthrie T, McGeary RP, Schafer K, Martin JL, Fairlie DP (2003) D-tyrosine as a chiral precusor to potent inhibitors of human non-pancreatic secretory phospholipase A_2 (IIa) with anti-inflammatory activity. *Chembiochem* 4: 181–185

101 Jain MK, Streb M, Rogers J, DeHaas GH (1984) Action of phospholipase A_2 on bilayers containing lysophosphatidylcholine analogs and the effect of inhibitors. *Biochem Pharmacol* 33: 2541–2551
102 Kokotos G, Kotsovolou S, Six DA, Constantinou-Kokotou V, Beltzner CC, Dennis EA (2002) Novel 2-oxoamide inhibitors of human group IVA phospholipase A_2. *J Med Chem* 45: 2891–2893
103 Kokotos G, Constantinou-Kokotou V, Noula C, Nicolaou A, Gibbons WA (1996) Synthesis of lipidic amino acid and dipeptide inhibitors of human platelet phospholipase A_2. *Int J Pept Protein Res* 48: 160–166
104 Lappas M, Munns MJ, King RG, Rice GE (2001) Antisense oligonucleotide inhibition of type II phospholipase A_2 expression, release and activity *in vitro*. *Placenta* 22: 418–424
105 Tanaka K, Arita H (1995) Secretory phospholipase A_2 inhibitors. Possible new anti-inflammatory agents. *Agents Actions* (Suppl) 46: 51–64
106 Kennedy BP, Payette P, Mudgett J, Vadas P, Pruzanski W, Kwan M, Tang C, Rancourt DE, Cromlish WA (1995) A natural disruption of the secretory group II phospholipase A_2 gene in inbred mouse strains. *J Biol Chem* 270: 22378–22385
107 Lilja I, Smedh K, Olaison G, Sjodahl R, Tagesson C, Gustafson-Svard C (1995)Phospholipase A_2 gene expression and activity in histologically normal ileal mucosa and in Crohn's ileitis. *Gut* 37: 380–385
108 MacPhee M, Chepenik KP, Liddell RA, Nelson KK, Siracusa LD, Buchberg AM (1995) The secretory phospholipase A_2 gene is a candidate for the Mom1 locus, a major modifier of ApcMin-induced intestinal neoplasia. *Cell* 81: 957–966
109 Fox N, Song M, Schrementi J, Sharp JD, White DL, Snyder DW, Hartley LW, Carlson DG, Bach NJ, Dillard RD et al (1996) Transgenic model for the discovery of novel human secretory non-pancreatic phospholipase A_2 inhibitors. *Eur J Pharmacol* 308: 195–203
110 Nevalainen TJ, Laine VJ, Grass DS (1997) Expression of human group II phospholipase A_2 in transgenic mice. *J Histochem Cytochem* 45: 1109–1119
111 Uozumi N, Kume K, Nagase T, Nakatani N, Ishii S, Tashiro F, Komagata Y, Maki K, Ikuta K, Ouchi Y et al (1997) Role of cytosolic phospholipase A_2 in allergic response and parturition. *Nature* 390: 618–622
112 Grass DS, Felkner RH, Chiang MY, Wallace RE, Nevalainen TJ, Bennett CF, Swanson ME (1996) Expression of human group II PLA_2 in transgenic mice results in epidermal hyperplasia in the absence of inflammatory infiltrate. *J Clin Invest* 97: 2233–2241
113 Bingham CO III, Murakami M, Fujishima H, Hunt JE, Austen KF, Arm JP (1996) A heparin-sensitive phospholipase A_2 and prostaglandin endoperoxide synthase-2 are functionally linked in the delayed phase of prostaglandin D_2 generation in mouse bone marrow-derived mast cells. *J Biol Chem* 271: 25936–25944
114 Balsinde J, Barbour SE, Bianco ID, Dennis EA (1994) Arachidonic acid mobilization in P388D1 macrophages is controlled by two distinct Ca^{2+}-dependent phospholipase A_2 enzymes. *Proc Natl Acad Sci USA* 91: 11060–11064

115 Fonteh AN, Samet JM, Surette M, Reed W, Chilton FH (1998) Mechanisms that account for the selective release of arachidonic acid from intact cells by secretory phospholipase A_2. *Biochim Biophys Acta* 1393: 253–266

116 Fonteh AN, Atsumi G, LaPorte T, Chilton FH (2000) Secretory phospholipase A_2 receptor-mediated activation of cytosolic phospholipase A_2 in murine bone marrow-derived mast cells. *J Immunol* 165: 2773–2782

117 Bingham CO III, Fijneman RJ, Friend DS, Goddeau RP, Rogers RA, Austen KF, Arm JP (1999) Low molecular weight group IIA and group V phospholipase A_2 enzymes have different intracellular locations in mouse bone marrow-derived mast cells. *J Biol Chem* 274: 31476–31484

118 Fujishima H, Sanchez Mejia RO, Bingham CO III, Lam BK, Sapirstein A, Bonventre JV, Austen KF, Arm JP (1999) Cytosolic phospholipase A_2 is essential for both the immediate and the delayed phases of eicosanoid generation in mouse bone marrow-derived mast cells. *Proc Natl Acad Sci USA* 96: 4803–4807

119 Murakami M, Austen KF, Arm JP (1995) The immediate phase of c-*sn*-ligand stimulation of mouse bone marrow-derived mast cells elicits rapid leukotriene C_4 generation through post-translational activation of cytosolic phospholipase A_2 and 5-lipoxygenase. *J Exp Med* 182: 197–206

120 Reddy ST, Winstead MV, Tischfield JA, Herschman HR (1997) Analysis of the secretory phospholipase A_2 that mediates prostaglandin production in mast cells. *J Biol Chem* 272: 13591–13596

121 Blom M, Tool AT, Wever PC, Wolbink GJ, Brouwer MC, Calafat J, Egesten A, Knol EF, Hack CE, Roos D et al (1998) Human eosinophils express, relative to other circulating leukocytes, large amounts of secretory 14-kD phospholipase A_2. *Blood* 91: 3037–3043

122 Hundley TR, Marshall LA, Hubbard WC, MacGlashan DW Jr (1998) Characteristics of arachidonic acid generation in human basophils: Relationship between the effects of inhibitors of secretory phospholipase A_2 activity and leukotriene C4 release. *J Pharmacol Exp Ther* 284: 847–857

123 Triggiani M, Granata F, Oriente A, De MV, Gentile M, Calabrese C, Palumbo C, Marone G (2000) Secretory phospholipases A_2 induce beta-glucuronidase release and IL-6 production from human lung macrophages. *J Immunol* 164: 4908–4915

124 Triggiani M, Granata F, Oriente A, Gentile M, Petraroli A, Balestrieri B, Marone G (2002) Secretory phospholipases A_2 induce cytokine release from blood and synovial fluid monocytes. *Eur J Immunol* 32: 67–76

125 Triggiani M, Granata F, Balestrieri B, Petraroli A, Scalia G, Del Vecchio L, Marone G (2003) Secretory phospholipases A_2 activate selective functions in human eosinophils. *J Immunol* 170: 3279–3288

126 Galli SJ (2000) Mast cells and basophils. *Curr Opin Hematol* 7: 32–39

127 Galli SJ (1993) New concepts about the mast cell. *N Engl J Med* 328: 257–265

128 Gordon JR, Galli SJ (1991) Release of both preformed and newly synthesized tumor necrosis factor alpha (TNF-α)/cachectin by mouse mast cells stimulated *via* the Fc

epsilon RI. A mechanism for the sustained action of mast cell-derived TNF-α during IgE-dependent biological responses. *J Exp Med* 174: 103–107
129 Wedemeyer J, Tsai M, Galli SJ (2000) Roles of mast cells and basophils in innate and acquired immunity. *Curr Opin Immunol* 12: 624–631
130 Fonteh AN, Marion CR, Barham BJ, Edens MB, Atsumi G, Samet JM, High KP, Chilton FH (2001) Enhancement of mast cell survival: A novel function of some secretory phospholipase A_2 isotypes. *J Immunol* 167: 4161–4171
131 Nair X, Nettleton D, Clever D, Tramposch KM, Ghosh S, Franson RC (1993) Swine as a model of skin inflammation. Phospholipase A_2-induced inflammation. *Inflammation* 17: 205–215
132 Babu AS, Gowda TV (1994) Dissociation of enzymatic activity from toxic properties of the most basic phospholipase A_2 from *Vipera russelli* snake venom by guanidination of lysine residues. *Toxicon* 32: 749–752
133 Arita H, Hanasaki K (1993) Physiological aspects of a high affinity binding site for pancreatic-type phospholipase A_2. *J Lipid Mediat* 6: 217–222
134 Hanasaki K, Arita H (2002) Phospholipase A_2 receptor: A regulator of biological functions of secretory phospholipase A_2. *Prostaglandins Other Lipid Mediat* 68–69: 71–82
135 Hanasaki K, Arita H (1996) Structure and function of phospholipase A_2 receptor. *Adv Exp Med Biol* 416: 315–319
136 Ishizaki J, Kishino J, Teraoka H, Ohara O, Arita H (1993) Receptor-binding capability of pancreatic phospholipase A_2 is separable from its enzymatic activity. *FEBS Lett* 324: 349–352
137 Lambeau G, Lazdunski M (1999) Receptors for a growing family of secreted phospholipases A_2. *Trends Pharmacol Sci* 20: 162–170
138 Lambeau G, Ancian P, Nicolas JP, Beiboer SH, Moinier D, Verheij H, Lazdunski M (1995) Structural elements of secretory phospholipases A_2 involved in the binding to M-type receptors. *J Biol Chem* 270: 5534–5540
139 Ancian P, Lambeau G, Mattei MG, Lazdunski M (1995) The human 180-kDa receptor for secretory phospholipases A_2. Molecular cloning, identification of a secreted soluble form, expression, and chromosomal localization. *J Biol Chem* 270: 8963–8970
140 Hanasaki K, Yokota Y, Ishizaki J, Itoh T, Arita H (1997) Resistance to endotoxic shock in phospholipase A_2 receptor-deficient mice. *J Biol Chem* 272: 32792–32797
141 Koduri RS, Gronroos JO, Laine VJ, Le Calvez C, Lambeau G, Nevalainen TJ, Gelb MH (2002) Bactericidal properties of human and murine groups I, II, V, X, and XII secreted phospholipases A_2. *J Biol Chem* 277: 5849–5857
142 Fourcade O, Simon MF, Viode C, Rugani N, Leballe F, Ragab A, Fournie B, Sarda L, Chap H (1995) Secretory phospholipase A_2 generates the novel lipid mediator lysophosphatidic acid in membrane microvesicles shed from activated cells. *Cell* 80: 919–927
143 Bevers EM, Comfurius P, Zwaal RF (1983) Changes in membrane phospholipid distribution during platelet activation. *Biochim Biophys Acta* 736: 57–66
144 Dekkers DW, Comfurius P, Bevers EM, Zwaal RF (2002) Comparison between Ca^{2+}-

induced scrambling of various fluorescently labelled lipid analogues in red blood cells. *Biochem J* 362: 741–747
145 Fonteh AN, Chilton FH (1993) Mobilization of different arachidonate pools and their roles in the generation of leukotrienes and free arachidonic acid during immunologic activation of mast cells. *J Immunol* 150: 563–570
146 Fonteh AN, Chilton FH (1992) Rapid remodeling of arachidonate from phosphatidylcholine to phosphatidylethanolamine pools during mast cell activation. *J Immunol* 148: 1784–1791
147 Benveniste J, Chignard M, Le Couedic JP, Vargaftig BB (1982) Biosynthesis of platelet-activating factor (PAF-ACETHER) II. Involvement of phospholipase A_2 in the formation of PAF-ACETHER and lyso-PAF-ACETHER from rabbit platelets. *Thromb Res* 25: 375–385
148 Blank ML, Fitzgerald V, Smith ZL, Snyder F (1995) Generation of the precursor (lyso-PAF) of platelet-activating factor via a CoA-dependent transacylase. *Biochem Biophys Res Commun* 210: 1052–1058
149 Snyder F, Lee TC, Blank M, Malone B, Woodard D, Robinson M (1985) Platelet-activating factor: Alternate pathways of biosynthesis, mechanism of inactivation, and reacylation of lyso-PAF with arachidonate. *Adv Prostaglandin Thromboxane Leukot Res* 15: 693–696
150 East L, Isacke CM (2002) The mannose receptor family. *Biochim Biophys Acta* 1572: 364–386
151 Durstin M, Durstin S, Molski TF, Becker EL, Sha'afi RI (1994) Cytoplasmic phospholipase A_2 translocates to membrane fraction in human neutrophils activated by stimuli that phosphorylate mitogen-activated protein kinase. *Proc Natl Acad Sci USA* 91: 3142–3146
152 Gordon RD, Leighton IA, Campbell DG, Cohen P, Creaney A, Wilton DC, Masters DJ, Ritchie GA, Mott R, Taylor IW et al (1996) Cloning and expression of cystolic phospholipase A_2 ($cPLA_2$) and a naturally occurring variant. Phosphorylation of Ser505 of recombinant $cPLA_2$ by p42 mitogen-activated protein kinase results in an increase in specific activity. *Eur J Biochem* 238: 690–697
153 Hazan-Halevy I, Levy R (2000) Activation of cytosolic phospholipase A_2 by opsonized zymosan in human neutrophils requires both ERK and p38 MAP-kinase. *Adv Exp Med Biol* 479: 115–123
154 Hefner Y, Borsch-Haubold AG, Murakami M, Wilde JI, Pasquet S, Schieltz D, Ghomashchi F, Yates JR III, Armstrong CG, Paterson A et al (2000) Serine 727 phosphorylation and activation of cytosolic phospholipase A_2 by MNK1-related protein kinases. *J Biol Chem* 275: 37542–37551
155 Kramer RM, Roberts EF, Hyslop PA, Utterback BG, Hui KY, Jakubowski JA (1995) Differential activation of cytosolic phospholipase A_2 ($cPLA_2$) by thrombin and thrombin receptor agonist peptide in human platelets. Evidence for activation of $cPLA_2$ independent of the mitogen-activated protein kinases ERK1/2. *J Biol Chem* 270: 14816–14823
156 Kramer RM, Roberts EF, Um SL, Borsch-Haubold AG, Watson SP, Fisher MJ,

Jakubowski JA (1996) p38 mitogen-activated protein kinase phosphorylates cytosolic phospholipase A_2 (cPLA2) in thrombin-stimulated platelets. Evidence that proline-directed phosphorylation is not required for mobilization of arachidonic acid by $cPLA_2$. *J Biol Chem* 271: 27723–27729

157 Kawakami T, Galli SJ (2002) Regulation of mast-cell and basophil function and survival by IgE. *Nat Rev Immunol* 2: 773–786

158 Hata D, Kawakami Y, Inagaki N, Lantz CS, Kitamura T, Khan WN, Maeda-Yamamoto M, Miura T, Han W, Hartman SE et al (1998) Involvement of Bruton's tyrosine kinase in Fc epsilon RI-dependent mast cell degranulation and cytokine production. *J Exp Med* 187: 1235–1247

159 Petro JB, Khan WN (2001) Phospholipase C-gamma 2 couples Bruton's tyrosine kinase to the NF-kappa B signaling pathway in B lymphocytes. *J Biol Chem* 276: 1715–1719

160 Tsai M, Chen RH, Tam SY, Blenis J, Galli SJ (1993) Activation of MAP kinases, pp90rsk and pp70-S6 kinases in mouse mast cells by signaling through the c-kit receptor tyrosine kinase or Fc epsilon RI: Rapamycin inhibits activation of pp70-S6 kinase and proliferation in mouse mast cells. *Eur J Immunol* 23: 3286–3291

161 Galli SJ, Tsai M, Wershil BK (1993) The c-kit receptor, stem cell factor, and mast cells. What each is teaching us about the others. *Am J Pathol* 142: 965–974

Control of arachidonic acid levels in resting and activated U937 phagocytic cells by Ca^{2+}-independent phospholipase A_2

Jesús Balsinde, Rebeca Pérez, Yolanda Sáez, and María A. Balboa

Institute of Molecular Biology and Genetics, University of Valladolid School of Medicine, Avenida Ramón y Cajal 7, E-47005 Valladolid, Spain

Phospholipase A_2 regulation of arachidonic acid levels

Availability of free arachidonic acid (AA) is widely recognized as a rate-limiting step in the formation of prostaglandins. This fatty acid is an intermediate of a reacylation/deacylation cycle of membrane phospholipids, the so-called Lands pathway, in which the fatty acid is cleaved from phospholipid by phospholipase A_2s (PLA_2) and reincorporated by acyltransferases. Whereas in resting cells reacylation dominates, in stimulated cells the dominant reaction is the PLA_2-mediated deacylation. Nevertheless, increased AA reacylation during cellular activation is still very significant, as manifested by the fact that only a minor portion of the free AA released by PLA_2 is converted into eicosanoids, the remainder being effectively incorporated back into phospholipids.

Phagocytic cells generally contain multiple PLA_2s [1, 2]. Thus the challenge in recent years has been both to identify these PLA_2s and to clarify their roles in AA metabolism. A general mechanism for PLA_2-regulated AA metabolism in resting and activated cells has emerged from the studies by several laboratories [3, 4], and involves participation of all three major classes of PLA_2, namely $cPLA_2$ (cytosolic PLA_2), $iPLA_2$ (Ca^{2+} independent PLA_2) and $sPLA_2$ (secreted PLA_2) (Fig. 1).

In resting conditions, $iPLA_2$ accounts for most of the PLA_2 activity of cells. $iPLA_2$ is therefore the dominant PLA_2 involved in the liberation of fatty acids, including AA, during the continuous recycling of membrane phospholipids that takes place under these conditions. Since, as indicated above, the rate of AA release by $iPLA_2$ is lesser than the rate of its reacylation back into phospholipids, no net accumulation of free fatty acid occurs. Stimulation of the cells by receptor agonists results in the activation of $cPLA_2$, which then becomes the dominant PLA_2 involved in AA release. Under these conditions, the rate of AA release clearly exceeds that of reincorporation into phospholipids; hence net accumulation of AA occurs that is followed by its conversion into different oxygenated compounds, collectively called the eicosanoids.

Arachidonate Remodeling and Inflammation, edited by Alfred N. Fonteh and Robert L. Wykle

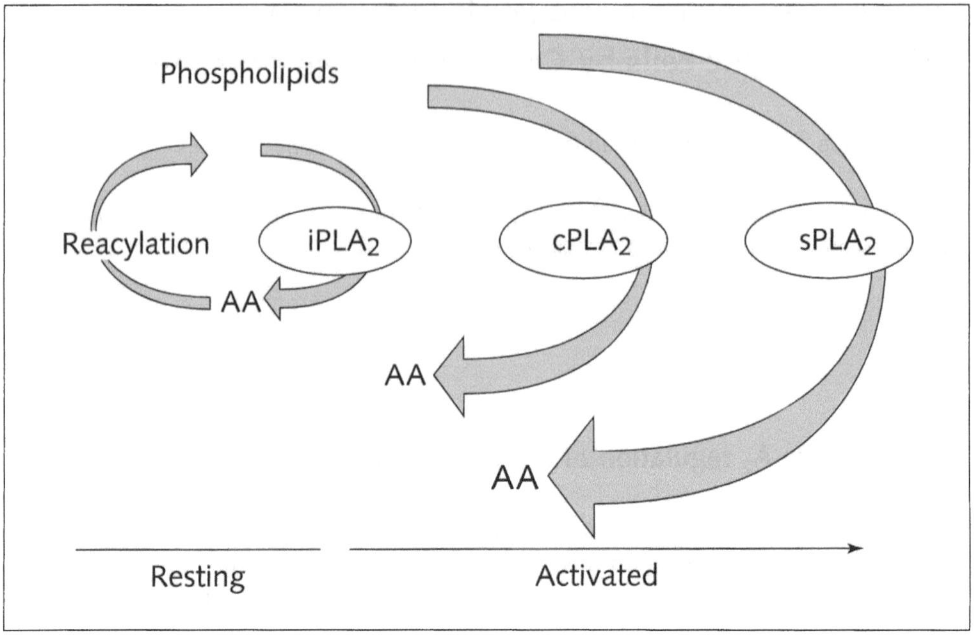

Figure 1
PLA_2 regulation of AA release.

During long-term AA mobilization responses, i.e., those that are typical from immunoinflammatory cells such as macrophages or mast cells, the inducible $sPLA_2$ also participates in the process, thereby creating an amplification loop that results in a greatly enhanced release of AA for eicosanoid synthesis. At these stages, the contribution of $sPLA_2$ to overall AA release often exceeds that of $cPLA_2$. The $sPLA_2$ acts not only upon the cells that synthesized it but also upon surrounding cells, which allows for efficient propagation of the inflammatory response (Fig. 1).

Within the framework of the model depicted in Figure 1, recent studies have further explored the involvement of other PLA_2s in addition to $iPLA_2$ in *housekeeping* phospholipid and fatty acid remodeling. Possible $iPLA_2$-roles in inflammatory signaling distinct from receptor-regulated AA metabolism have also been investigated. These are discussed in the following sections.

Production of lysophospholipid acceptors for AA incorporation and remodeling into phospholipids

Unlike saturated fatty acids, AA at physiologically relevant nanomolar levels does not generally enter cellular phospholipids *via* direct acylation of glycerol phosphate/dihydroxyacetone phosphate or lysophosphatidic acid (i.e., the *de novo* pathway) but rather does it at later stage, *via* direct acylation of pre-existing lysophospholipid acceptors (the Lands pathway). Since lysophospholipids are produced by the hydrolytic action of (PLA$_2$ on phospholipids, this class of enzymes necessarily plays a key role in AA incorporation into phospholipids. Initial incorporation of AA into phospholipids takes place primarily into phosphatidylcholine (PC); hence the lysophospholipid acceptor utilized is lysophosphatidylcholine (lysoPC) [5]. In many cells, the steady-state levels of lysoPC appear to be maintained by the continuing action of Ca^{2+}-independent group VI phospholipase A$_2$ (iPLA$_2$) on cellular phospholipids [6]. Thus, a decrease in the activity of the iPLA$_2$ frequently results in the diminished production of lysoPC and hence in the inhibition of AA incorporation into phospholipids.

Once the AA has been incorporated into PC by the action of CoA-dependent acyltransferases, it is then transferred to certain lysophospholipids, particularly the ethanolamine lysophospholipids (lysoPE) in a process that generally takes several hours, and is governed by the enzyme CoA-independent transacylase (CoA-IT) [5]. Thus, for the efficient incorporation of AA into phospholipids, two kinds of lysophospholipid acceptors should be readily available in the cell (Fig. 2).

Earlier studies on the initial incorporation of AA into the phospholipids of murine macrophages demonstrated that the process was essentially Ca^{2+}-independent [7]. This suggested that the PLA$_2$ activity responsible for generating lysophospholipid acceptors for AA incorporation would correspond to that of an iPLA$_2$-like enzyme [7]. Such an activity was later identified to belong to the group VI PLA$_2$ in studies carried out with murine P388D$_1$ macrophages [8, 9]. These findings were later extended by other authors to different cellular systems, such as human neutrophils [10], rat submandibular ductal cells [11], and rat uterine stromal cells [12].

Importantly, the contribution of group VI iPLA$_2$ to maintaining the lysophospholipid pool that facilitates AA incorporation appears to largely depend on cell type. Based on studies of iPLA$_2$ inhibition by the selective inhibitor bromoenol lactone (BEL), the iPLA$_2$ contribution ranges from ~90% in rat submandibular ductal cells [11], to 50–60 in phagocytic cells [8–10], and to only 20–25% in rat uterine stromal cells [12]. It follows from these findings that, in addition to the contribution of group VI PLA$_2$ to the cellular lysophospholipid pools, the cells may possess other mechanisms to generate and maintain the appropriate levels of lysophospholipid acceptors.

The above notion was highlighted by a recent report in rat pancreatic islets [13], where iPLA$_2$-inhibition by BEL does not result in diminished AA incorporation into phospholipids. Nevertheless, group VI PLA$_2$ is estimated to contribute to at least 20% of the very high steady-state lysophospholipid levels in pancreatic islets, indicating that the enzyme possesses significant *housekeeping* activity in islets as well. Since rat pancreatic islets maintain cellular lysophospholipid levels at high levels, it is very likely that the amount of lysophospholipid present in these cells after BEL

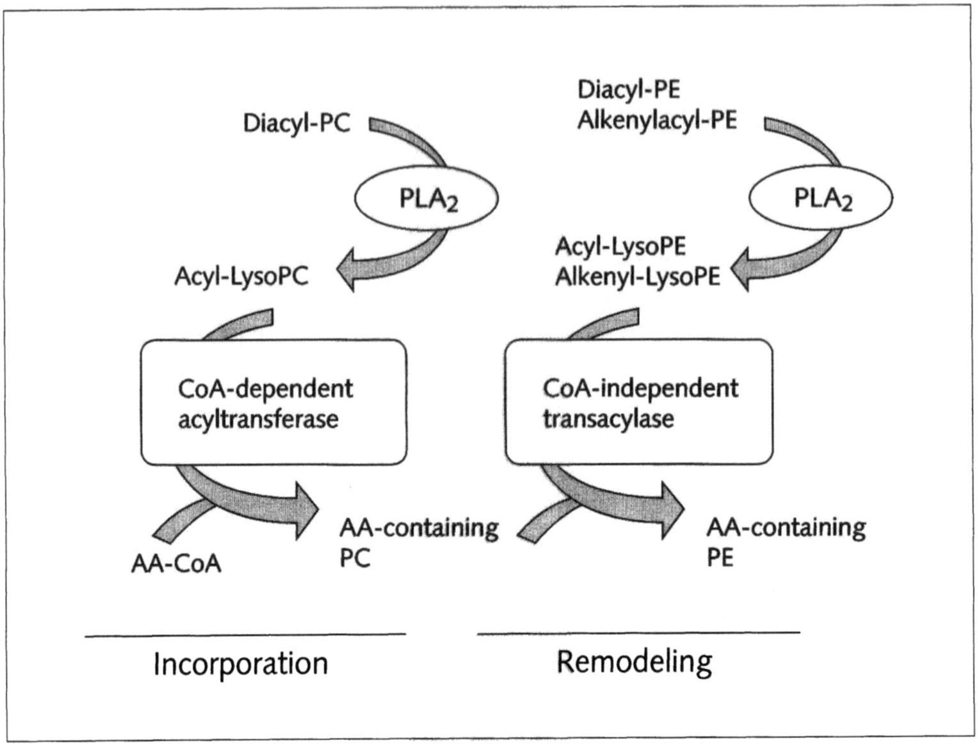

Figure 2
AA incorporation into and remodeling within phospholipids.

treatment is still high enough to account for a normal rate of AA incorporation into phospholipids. In support of this contention, recent studies on AA incorporation utilizing cells over-expressing group VI PLA_2 have demonstrated that the excess amount of lysophospholipid produced under those conditions does not serve to increase the rate of fatty acid incorporation [14]. Collectively, these findings have suggested that, while a threshold level of lysophospholipid seems necessary to support AA incorporation into phospholipids, increasing cellular lysophospholipid above that threshold level does not increase the rate of AA incorporation. Thus other factor(s) in addition to lysophospholipid availability limit AA incorporation into phospholipid.

We have recently employed the U937 cell line as a system model to study the enzymes involved in generating and maintaining the lysophospholipid threshold level that is necessary for fatty acid incorporation into phospholipids. These cells differentiate into macrophage-like cells when treated with phorbol esters such as phorbol myristate acetate (PMA). U937 cells express group IV PLA_2 (also known as

$cPLA_2$) and group VI PLA_2 (generally referred to as $iPLA_2$) but not any of the AA-releasing $sPLA_2$s, including group IIA, IID, V and X (Balboa, Sáez, Pérez and Balsinde, unpublished observations).

AA incorporation into phospholipids of both resting and ConA-activated PMA-differentiated U937 cells exhibits features that are fully similar to those of other phagocytic cells such as $P388D_1$ macrophages and neutrophils [8–10], i.e., it is Ca^{2+}-independent and BEL-inhibitable. Interestingly, BEL reduces AA incorporation only partially, suggesting again the existence of Ca^{2+}-independent pathway(s) for AA incorporation into phospholipids that do not involve the BEL-sensitive group VI PLA_2 [15].

Parallel studies comparing the features of AA incorporation to those of eicosapentaenoic acid (C20:5, ω-3; EPA), show that the two fatty acids compete with each other for incorporation into the phospholipids of resting cells. Thus, AA and EPA share a common pathway for incorporation into phospholipids in resting cells. Since EPA incorporation is not affected by BEL, this common pathway must also be insensitive to BEL [15]. This situation contrasts with the one seen in ConA-activated cells, where AA and EPA do not compete with each other. Unlike AA, EPA incorporation into phospholipid is sensitive to $cPLA_2$ inhibitors in ConA-activated cells [15]. Thus, it seems that the increased lysophospholipid availability produced as a consequence of stimulus-induced $cPLA_2$ activation plays a role in EPA incorporation. In accord with this idea, the PE pools appear to be preferential targets for $cPLA_2$ phospholipolysis in stimulated phagocytes [16, 17], and PE is the phospholipid class to which EPA preferentially incorporates [15]. Thus, $cPLA_2$ would provide an additional supply of lysoPE acceptors to be used by EPA but not by AA, and the existence of such an alternative route for EPA incorporation may explain why in the activated cells EPA does not compete with AA.

After the initial incorporation of AA mostly into PC, a slow transfer of the fatty acid occurs toward PE in most cell types. These changes have been well documented in other cell systems and reflect the remodeling action of the enzyme CoA-IT on cellular phospholipids [5]. A PLA_2 is strikingly involved in the CoA-IT-driven remodeling reactions by providing the lysoPE acceptors utilized in the transacylation reaction [5].

The nature of such a PLA_2 has recently been investigated in U937 macrophages [15] and peripheral T lymphocytes [18] by measuring the transfer of AA from PC to PE in the presence of different PLA_2 inhibitors. Inhibitors of $cPLA_2$ (methyl arachidonyl fluorophosphonate), $iPLA_2$ (BEL) and $sPLA_2$ (LY311727), all fail to exert any detectable effect on the transfer of AA from PE to PC in either cell type. These findings suggest that the PLA_2 involved in this pathway might not be any of the previously identified PLA_2s. The Ca^{2+}-independent nature of the response suggests the involvement of an $iPLA_2$-like activity different from the group VI enzyme. It should be noted that the aforementioned experiments are reported to be carried out at inhibitor doses that completely ablate the corresponding PLA_2 activity [15,

18]. That excludes the possibility that residual PLA_2 activity in the presence of any of these inhibitors might provide enough lysoPE to accommodate substrate needs of CoA-IT.

Unlike other macrophage-like cells such as $P388D_1$ cells [8], PMA-differentiated U937 cells exhibit a significant $iPLA_2$ activity component that is resistant to BEL [15, 19]. This component is more evident when PE is used as a substrate for the assay [15]. It is very intriguing to speculate with the possibility that such an activity is the one generating lysoPC and lysoPE acceptors for incorporation into and remodeling among phospholipids under conditions where all other known PLA_2s are not involved. Purification and characterization of this novel $iPLA_2$ activity seems important to further clarify the mechanisms responsible for lysophospholipid level maintenance in resting and activated phagocytes.

The intracellular level of lysophosphatidylcholine as a regulatory signal for phagocyte secretion

As a part of their surveillance functions in the immune system, monocytes/macrophages secrete large amounts of the bactericidal enzyme lysozyme to the extracellular medium. Lysozyme degrades bacterial cell walls of Gram-positive bacteria and the chitinous components of fungal cell walls. The enzyme occurs in many body fluids such as tears, saliva or mucus, and is produced and secreted by phagocytic cells and a variety of cells of epithelial origin. Stimuli that induce lysozyme secretion from phagocytic cells also induce the PLA_2-mediated mobilization of free AA. Thus the question arises as to whether these two phenomena are causally related.

Studies with $cPLA_2$ knock-out mice have unambiguously demonstrated that this is the key enzyme in stimulus-induced AA mobilization [20, 21]. In keeping with these data, countless articles in the scientific literature have shown that inhibition of $cPLA_2$ results in greatly diminished AA release responses. The classical method for studying the involvement of $cPLA_2$ in a given cellular response is to use chemical inhibitors that are reasonably selective for this enzyme. While some of these studies may be flawed by the use of inhibitors that are actually not that selective for $cPLA_2$ [22], recent studies have described new chemical inhibitors of $cPLA_2$ with improved potency and specificity. These are pyrrophenone [23] and related compounds [24]. Inhibition by pyrrophenone of pure $cPLA_2$, as measured in an *in vitro* assay, is over two orders of magnitude more potent than that for other PLA_2 types [23]. Importantly, much of this selectivity is retained in assays utilizing cell homogenates as the source of enzyme [24, 25], implying that pyrrophenone may indeed constitute an excellent tool to study the involvement of $cPLA_2$ in cell function. For example, in U937 cells pyrrophenone completely inhibits stimulus-induced AA release at concentrations well below 1 μM [25], and at this concentration no effect is seen on either $iPLA_2$ or $sPLA_2$ [26].

Despite its potent effects on AA mobilization, pyrrophenone does not exert any effect on lysozyme secretion in activated U937 cells, indicating that $cPLA_2$ has no role in this process. However, lysozyme secretion is impaired by BEL, which suggests the involvement of $iPLA_2$ [27]. Such an involvement has been confirmed by antisense oligonucleotide inhibition experiments, which selectively target the $iPLA_2$ [27]

The low lysoPC level found in $iPLA_2$-deficient cells appears to be related to the diminished capacity of these cells to secrete lysozyme, since exogenous supplementation of lysoPC fully restores the response. Other putative PLA_2 products such as exogenous AA and other fatty acids fail to restore lysozyme secretion in the U937 cells deficient in $iPLA_2$ activity [27]. This suggests that stimulus-triggered increases in free AA levels have no role in regulating lysozyme secretion. Since in activated U937 cells AA mobilization appears to be under the control of $cPLA_2$, these results are fully consistent with the aforementioned studies showing that $cPLA_2$ inhibition by pyrrophenone has no effect on lysozyme release.

$cPLA_2$ activation transiently elevates cellular lysoPC levels in activated phagocytic cells [28]. Since $cPLA_2$ plays no discernible role in lysozyme release, it is the steady-state level of lysoPC ($iPLA_2$-mediated), not the transient increase in lysoPC that occurs as a consequence of cellular activation ($cPLA_2$-mediated), which is important for lysozyme secretion. In agreement with these observations, exogenous lysoPC in the absence of PMA neither triggers secretion on its own nor increases the secretory response of cells containing normal $iPLA_2$ levels (and hence, exhibiting normal steady-state lysoPC levels) [27].

Recently two important cellular functions that, like enzyme secretion, require profound membrane rearrangement have also been suggested to involve participation of the $iPLA_2$. These are chemotaxis [29], and cell spreading [30]. Coincident with the aforementioned studies on lysozyme secretion, it is the constitutive activity of the $iPLA_2$ that was found to be necessary to sustain both of these functions. In addition, in these studies the contribution of the $iPLA_2$ was dissociated from $cPLA_2$ activation [29, 30]. Altogether these findings underscore the importance of $iPLA_2$ in regulating processes that require changes in membrane phospholipid homeostasis and provide support to the idea that $iPLA_2$ and $cPLA_2$ play separate and often unique roles in inflammatory cell signaling.

$iPLA_2$-derived products and oxidative stress

Phagocytic cells produce reactive oxygen intermediates such as superoxide anion and hydrogen peroxide in response to a variety of agonists. While the production of these oxygen metabolites plays an important role in cellular signaling and host defense, their uncontrolled production may constitute a serious pathophysiological factor for a wide variety of vascular-based disorders. Oxidative damage is often associated with AA mobilization from cells from the vascular system, such as endothelial cells, smooth muscle cells, platelets and phagocytes. Reactive oxygen intermediates enhance AA release and prostaglandin production in different cell systems, but the molecular mechanism responsible for these effects appears to vary from cell to cell. PLA_2 activation has been pointed out as the most likely mechanism for AA mobilization in vascular smooth muscle cells, stromal cells, and striatal neurons exposed to H_2O_2 [31–34]. In other systems however, diminished AA incorporation into phospholipids, not PLA_2 activation, has been suggested to be the event responsible for free AA accumulation [35, 36].

Our recent data in U937 cells have established a novel mechanism for AA mobilization in phagocytic cells under an oxidative stress that involves participation of $iPLA_2$ rather than of $cPLA_2$ [25]. In these cells, H_2O_2 is able to induce a delayed AA mobilization response with a kinetics that strongly contrasts with the response of the cells to Con A, which is mediated by $cPLA_2$ and hence shows the typical saturation kinetics that is expected from a highly regulated cellular response such as AA release. The response to H_2O_2 however, is inhibited by BEL and by specific $iPLA_2$ antisense oligonucleotides [25].

Importantly, H_2O_2 treatment of the U937 cells does not increase the $iPLA_2$ specific activity of the cells, as measured by different *in vitro* assays, indicating that a stable activation of the $iPLA_2$ (e.g., phosphorylation) is not the mechanism for H_2O_2-mediated AA release in U937 cells. When membranes from H_2O_2-treated cells are used in the assay, the $iPLA_2$ activity measured is significantly higher than that found in membranes from otherwise unstimulated cells. Therefore, treating the cells with H_2O_2 results in facilitated $iPLA_2$ attack on membrane phospholipids. Since membranes from H_2O_2-treated cells contain significantly higher amounts of lipid peroxides than membranes from untreated cells [25], these findings suggest that lipid hydrolysis by $iPLA_2$ occurs more readily in H_2O_2-treated cells because of changes in the physical state of membrane substrates, which may result, at least in part, from lipid peroxide accumulation.

Taken together, the above findings have suggested a model for fatty acid mobilization in H_2O_2-treated cells whereby the oxidant induces lipid oxidation, which results in accumulation of lipid peroxides at the membrane. These lipid peroxides destabilize the membrane and render it more susceptible to $iPLA_2$ attack, which results in increased liberation of fatty acids. An important aspect of the above model is that this fatty acid release appears to occur in the absence of $cPLA_2$ activation, which underscores the apparent lack of a regulated signaling component in the process. Still, a mechanism such as this one may be very relevant under pathophysiological conditions such as oxidative stress, where increased $iPLA_2$ activity may account for a significant phospholipid hydrolysis before cellular homeostasis is reestablished.

Analysis of the AA metabolites produced after exposure of the cells to H_2O_2 reveals a significant production of prostaglandins, particularly the pro-inflammatory prostaglandins E_2 and D_2. This may suggest that an immediate biological consequence of H_2O_2-induced AA release is to generate mediators that propagate and/or amplify the oxidative injury. However, unmetabolized free AA is, by far and large, the main compound released into the medium after exposure to H_2O_2, which raises the possibility that its metabolism to eicosanoid mediators might not be its actual fate or, at least, not the only one. Thus, further studies will be needed to address the biological roles of H_2O_2-mediated release of unmetabolized AA in phagocytic cells.

Another aspect that remains unknown is whether $iPLA_2$, in addition to its housekeeping role in U937 cells and phagocytic cells in general, also plays some role in regulated phospholipid hydrolysis in phagocytic cells. The fact that multiple splice variants of $iPLA_2$ exist in some cells and that other $iPLA_2$s distinct from the "classical" group VI enzyme have recently been described [6], suggest the possibility that $iPLA_2$ may be subject to complex regulatory mechanisms that differ among cell types. Two recent reports utilizing cells over-expressing group VI PLA_2 have shown the enzyme to be responsive to Ca^{2+} ionophore in HEK293 cells [37] and to glucose plus cAMP-elevating agents in INS-1 insulinoma cells [38], thus suggesting that the enzyme might be capable of playing some signaling roles in cells.

References

1 Six DA, Dennis EA (2000) The expanding superfamily of phospholipase A_2 enzymes: Classification and characterization. *Biochim Biophys Acta* 31: 1–19

2 Kudo I, Murakami M (2002) Phospholipase A_2 enzymes. *Prostaglandins* 68–69, 3–58

3 Balsinde J, Balboa MA, Insel PA, Dennis EA (1999) Regulation and inhibition of phospholipase A_2. *Annu Rev Pharmacol Toxicol* 39: 175–189

4 Balsinde J, Winstead MV, Dennis EA (2002) Phospholipase A_2 regulation of arachidonic acid mobilization. *FEBS Lett* 531: 2–6

5 Chilton FH, Fonteh AN, Surette ME, Triggiani M, Winkler JD (1996) Control of arachidonate levels within inflammatory cells. *Biochim Biophys Acta* 1299: 1–15

6 Winstead MV, Balsinde J, Dennis EA (2000) Calcium-independent phospholipase A_2: Structure and function. *Biochim Biophys Acta* 1488: 28–39

7 Balsinde J, Fernández B, Solís-Herruzo JA (1994) Increased incorporation of arachidonic acid into phospholipids in zymosan-stimulated mouse peritoneal macrophages. *Eur J Biochem* 221: 1013–1018

8 Balsinde J, Bianco ID, Ackermann EJ, Conde-Frieboes K, Dennis EA (1995) Inhibition of calcium-independent phospholipase A_2 prevents arachidonic acid incorporation and phospholipid remodeling in $P388D_1$ macrophages. *Proc Natl Acad Sci USA* 92: 8527–8531

9 Balsinde J, Balboa MA, Dennis EA (1997) Antisense inhibition of group VI Ca^{2+}-independent phospholipase A_2 blocks phospholipid fatty acid remodeling in murine $P388D_1$ macrophages. *J Biol Chem* 272: 29317–29321

10 Daniele JJ, Fidelio GD, Bianco ID (1999) Calcium dependency of arachidonic acid incorporation into cellular phospholipids of different cell types. *Prostaglandins* 57: 341–350

11 Alzola E, Perez-Etxebarria A, Kabre E, Fogarty DJ, Metioui M, Chaib N, Macarulla JM, Matute C, Dehaye JP, Marino A (1998) Activation by P2X7 agonists of two phospholipases A_2 (PLA_2) in ductal cells of rat submandibular gland. Coupling of the calcium-independent PLA_2 with kallikrein secretion. *J Biol Chem* 273: 30208–30217

12 Birbes H, Drevet S, Pageaux JF, Lagarde M, Laugier C (2000) Involvement of calcium-independent phospholipase A_2 in uterine stromal cell phospholipid remodeling. *Eur J Biochem* 267: 7118–7127

13 Ramanadham S, Hsu FF, Bohrer A, Ma Z, Turk J (1999) Studies of the role of group VI phospholipase A_2 in fatty acid incorporation, phospholipid remodeling, lysophosphatidylcholine generation, and secretagogue-induced arachidonic acid release in pancreatic islets and insulinoma cells. *J Biol Chem* 274: 13915–13927

14 Chiu CH, Jackowski S (2001) Role of calcium-independent phospholipases ($iPLA_2$) in phosphatidylcholine metabolism. *Biochem Biophys Res Commun* 287: 600–606

15 Balsinde J (2002) Roles of various phospholipases A_2 in providing lysophospholipid acceptors for fatty acid phospholipid incorporation and remodelling. *Biochem J* 364: 695–702

16 Balsinde J, Dennis EA (1996) Distinct roles in signal transduction for each of the phospholipase A_2 enzymes present in $P388D_1$ macrophages. *J Biol Chem* 271: 6758–6765

17 Chilton FH, Connell TR (1988) 1-Ether-linked phosphoglycerides. Major endogenous sources of arachidonate in the human neutrophil. *J Biol Chem* 263: 5260–5265

18 Boilard E, Surette ME (2001) Anti-CD3 and concanavalin A-induced human T cell proliferation is associated with an increased rate of arachidonate-phospholipid remodeling. Lack of involvement of group IV and group VI phospholipase A_2 in remodeling and increased susceptibility of proliferating T cells to CoA-independent transacyclase inhibitor-induced apoptosis. *J Biol Chem* 276: 17568–17575

19 Hsu FF, Ma Z, Wohltmann M, Bohrer A, Nowatzke W, Ramanadham S, Turk J (2000) Electrospray ionization/mass spectrometric analyses of human pro-monocytic U937 cell glycerolipids and evidence that differentiation is associated with membrane lipid composition changes that facilitate phospholipase A_2 activation. *J Biol Chem* 275: 16579–16589

20 Bonventre JV, Huang Z, Taheri MR, O'Leary E, Li E, Moskowitz MA, Sapirstein, A (1997) Reduced fertility and post-ischaemic brain injury in mice deficient in cytosolic phospholipase A_2. *Nature* 390: 622–625

21 Uozumi N, Kume K, Nagase T, Nakatani N, Ishii S, Tashiro F, Komagata Y, Maki K, Ikuta K, Ouchi Y, Miyazaki J, Shimizu T (1997) Role of cytosolic phospholipase A_2 in allergic response and parturition. *Nature* 390: 618–622

22 Fonteh AN (2002) Differential effects of arachidonoyl trifluoromethyl ketone on arachidonic acid release and lipid mediator biosynthesis by human neutrophils. Evidence for different arachidonate pools. *Eur J Biochem* 269: 3760–3770

23 Ono T, Yamada K, Chikazawa Y, Ueno M, Nakamoto S, Okuno T, Seno K (2002) Characterization of a novel inhibitor of cytosolic phospholipase $A_2\alpha$. *Biochem J* 363: 727–735

24 Ghomashchi F, Stewart A, Hefner Y, Ramanadham S, Turk J, Leslie CC, Gelb MH (2001) Pyrrolidine-based specific inhibitor of cytosolic phospholipase $A_2\alpha$ blocks arachidonic acid release in a variety of mammalian cells. *Biochim Biophys Acta* 1513: 160–166

25 Balboa MA, Balsinde J (2002) Involvement of calcium-dependent phospholipase A_2 in hydrogen peroxide-induced accumulation of free fatty acids in human U937 cells. *J Biol Chem* 277: 40384–40389

26 Balboa MA, Pérez R, Balsinde J (2003) Amplification mechanisms of inflammation: paracrine stimulation of arachidonic acid mobilization by secreted phospholipase A_2 is regulated by cytosolic phospholipase A_2-derived hydroperoxyeicosateatraenoic acid. *J Immunol* 171: 989–994

27 Balboa MA, Sáez Y, Balsinde J (2003) Calcium-independent phospholipase A_2 is required for lysozyme secretion in U937 promonocytes. *J Immunol* 170: 5276-5280

28 Balsinde J, Dennis EA (1996) Distinct roles in signal transduction for each of the phospholipase A_2 enzymes present in $P388D_1$ macrophages. *J Biol Chem* 271: 6758–6765

29 Carnevale KA, Cathcart MK (2001) Calcium-independent phospholipase A_2 is required for human monocyte chemotaxis to monocyte chemo-attractant protein 1. *J Immunol* 167: 3414–3421

30 Teslenko V, Rogers M, Lefkowith JB (1997) Macrophage arachidonate release *via* both the cytosolic Ca^{2+}-dependent and -independent phospholipases is necessary for cell spreading. *Biochim Biophys Acta* 1344, 189–199

31 Birbes H, Gothié E, Pageaux JF, Lagarde M, Laugier C (2000) Hydrogen peroxide activation of Ca^{2+}-independent phospholipase A_2 in uterine stromal cells. *Biochem Biophys Res Commun* 276: 613–618

32 Rao GN, Runge MS, Alexander RW (1995) Hydrogen peroxide activation of cytosolic phospholipase A_2 in vascular smooth muscle cells. *Biochim Biophys Acta* 1265: 67–72

33 Boyer CS, Bannenberg GL, Neve EP, Ryrfeldt A, Moldeus P (1995) Evidence for the activation of the signal-responsive phospholipase A_2 by exogenous hydrogen peroxide. *Biochem Pharmacol* 50: 753–761

34 Samanta S, Perkington MS, Morgan M, Williams RJ (1998) Hydrogen peroxide enhances signal-responsive arachidonic acid release from neurons: role of mitogen-activated protein kinase. *J Neurochem* 70: 2082–2090

35 Sporn PH, Marshall TM, Peters-Golden M (1992) Hydrogen peroxide increases the availability of arachidonic acid for oxidative metabolism by inhibiting acylation into phospholipids in the alveolar macrophage. *Am J Resp Cell Mol Biol* 7: 307–316

36 Cane A, Breton M, Koumanov K, Bereziat G, Colard O (1998) Oxidant-induced arachidonic acid release and impairment of fatty acid acylation in vascular smooth muscle cells. *Am J Physiol* 274: C1040–C1046

37 Murakami M, Kambe T, Shimbara S, Kudo I (1999) Functional coupling between vari-

ous phospholipase A_2s and cyclo-oxygenases in immediate and delayed prostanoid biosynthetic pathways. *J Biol Chem* 274: 3103–3115
38 Ma Z, Zhang S, Turk J, Ramanadham S (2002) Stimulation of insulin secretion and associated nuclear accumulation of iPLA$_2\beta$ in INS-1 insulinoma cells. *Am J Physiol* 282: E820–E833

Arachidonate remodeling and PAF synthesis in human neutrophils

Robert L. Wykle

Department of Biochemistry, Wake Forest University School of Medicine, Medical Center Boulevard, Winston-Salem, NC 27157, USA

Introduction

Both AA-derived eicosanoids and platelet activating factor (PAF) serve important physiological functions, but also participate in pathological developments [1–4]. The metabolism and actions of PAF, arachidonic acid (AA), and AA-derived eicosanoids are closely linked in neutrophils, indeed they can be derived from the same phospholipid precursor, 1-O-alkyl-2-AA-GPC (sn-glycero-3-phosphocholine). Neutrophils play a major role in host defense and inflammation and many studies have focused on neutrophil signaling systems and pharmacological intervention in these systems. Since neutrophils are terminally differentiated cells and can be maintained for only short periods of time after their isolation, molecular approaches for their study have been limited. In addition, many of the enzymes that are responsible for PAF and AA metabolism, except for $cPLA_2$ (85 kDa cytosolic phospholipase A_2) and 5-lipoxygenase, are membrane proteins and have not been isolated. Thus much of our knowledge of the lipid metabolism and lipid-mediated signaling has been obtained through the study of crude systems and intact cells. This review attempts to summarize the work of my colleagues and me in this area and closely related work of others. Our understanding of neutrophil signaling draws on numerous studies of other cells and tissues largely beyond the scope of this review but generally acknowledged in the primary literature cited.

Human neutrophils contain high levels of ether-linked phospholipids enriched in arachidonate

At the time we began our studies of AA and PAF metabolism in neutrophils, two biosynthetic pathways for the formation of PAF had been found, a remodeling pathway in which lyso PAF derived from membrane lipids could be acetylated by an acetyltransferase to form PAF [5] and a *de novo* route which converted an acetylated alkyl diglyceride to PAF *via* a CDP-choline cholinephosphotransferase-catalyzed reaction [6]. We examined human neutrophils to determine if they contained enough 1-O-

Arachidonate Remodeling and Inflammation, edited by Alfred N. Fonteh and Robert L. Wykle

alkyl-2-acyl-GPC to support PAF synthesis by the remodeling pathway and conducted experiments to see if we could find evidence for the *de novo* pathway.

We were surprised to find that almost half of the cholined-containing phosphoglyceride class (PC) was comprised of the 1-O-alkyl ether-linked subclass [7]. In addition, we found that two-thirds of the ethanolamine-containing phosphoglyceride class (PE) was 1-O-alk-1'-enyl linked (plasmalogen). Thus the ether-linked phospholipids are major components of the neutrophil membranes. Eosinophils were found to contain even higher levels of the 1-O-alkyl and plasmalogen subclasses comprising 70 mol% of PC and 75 mol% of PE, respectively [8]. It is clear these cells contain far more 1-O-alkyl-2-acyl-GPC than needed to support PAF synthesis by the remodeling pathway. Since the ether bonds are not hydrolyzed by phospholipases it is possible the ether-linked subclasses help stabilize neutrophil membranes when the cells leave the blood stream to invade other tissues. The high plasmalogen content might also help protect against reactive oxygen species [9], which are actively produced by the cells particularly during the oxidative burst that is initiated to kill invading organisms.

In studies of the ether lipid content of neutrophil lipids, Mueller and co-workers [10] found that although AA comprised only 7% of all fatty acyl chains found in the PC fraction, 63% of this amount was found as 1-O-alkyl-2-AA-GPC. The PE contained much greater levels of AA comprising 27% of the acyl chains; 80% of this amount was found as 1-O-alk-1'-enyl-2-AA-GPE (1-O-alk-l'-enyl-2-arachidonoyl-sn-glycero-3-phosphoethanolamine). Chilton and Connell [11] observed a similar distribution of AA in human neutrophils based on mass spectral analysis. Similar enrichment of AA in the ether linked subclasses of rat neutrophils was reported by Ramesha and Pickett [12]. The subcellular distribution of the ether-linked phospholipid classes and AA distribution was examined by MacDonald and Sprecher [13] who found that the subclass composition was markedly uniform throughout the subcellular membrane fractions yet the major pool of AA was localized in intracellular membranes. Our laboratory made very similar observations (Venable and Wykle, unpublished).

A number of studies have shown that when the diets of rats [14], monkeys [15] or humans [16–18] are supplemented with a fish oil diet containing eicosapentaenoic acid (EPA), the EPA is incorporated in a pattern almost identical to that of AA. Thus it is enriched in the ether-linked subclasses and is released form these pools upon stimulation of the cells. The incorporation of gammalinolenic acid into human neutrophils has also been examined [19]. Since the n-3 fatty acids yield less active lipoxygenase products, the substitution of EPA or other polyunsaturated fatty acids could importantly affect eicosanoid responses in inflammation [16].

These studies have revealed that human neutrophils contain high levels of ether-linked phospholipids in their membranes and that these subclasses are enriched in AA. The ethanolamine plasmalogen fraction contains by far the greatest pool of AA in the cells and donate the most free AA upon stimulation [11].

Co-synthesis of PAF and eicosanoids

It was early recognized that the same stimuli which elicit eicosanoids also elicit synthesis of PAF. PAF and the eicosanoids have overlapping activities and act synergistically to promote cell function [20, 21]. The formation of lyso PAF from 1-O-alkyl-2-acyl-GPC in the remodeling pathway of PAF synthesis requires the removal of the 2-acyl chain by a phospholipase A_2 (PLA_2) or transacylase reaction, a reaction which also requires formation of a lysophospholipid acceptor by a PLA_2 reaction. It was at first puzzling why no [^{14}C] lysophospholipid was observed in stimulated neutrophils labeled with [^{3}H] AA in the *sn*-2 position and [^{14}C] stearate in the sn-1 position of PC, even though [^{3}H] AA release was seen [22]. This was subsequently explained by the finding from the labeling studies that [^{3}H] AA and [^{14}C] stearate are not found in the same molecules; [^{3}H] AA is in molecules containing unlabeled stearate in the *sn*-1 position while stearate-labeled PC species contain oleate and linoleate but not AA in the *sn*-2 position. Thus the specific release of [^{3}H] AA releases only unlabeled lyso PC. In further studies [^{3}H] AA was shown to be incorporated into seven molecular species of PC, 1-O-alkyl ether-linked species and 1-acyl-linked species [23]. A similar pattern of incorporation was observed employing mass spectroscopy [24, 25]. These studies revealed that 1-O-alkyl-2-AA-GPC is actively formed in neutrophils and provided evidence that it could yield 1-O-alkyl-2-lyso-GPC (lyso PAF) through the AA-specific action of PLA_2.

One of our most striking discoveries was the finding that when human neutrophils are exposed to PAF labeled in the 1-O-alkyl chain, 1-O[^{3}H]hexadecyl-2-acetyl-GPC, the cells rapidly remove the acetate and reacylate the [^{3}H]lyso PAF in a highly specific manner with AA accounting for approximately 90% of the added acyl chains [26, 27]. Similarly exogenous [^{3}H]lyso PAF was also rapidly taken up and acylated with AA [26–28]. This observation provided a labeling approach to specifically label the neutrophils with 1-O-[^{3}H]alkyl-2-AA-GPC and only traces of other labeled products. Using neutrophils labeled in such a manner we were able to show directly that the arachidonate-containing species, 1-O-alkyl-2-AA-GPC is readily converted to PAF upon stimulation [28]. Thus the same precursor can yield both PAF and AA products. Similar conclusions were drawn from studies of macrophages by Albert and Synder [29]. In the acylation of lyso PAF, we found that neutrophils from monkeys on a fish oil diet substituted EPA for AA in the acylation of lyso-PAF in the same ratio of EPA:AA as found in the cells [15]. Again, it appears EPA from fish oil may interact in a centrally important pathway.

Evidence that the arachidonate-containing molecular species, 1-O-alkyl-2-AA-GPC is the obligate precursor of PAF was obtained from studies of HL60 cells grown in essential fatty acid-deficient media and differentiated into neutrophil-like granulocytes [30]. The cells grown in essential fatty acid-deficient media could not make PAF unless they were supplemented with AA or other polyunsaturated fatty acids [30]. Rat neutrophils that were 90% depleted of AA synthesized 85% less PAF than control cells containing normal levels of AA [31, 32]. Much of the AA was replaced by eicosatrienoate (20:3) but the 20:3-containing 1-O-alkyl-2-acyl-GPC did not serve as a precursor of PAF, indicating a highly specific deacylation reaction is required [33].

Molecular species of PAF and related products synthesized by stimulated neutrophils

Although there are reports of the production of highly selective molecular species of PAF by human neutrophils, our laboratory observed several molecular species of PAF labeled by acetate formed upon fMLP stimulation, with the 16:0 (39%) species being the major product followed by 18:1 (22%), 18:0 (15%) and other minor alkyl-linked species [34]. In addition, 1-acyl-2-acetyl-GPC was produced and accounted for approximately 12–15% of the acetate-labeled products; the 1-acyl-2-acetyl-GPC also accounted for the major PAF-like products synthesized by other cells [35]. The molecular species of acetate-labeled products was significantly different from the alkyl chain composition of the PC fraction and more closely followed the AA-linked species. Pinckard and co-workers [36] also observed five molecular species of PAF as detected by reverse phase high performance liquid chromatography coupled with rabbit platelet aggregation assays; only one species was identified by mass spectroscopy, the 16:0 species.

1-Acyl-2-acetyl-GPC containing a 1-acyl linkage rather than ether linkage, which is at least a hundred-fold less active than the ether-linked species [37] is formed in a number of cells. Its predominance appears to reflect the arachidonate-containing species of PC. If cells, such as endothelial cells, contain low levels of 1-O-alkyl-2-AA-GPC and more 1-acyl-2-AA-GPC the acetylated product will be predominantly 1-acyl-2-acetyl-GPC as shown by Chilton and co-workers. The function of this product is not clear but it may be able to act synergistically with other agonists to elicit responses.

In addition to choline-containing acetate species, neutrophils produce an acetate-containing 1-O-alk-1'-enyl-2-acetyl-GPE (ethanolamine plasmalogen) upon stimulation [38]. The production of this compound is not surprising since large amounts of 1-O-alk-1'-enyl-2-lyso-GPE accumulates in stimulated neutrophils and reflects the high levels of AA found as 1-O-alk-1-enyl-2-AA-GPE. Upon stimulation of human neutrophils, the arachidonate- containing species of PE are selectively hydrolyzed [39].

Arachidonic acid incorporation into phospholipids and remodeling

As in many other cells and tissues [40–42], radiolabeled AA given to neutrophils is first incorporated primarily into diacyl-GPC and phosphatidylinositol (PI) and gradually transferred to 1-O-alkyl-2-AA-GPC and the ethanolamine plasmalogen, 1-O-alk-1'-enyl-2-AA-GPE [24, 43]. At higher concentrations, much of the exogenous AA is initially incorporated into triglyceride. It is generally agreed that exogenous AA is initially incorporated by a Land's cycle-type mechanism by which arachidonic acid is first converted to CoA-AA and then incorporated into 1-acyl-2-lyso-GPC and PI by a CoA-dependent acyltransferase reaction. However, the ultimate disposition of the AA into the ether-linked pools is catalyzed by a CoA-independent transacylase (CoA-IT)-catalyzed reaction [41, 44–48]. The CoA-IT is an intrinsic membrane protein that is highly specific for polyunsaturated fatty chains especially AA and EPA [15, 41, 42, 44, 48, 49]. It has no metal ion requirements or requirement for ATP. The enzyme was concluded in one study to be activated by TNF-α, but we were unable to confirm the activation of the enzyme in neutrophils [50]. The assay systems for measuring CoA-IT activity are quite complex and may explain these discrepant results; we now believe the CoA-IT is constitutatively active and specifically transfers AA to lysophospholipids and that its activity is dependent only on the appearance of an appropriate lysophospholipid acceptor. We found that the CoA-IT has a preference for 1-O-alk'-enyl-2-lyso-GPE, which could explain the enrichment of AA found in the ethanolamine-containing plasmalogen subclass [51]. However the enzyme also readily transfers AA to 1-acyl-2-lyso-GPC and 1-O-alkyl-2-lyso-GPC. It seems likely that the CoA-IT acts in synchrony with the Ca^{2+}-independent PLA_2 and the low molecular weight secreted PLA_2 to achieve the pattern of AA distribution observed in intact neutrophils.

A number of intriguing questions surround the role and action of CoA-IT. The high specificity of the enzyme for AA is maintained in isolated membrane fractions which readily convert trace amounts of lyso PAF, 1-O-alkyl-2-lyso-GPC, to 1-O-alkyl-2-AA-GPC. The membrane fractions even more readily convert 1-O-alk-1'-enyl-2-lyso-GPE to 1-O-alkyl-1'-enyl-2-AA-GPE. In these reactions the membranes supply the AA but what is the origin of the AA transferred? Also the transfer of AA from one molecular species of phospholipid to a lysophospholipid only generates another lysophospholipid. When 1-O-alkyl-2-lyso-GPC is formed by CoA-IT, some of it is converted to PAF by acetylation; more is likely reacylated by acyl-CoA-dependent mechanisms. Many PLA_2's and lysophospholipases exhibit transacylase activity. This raises the possibility that CoA-IT with the help of chaperon proteins, or under certain conditions could act as a PLA_2 and release free AA by transferring the enzyme-bound AA to water, rather than to a lysophospholipid acceptor; to our knowledge no evidence for such an action has been found. However, the CoA-IT activity observed under optimal enzymatic assay conditions is much greater than any cellular PLA_2 activity we have observed. The CoA-IT clearly has the ability to cleave AA from phospholipids donors. Are the transferase and acylase sites of the enzyme the same? It should be possible to answer many of these questions once the structure of the enzyme is determined. Its purification from membranes has been problematic, in part because both donor and acceptor phospholipid substrates are required and the effects of detergents on the interactions of the enzyme and substrates are unknown. We have speculated that CoA-IT might be related to lysolecthin:cholesterol acyltransferase (LCAT).

CoA-IT is clearly one of the most important enzymes in the remodeling of AA from diacyl-linked species to the ether-linked phospholipids and PE, and may also be a key enzyme in the synthesis of PAF as discussed below. Since the equilibration of labeled AA with endogenous pools requires many hours, studies of AA release based on short-period labeling can yield very misleading results that do not agree even closely with results based on mass. Thus the PE fraction of neutrophil phospholipids is poorly labeled by exogenous AA within the time constraints of these labile cells and most observed loss of AA-label upon stimulation of prelabeled cells is from the highly labeled PC and phosphatidylinositol classes, whereas mass determinations reveal that far more AA is lost from the PE plasmalogen fraction [11], which contains little labeled AA.

The shift over time of AA from the diacyl-GPC into 1-O-alkyl-2-AA-GPC and 1-O-alk-1'-enl-2-AA-GPE appears to provide an exciting tool for examining the subcellular movement of AA among organelles as well as the metabolic pools giving rise to 5-HETE, LTB_4 and other eicosanoids. Thus by double labeling the cells by first incubating them for 30 to 60 minutes with [^{14}C] AA, then washing the cells followed by a five minute pulse labeling with [^{3}H] AA, one should be able to observe a higher ^{3}H:^{14}C ratio in the more rapidly labeled pools and a lower ratio in the more slowly labeled pools such as 1-O-alkyl-2-AA-GPC and PE. Upon stimulation of the doubly labeled cells the ^{3}H:^{14}C ratio of products such as 5-HETE and LTB_4 might reflect and reveal the pool of origin. By cooling the cells, isolating subcellular fractions, and determining their ^{3}H:^{14}C ratio it might be possible to follow the movement of the exogenous AA among the subcellular fractions. Although we obtained promising results using this approach, a number of problems make interpretation of such data difficult. One of the problems is the rapid labeling of phosphatidylinositol and release of labeled AA from that pool upon stimulation. In addition the subcellular pools appear to be labeled very rapidly. Further, refinement of this approach might yet shed light on the role of various AA pools of neutrophils. The approach has been used in other cells.

Stimulated release of AA and PAF synthesis

Synthesis of PAF and release of AA

The *de novo* pathway of PAF synthesis, which is dependent in the final step on a specific dithiotheitol insensitive CDP-choline cholinephospho-photransferase to convert 1-O-alkyl-2-acetyl-*sn*-glyceryl to PAF [6]. This pathway does not appear to be a significant route of PAF synthesis in neutrophils or other inflammatory cells.

Since the ether bond is derived from a fatty alcohol, hexadecanol e.g., should be incorporated into PAF in a 1:1 molar ratio with acetate to form 16:0 PAF. We observed the synthesis of 1-O-alkyl-2-acyl-*sn*-glycerophosphate from [^{14}C] hexadecanol but no species containing a 2-acetyl moiety [52]. One study concluded that holine from CDP-choline can be incorporated into PAF upon treatment with PMA [53], but further confirmation of this finding and its mechanism of incorporation have not been reported. Since studies have not shown the direct incorporation of fatty alcohol into PAF, it is possible that the *de novo* pathway could be initiated by PLA_2 and lysophospholipase D acting on 1-O-alkyl-2-lyso-GPC to generate 1-O-alkyl-2-lyso-*sn*-glycero-3-phosphate that could then be acetylated and yield PAF. However, we have been unable to detect lysophospholipase D in neutrophils. Overall, it does not appear the *de novo* pathway is responsible for PAF synthesis in stimulated neutrophils. On the other hand, the remodeling pathway has been conclusively demonstrated.

Although the remodeling pathway of PAF synthesis is firmly established, what is less clear is the relative contributions of a direct route of synthesis initiated by PLA_2 acting directly on 1-O-alkyl-2-AA-GPC to generate lyso PAF, and an indirect route in which PLA_2 first acts on ethanolamine plasmalogen to generate 1-O-alk-1'-enyl-2-lyso-GPE which then serves as an acceptor for the transfer of AA from 1-O-alkyl-2-AA-GPC by the CoA-independent transacylase (CoA-IT) to generate lyso PAF. We believe both pathways are likely operative in stimulated neutrophils. The acetyl-CoA:lyso PAF acetyltransferase responsible for the conversion of lyso PAF to PAF prefers 1-O-alkyl-2-lyso-GPC as a substrate but can also accept 1-acyl-2-lyso-GPC [5]; the same enzyme may be able to form the 1-O-alk-1'-enyl-2-acetyl-GPE observed [38]; however, Lee and coworkers [54] found evidence that a transacetylase which transfers acetate from PAF to 1-O-alk-1'-enyl-2-lyso-GPE is more active in the synthesis of the PE-derived product. In support of an indirect route of PAF synthesis:

1) CoA-IT is highly specific for AA
2) an accumulation of 1-O-alk-1'-enyl-2-lyso-GPE and its acetylated derivative is observed upon stimulation of the cells;
3) the bulk of AA released upon stimulation is derived from the ethanolamine plasmalogen;
4) a loss of specificity for AA in the reacylation of lyso PAF is observed and explained by the buildup of 1-O-alk-1'-enyl-2-lyso-GPE which competes as an acceptor for AA in the CoA-IT reaction [3];
5) and the indirect route can be demonstrated in an enzymatic system or permeabilized neutrophils [55].

In support of the direct route, the 85 kDa cytosolic PLA_2, $cPLA_2$ can readily be shown to act directly on 1-O-alkyl-2-AA-GPC to generate lyso PAF and is also highly specific for AA. The direct route thus yields the concomitant release of AA and lyso PAF, which can be converted to PAF. $cPLA_2$ is widely believed to be the PLA_2 responsible for initiating PAF synthesis in neutrophils. Some of the strongest evidence for such a role comes from studies of cells derived from $cPLA_2$-knockout mice, which lose the ability to synthesize PAF [56]. Low molecular weight secreted PLA_2, $sPLA_2$, could also play a role [57, 58] but might be acting through its receptors to activate $cPLA_2$ [59]. The relative importance of the direct and indirect routes has been difficult to determine; it should be noted that both routes require the action of a PLA_2. The $sPLA_2$ is not specific for AA, but by generating an ethanolamine acceptor for CoA-IT could lead to the specific release of AA from the PC class. We observed a selective loss of AA-containing species of PE based on measuring the molecular species of PE in stimulated and unstimulated neutrophils, a finding that appears to favor hydrolysis by an AA-selective PLA_2 such as $cPLA_2$ [39].

The $cPLA_2$ has been extensively studied in many cells [60]. Its activity in neutrophils was earlier difficult to demonstrate since the cells contain high levels of proteases that rapidly destroy its activity when the cells are disrupted. The activity of the enzyme is recovered when cells are disrupted in buffer containing a mixture of protease inhibitors [61]. The properties of $cPLA_2$ and its movement from the cytosol to membranes in the presence of Ca^{2+} and upon phosphorylation has been well documented; see [60] for review. Phosphorylation has recently been shown not to be required for movement to the membranes [60]. The enzyme was thought earlier to migrate largely to the nuclear membrane but is now found to move to membranes throughout the cells based on studies of Leslie and co-workers [62].

Once AA is released, much remains as free arachidonic acid, which likely acts as a mediator itself, e.g., possibly serving to translocate protein kinase C to membranes and assist its activation. The free AA is also converted to eicosanoids largely leukotriene B_4 (LTB_4) and 5-hydroxyeicosatetraenoic acid (5-HETE), both important mediators of inflammation. The 5-HETE can be oxidized to the 5-keto compound, 5-oxo-ETE [63]. The 5-HETE can also be reincorporated into phospholipids.

PAF and the eicosanoids are not only synthesized in concert, they act synergistically to elicit responses. The mediators serve to signal both other neutrophils and an array of other cells including endothelial cells that participate in the inflammatory process. Early studies of PAF revealed that although PAF alone can elicit the degranulation response of neutrophils, the same response can be obtained with 100-fold lower concentrations of PAF in the presence of 5-HETE. In more recent studies, 5-oxo-ETE was found to profoundly potentiate the activity of cytokines in neutrophils [64] and of PAF in eosinophils [65]. The findings suggest that 5-oxo-ETE produced by neutrophils may increase the responses of eosinophils to other stimuli and thus plays a major role in inflammation.

Neutrophils exposed to certain agents that do not themselves elicit AA release become more responsive (primed) to subsequent exposure to agonists that do. The mechanisms of priming are still poorly understood. In studies to compare the priming ability of diacyl diglycerides, which activate protein kinase C and 1-O-alkyl-2-acyl diglycerides, which do not activate protein kinase C, both diglycerides primed neutrophils to release markedly more AA in response to the chemotactic peptide fMLP [66, 67]. However, one of the most interesting findings was that priming with the diacyl diglyceride resulted in the further conversion of the released AA to 5-HETE and LTB_4 whereas no conversion of the released AA to lipoxygenase products was observed when the cells were primed with the alkyl acyl diglyceride. Subsequent studies revealed that while the diacyl diglyceride activates $cPLA_2$, the ether species does not [68] but rather involves the mobilization of low molecular weight $sPLA_2$ [57]. The lack of conversion to 5-HETE and LTB_4 in the cells primed by the ether-linked diglyceride might result from lack of activation of the lipoxygenase, or release of AA from a pool not accessible to the lipoxygenase, or from other parameters.

Activation of PAF synthesizing enzymes

The key enzymes in the synthesis of PAF and the mobilization of arachidonate are the acetyltransferase, which converts lyso PAF to PAF by transferring an acetyl group from acetyl-CoA, the 85 kDa $cPLA_2$, which specifically hydrolyzes AA from phospholipids, and the CoA-independent transacylase (CoA-IT), which allows the formation of 1-O-alk-1'-enyl-2-lyso-GPE to trigger lyso PAF formation by transferring AA from 1-O-alkyl-2-AA-GPC to the lyso plasmalogen acceptor, a reaction that is highly specific for AA and can generate lyso PAF without releasing free arachidonic acid. A role for $sPLA_2$ or the Ca^{2+}-independent phospholipase A_2 ($iPLA_2$) has not been ruled out. We have been unable to demonstrate activation of CoA-IT as discussed above; the appearance of a lysophospholipid acceptor apparently is sufficient to trigger action of the enzyme to remove and transfer AA. On the other hand, activation of $cPLA_2$ has long been recognized and widely studied as recently reviewed [60, 69] while the rapid activation (30 seconds) of the acetyltransferase (four- to ten-fold basal activity) upon stimulation was early recognized in a number of cell types [70–74].

The exact signaling pathways responsible for activation of the $cPLA_2$ and the acetyltransferase have been difficult to pinpoint because of the extensive cross-talk between systems; e.g., protein kinase C activation can result in activation of ERK and p38 MAP kinases. Thus activation responses due to a protein kinase C activator such as PMA yet dependent on the direct action of a MAP kinase may be blocked by protein kinase C inhibitors due to the cross-talk.

We recently carried out a series of studies that indicate the ERK's and p38 kinase play a central role in the regulation of PAF synthesis in human neutrophils [50, 75]. We further examined the priming of the cells by alkylacylglycerol and diacylglycerol

[68] and found that the diacyl diglyceride activated the p42 and p44 MAP kinases along with $cPLA_2$, whereas the alkyl diglyceride, which did not activate protein kinase C, did not activate MAP kinases on $cPLA_2$. These studies provided support for the phosphorylation of $cPLA_2$ by ERK's linked to activation by protein kinase C. Many studies of neutrophil stimulation have employed the Ca^{2+} ionophore A23187; we recently found that the ionophore closely mimics physiological stimuli in the activation of MAP kinase and enzyme activities [50]. In further studies [50], the stimulated phosphorylation and activation of cPLA2 was found to be reduced both by SB203580, a p38 MAP kinase inhibitor and by the MEK inhibitor PD98059 which blocks activation of the ERK's; $cPLA_2$ activity was suppressed below unstimulated levels when a combination of both inhibitors was used. On the other hand the acetyltransferase activation was blocked by the p38 inhibitor but not by the ERK inhibitor [75]. We also demonstrated using membrane fractions that active recombinant p38 increased the acetyltransferase to the maximal level observed with TNF-α. However, recombinant ERK's did not activate the enzyme [75]. We have found as reported by others [76, 77] that both the recombinant p38 and ERK's can phosphorylate and activate recombinant $cPLA_2$.

These findings support a scheme by which both ERK's and p38 MAP kinase phosphorylate and activate $cPLA_2$ whereas only p38 activates the acetyltransferase.

Overall, a complex network of signaling pathways have now been revealed in human neutrophils linking PAF and AA metabolism through signaling by ERK's, p38 kinase and protein kinase C. More definitive studies can be carried out once the structures and genes for the acetyltransferase and CoA-IT have been determined. The earlier studies have shown that the metabolism of PAF and AA and their actions are very closely linked and constitute an important signaling system for neutrophils and other inflammatory cells as well as endothelial cells.

References

1 Snyder F (1989) Biochemistry of platelet-activating factor: A unique class of biologically active phospholipids. *Proc Soc Exp Biol Med* 190: 125–135

2 Prescott SM, Zimmerman GA, McIntyre TM (1990) Platelet-activating factor. *J Biol Chem* 265: 17381–17384

3 Venable ME, Olson SC, Nieto ML, Wykle RL (1993) Enzymatic studies of lyso platelet-activating factor acylation in human neutrophils and changes upon stimulation. *J Biol Chem* 268: 7965–7975

4 Snyder F (1995) Platelet-activating factor: the biosynthetic and catabolic enzymes. *Biochem J* 305: 689–705

5 Wykle RL, Malone B, Snyder F (1980) Enzymatic synthesis of 1-alkyl-2-acetyl-*sn*-glycero-3-phosphocholine, a hypotensive and platelet-aggregating lipid. *J Biol Chem* 255: 10256–10260

6 Renooij W, Snyder F (1981) Biosynthesis of 1-alkyl-2-acetyl-*sn*-glycero-3-phosphocholine (platelet activating factor and a hypotensive lipid) by cholinephosphotransferase in various rat tissues. *Biochim Biophys Acta* 663: 545–556
7 Mueller HW, O'Flaherty JT, Wykle RL (1982) Ether lipid content and fatty acid distribution in rabbit polymorphonuclear neutrophil phospholipids. *Lipids* 17: 72–77
8 Ojima-Uchiyama A, Masuzawa Y, Sugiura T, Waku K, Saito H, Yui Y, Tomioka H (1988) Phospholipid analysis of human eosinophils: High levels of alkylacylglycerophosphocholine (PAF precursor). *Lipids* 23: 815–817
9 Khaselev N, Murphy RC (2000) Structural characterization of oxidized phospholipid products derived from arachidonate-containing plasmenyl glycerophosphocholine. *J Lipid Res* 41: 564–572
10 Mueller HW, O'Flaherty JT, Greene DG, Samuel MP, Wykle RL (1984) 1-O-alkyl-linked glycerophospholipids of human neutrophils: Distribution of arachidonate and other acyl residues in the ether-linked and diacyl species. *J Lipid Res* 25: 383–388
11 Chilton FH, Connell TR (1988) 1-ether-linked phosphoglycerides. Major endogenous sources of arachidonate in the human neutrophil. *J Biol Chem* 263: 5260–5265
12 Ramesha CS, Pickett WC (1987) Fatty acid composition of diacyl, alkylacyl, and alkenylacyl phospholipids of control and arachidonate-depleted rat polymorphonuclear leukocytes. *J Lipid Res* 28: 326–331
13 MacDonald JI, Sprecher H (1989) Distribution of arachidonic acid in choline- and ethanolamine-containing phosphoglycerides in subfractionated human neutrophils. *J Biol Chem* 264: 17718–17726
14 Careaga-Houck M, Sprecher H (1989) Effect of a fish oil diet on the composition of rat neutrophil lipids and the molecular species of choline and ethanolamine glycerophospholipids. *J Lipid Res* 30: 77–87
15 Chabot MC, Schmitt J D, Bullock BC, Wykle RL (1987) Reacylation of platelet activating factor with eicosapentaenoic acid in fish-oil-enriched monkey neutrophils. *Biochim Biophys Acta* 922: 214–220
16 Prescott SM (1984) The effect of eicosapentaenoic acid on leukotriene B production by human neutrophils. *J Biol Chem* 259: 7615–7621
17 Triggiani M, Connell TR, Chilton FH (1990) Evidence that increasing the cellular content of eicosapentaenoic acid does not reduce the biosynthesis of platelet-activating factor. *J Immunol* 145: 2241–2248
18 Chilton FH, Patel M, Fonteh AN, Hubbard WC, Triggiani M (1993) Dietary n-3 fatty acid effects on neutrophil lipid composition and mediator production. Influence of duration and dosage. *J Clin Invest* 91: 115–122
19 Chilton L, Surette ME, Swan DD, Fonteh AN, Johnson MM, Chilton FH (1996) Metabolism of gammalinolenic acid in human neutrophils. *J Immunol* 156: 2941–2947
20 O'Flaherty JT, Thomas MJ, Hammett MJ, Carroll C, McCall CE, Wykle RL (1983) 5-L-hydroxy-6,8,11,14-eicosatetraenoate potentiates the human neutrophil degranulating action of platelet-activating factor. *Biochem Biophys Res Commun* 111: 1–7
21 O'Flaherty JT, Thomas MJ, McCall CE, Wykle RL (1983) Potentiating actions of

hydroxyeicosatetraenoates on human neutrophil degranulation responses to leukotriene B4 and phorbol myristate acetate. *Res Commun Chem Pathol Pharmacol* 40: 475–487

22 Walsh CE, Dechatelet LR, Chilton FH, Wykle RL, Waite M (1983) Mechanism of arachidonic acid release in human polymorphonuclear leukocytes. *Biochim Biophys Acta* 750: 32–40

23 Swendsen CL, Chilton FH, O'Flaherty JT, Surles JR, Piantadosi C, Waite M, Wykle RL (1987) Human neutrophils incorporate arachidonic acid and saturated fatty acids into separate molecular species of phospholipids. *Biochim Biophys Acta* 919: 79–89

24 Chilton FH, Murphy RC (1986) Remodeling of arachidonate-containing phosphoglycerides within the human neutrophil. *J Biol Chem* 261: 7771–7777

25 Chilton FH, Murphy RC (1986) Characterization of the arachidonate-containing molecular species of phosphoglycerides in the human neutrophil. *Prostaglandins Leukot Med* 23: 141–148

26 Chilton FH, O'Flaherty JT, Ellis JM, Swendsen CL, Wykle RL (1983) Selective acylation of lyso platelet activating factor by arachidonate in human neutrophils. *J Biol Chem* 258: 7268–7271

27 Chilton FH, O'Flaherty JT, Ellis JM, Swendsen CL, Wykle RL (1983) Metabolic fate of platelet-activating factor in neutrophils. *J Biol Chem* 258: 6357–6361

28 Chilton FH, Ellis JM, Olson SC, Wykle RL (1984) 1-O-alkyl-2-arachidonoyl-*sn*-glycero-3-phosphocholine. A common source of platelet-activating factor and arachidonate in human polymorphonuclear leukocytes. *J Biol Chem* 259: 12014–12019

29 Albert DH, Snyder F (1984) Release of arachidonic acid from 1-alkyl-2-acyl-*sn*-glycero-3-phosphocholine, a precursor of platelet-activating factor, in rat alveolar macrophages. *Biochim Biophys Acta* 796: 92–101

30 Suga K, Kawasaki T, Blank ML, Snyder F (1990) An arachidonoyl (polyenoic)-specific phospholipase A_2 activity regulates the synthesis of platelet-activating factor in granulocytic HL-60 cells. *J Biol Chem* 265: 12363–12371

31 Ramesha CS, Pickett WC (1986) Metabolism of platelet-activating factor by arachidonic acid-depleted rat polymorphonuclear leukocytes. *J Biol Chem* 261: 15519–15523

32 Ramesha CS, Pickett WC (1986) Platelet-activating factor and leukotriene biosynthesis is inhibited in polymorphonuclear leukocytes depleted of arachidonic acid. *J Biol Chem* 261: 7592–7595

33 Pickett WC, Nytko D, Dondero-Zahn C, Ramesha C (1986) The effect of endogenous eicosapentaenoic acid on PMN leukotriene and PAF biosynthesis. *Prostaglandins Leukot Med* 23: 135–140

34 Mueller HW, O'Flaherty JT, Wykle RL (1984) The molecular species distribution of platelet-activating factor synthesized by rabbit and human neutrophils. *J Biol Chem* 259: 14554–14559

35 Mueller HW, Nollert MU, Eskin SG (1991) Synthesis of 1-acyl-2-[^{3}H]acetyl-SN-glycero-3-phosphocholine, a structural analog of platelet activating factor, by vascular endothelial cells. *Biochem Biophys Res Commun* 176: 1557–1564

36 Pinckard RN, Jackson EM, Hoppens C, Weintraub ST, Ludwig JC, McManus LM, Mott

G (1984) Molecular heterogeneity of platelet-activating factor produced by stimulated human polymorphonuclear leukocytes. *Biochem Biophys Res Commun* 122: 325–332

37 Demopoulos CA, Pinckard RN, Hanahan DJ (1979) Platelet-activating factor. Evidence for 1-O-alkyl-2-acetyl-*sn*-glyceryl-3-phosphorylcholine as the active component (a new class of lipid chemical mediators). *J Biol Chem* 254: 9355–9358

38 Tessner TG, Wykle RL (1987) Stimulated neutrophils produce an ethanolamine plasmalogen analog of platelet-activating factor. *J Biol Chem* 262: 12660–12664

39 Tessner TG, Greene DG, Wykle RL (1990) Selective deacylation of arachidonate-containing ethanolamine-linked phosphoglycerides in stimulated human neutrophils. *J Biol Chem* 265: 21032–21038

40 Blank ML, Wykle RL, Snyder F (1973) The retention of arachidonic acid in ethanolamine plasmalogens of rat testes during essential fatty acid deficiency. *Biochim Biophys Acta* 316: 28–34

41 Kramer RM, Deykin D (1983) Arachidonoyl transacylase in human platelets. Coenzyme A-independent transfer of arachidonate from phosphatidylcholine to lysoplasmenylethanolamine. *J Biol Chem* 258: 13806–13811

42 Sugiura T, Katayama O, Fukui J, Nakagawa Y, Waku K (1984) Mobilization of arachidonic acid between diacyl and ether phospholipids in rabbit alveolar macrophages. *FEBS Lett* 165: 273–276

43 Chilton FH, Hadley JS, Murphy RC (1987) Incorporation of arachidonic acid into 1-acyl-2-lyso-*sn*-glycero-3-phosphocholine of the human neutrophil. *Biochim Biophys Acta* 917: 48–56

44 Kramer RE, Simpson ER, Waterman MR (1983) Induction of 11 beta-hydroxylase by corticotropin in primary cultures of bovine adrenocortical cells. *J Biol Chem* 258: 3000–3005

45 Kramer RM, Patton GM, Pritzker CR, Deykin D (1984) Metabolism of platelet-activating factor in human platelets. Transacylase-mediated synthesis of 1-O-alkyl-2-arachidonoyl-*sn*-glycero-3-phosphocholine. *J Biol Chem* 259: 13316–13320

46 Kramer RM, Pritzker CR, Deykin D (1984) Coenzyme A-mediated arachidonic acid transacylation in human platelets. *J Biol Chem* 259: 2403–2406

47 Kramer RE, Rainey WE, Funkenstein B, Dee A, Simpson ER, Waterman MR (1984) Induction of synthesis of mitochondrial steroidogenic enzymes of bovine adrenocortical cells by analogs of cyclic AMP. *J Biol Chem* 259: 707–13

48 Snyder F, Lee TC, Blank ML (1992) The role of transacylases in the metabolism of arachidonate and platelet activating factor. *Prog Lipid Res* 31: 65–86

49 Sugiura T, Waku K (1985) CoA-independent transfer of arachidonic acid from 1,2-diacyl-*sn*-glycero-3-phosphocholine to 1-O-alkyl-*sn*-glycero-3-phosphocholine (lyso platelet-activating factor) by macrophage microsomes. *Biochem Biophys Res Commun* 127: 384–390

50 Baker PR, Owen JS, Nixon AB, Thomas LN, Wooten R, Daniel LW, O'Flaherty JT, Wykle RL (2002) Regulation of platelet-activating factor synthesis in human neutrophils by MAP kinases. *Biochim Biophys Acta* 1592: 175–184

51 Nixon AB, Greene DG, Wykle RL (1996) Comparison of acceptor and donor substrates in the CoA-independent transacylase reaction in human neutrophils. *Biochim Biophys Acta* 1300: 187–196
52 Mueller HW, O'Flaherty JT, Wykle RL (1983) Biosynthesis of platelet activating factor in rabbit polymorphonuclear neutrophils. *J Biol Chem* 258: 6213–6218
53 Sanchez Crespo M, Nieto ML (1989) Modulation of PAF biosynthesis in human polymorphonuclear leukocytes. The role of phorbol esters and protein kinases. *Agents Actions* 26: 111–112
54 Lee TC, Uemura Y, Snyder F (1992) A novel CoA-independent transacetylase produces the ethanolamine plasmalogen and acyl analogs of platelet-activating factor (PAF) with PAF as the acetate donor in HL-60 cells. *J Biol Chem* 267: 19992–20001
55 Nieto ML, Venable ME, Bauldry SA, Greene DG, Kennedy M, Bass DA, Wykle RL (1991) Evidence that hydrolysis of ethanolamine plasmalogens triggers synthesis of platelet-activating factor *via* a transacylation reaction. *J Biol Chem* 266: 18699–18706
56 Bonventre JV, Huang Z, Taheri MR, O'Leary E, Li E, Moskowitz MA, Sapirstein A (1997) Reduced fertility and postischaemic brain injury in mice deficient in cytosolic phospholipase A_2. *Nature* 390: 622–625
57 Seeds MC, Nixon AB, Wykle RL, Bass DA (1998) Differential activation of human neutrophil cytosolic phospholipase A_2 and secretory phospholipase A_2 during priming by 1,2-diacyl- and 1-O-alkyl-2-acylglycerols. *Biochim Biophys Acta* 1394: 224–234
58 Seeds MC, Jones DF, Chilton FH, Bass DA (1998) Secretory and cytosolic phospholipases A2 are activated during TNF priming of human neutrophils. *Biochim Biophys Acta* 1389: 273–84
59 Fonteh AN, Samet JM, Surette M, Reed W, Chilton FH (1998) Mechanisms that account for the selective release of arachidonic acid from intact cells by secretory phospholipase A_2. *Biochim Biophys Acta* 1393: 253–266
60 Leslie CC (1997) Properties and regulation of cytosolic phospholipase A_2. *J Biol Chem* 272: 16709–16712
61 Wijkander J, O'Flaherty JT, Nixon AB, Wykle RL (1995) 5-Lipoxygenase products modulate the activity of the 85-kDa phospholipase A_2 in human neutrophils. *J Biol Chem* 270: 26543–26549
62 Evans JH, Fergus DJ, Leslie CC (2003) Regulation of cytosolic phospholipase A(2) translocation. *Adv Enzyme Regul* 43: 229–244
63 Powell WS, Gravelle F, Gravel S (1992) Metabolism of 5(S)-hydroxy-6,8,11,14-eicosatetraenoic acid and other 5(S)-hydroxyeicosanoids by a specific dehydrogenase in human polymorphonuclear leukocytes. *J Biol Chem* 267: 19233–19241
64 O'Flaherty JT, Kuroki M, Nixon AB, Wijkander J, Yee E, Lee SL, Smitherman PK, Wykle RL, Daniel LW (1996) 5-Oxo-eicosanoids and hematopoietic cytokines cooperate in stimulating neutrophil function and the mitogen-activated protein kinase pathway. *J Biol Chem* 271: 17821–17828
65 O'Flaherty JT, Kuroki M, Nixon AB, Wijkander J, Yee E, Lee SL, Smitherman PK,

Wykle RL, Daniel LW (1996) 5-Oxo-eicosatetraenoate is a broadly active, eosinophil-selective stimulus for human granulocytes. *J Immunol* 157: 336–342
66 Bauldry SA, McCall CE, Cousart SL, Bass DA (1991) Tumor necrosis factor-alpha priming of phospholipase A_2 activation in human neutrophils. An alternative mechanism of priming. *J Immunol* 146: 1277–1285
67 Bauldry SA, Wykle RL, Bass DA (1991) Differential actions of diacyl- and alkylacylglycerols in priming phospholipase A_2, 5-lipoxygenase and acetyltransferase activation in human neutrophils. *Biochim Biophys Acta* 1084: 178–184
68 Nixon AB, Seeds MC, Bass DA, Smitherman PK, O'Flaherty JT, Daniel LW, Wykle RL (1997) Comparison of alkylacylglycerol versus diacylglycerol as activators of mitogen-activated protein kinase and cytosolic phospholipase A_2 in human neutrophil priming. *Biochim Biophys Acta* 1347: 219–230
69 Murakami M, Nakatani Y, Atsumi G, Inoue K, Kudo I (1997) Regulatory functions of phospholipase A_2. *Crit Rev Immunol* 17: 225–283
70 Lenihan DJ, Lee TC (1984) Regulation of platelet activating factor synthesis: modulation of 1-alkyl-2-lyso-*sn*-glycero-3-phosphocholine:acetyl-CoA acetyltransferase by phosphorylation and dephosphorylation in rat spleen microsomes. *Biochem Biophys Res Commun* 120: 834–839
71 Nieto ML, Velasco S, Sanchez Crespo M (1988) Modulation of acetyl-CoA:1-alkyl-2-lyso-*sn*-glycero-3-phosphocholine (lyso-PAF) acetyltransferase in human polymorphonuclears. The role of cyclic AMP-dependent and phospholipid sensitive, calcium-dependent protein kinases. *J Biol Chem* 263: 4607–4611
72 Nieto ML, Velasco S, Sanchez Crespo M (1988) Biosynthesis of platelet-activating factor in human polymorphonuclear leukocytes. Involvement of the cholinephosphotransferase pathway in response to the phorbol esters. *J Biol Chem* 263: 2217–2222
73 Ninio E, Joly F, Bessou G (1988) Biosynthesis of paf-acether. XI. Regulation of acetyltransferase by enzyme-substrate imbalance. *Biochim Biophys Acta* 963: 288–294
74 Alonso F, Gil MG, Sanchez-Crespo M, Mato JM (1982) Activation of 1-alkyl-2-lyso-glycero-3-phosphocholine. Acetyl-CoA transferase during phagocytosis in human polymorphonuclear leukocytes. *J Biol Chem* 257: 3376–3378
75 Nixon AB, O'Flaherty JT, Salyer JK, Wykle RL (1999) Acetyl-CoA:1-O-alkyl-2-lyso-*sn*-glycero-3-phosphocholine acetyltransferase is directly activated by p38 kinase. *J Biol Chem* 274: 5469–5473
76 Borsch-Haubold AG, Kramer RM, Watson SP (197) Phosphorylation and activation of cytosolic phospholipase A2 by 38-kDa mitogen-activated protein kinase in collagen-stimulated human platelets. *Eur J Biochem* 245: 751–759
77 Kramer RM, Roberts EF, Um SL, Borsch-Haubold AG, Watson SP, Fisher MJ, Jakubowski JA (1996) p38 mitogen-activated protein kinase phosphorylates cytosolic phospholipase A_2 ($cPLA_2$) in thrombin-stimulated platelets. Evidence that proline-directed phosphorylation is not required for mobilization of arachidonic acid by $cPLA_2$. *J Biol Chem* 271: 27723–27729

Control of long chain polyunsaturated fatty acid levels and the role of inhibitors of incorporation and remodeling on the biosynthesis of lipid mediators

Allen M. McAlexander[1], Brooke J. Barham[2], Margaret Johnson[3] and Alfred N. Fonteh[4]

[1]Department of Biology and Health Sciences, Meridith College, Raleigh, North Carolina, USA; [2]Comprehensive Center for Inflammatory Diseases, Department of Dentistry, University of North Carolina, USA; [3]Department of Medicine, Mayo Clinic, Jacksonville, Florida, USA; [4]Molecular Neurology Program, Huntington Medical Research Institutes, 99 N El Molino Avenue, Pasadena, CA 91101, USA

Overview of AA and PUFA metabolism

AA and 20–22 carbon PUFAs play a number of roles in mammalian physiology [1–7]. These fatty acids serve as structural components of cellular membranes; as signaling molecules; and as precursors of mediators of inflammation, pain, cellular differentiation and cell growth [8–13]. Thus, their levels in mammalian systems are more tightly controlled than those of saturated fatty acids, which appear to primarily serve a structural role.

There are several levels of control of long chain PUFA levels. Mammalian cells lack Δ-9 desaturase and cannot readily convert the more readily available oleic acid to PUFAs. PUFAs of the n-6 or n-3 families are obtained from the diet, mainly from plant oils and marine oils and are thus considered essential fatty acids (EFA) [14, 15]. In addition to dietary sources, elongation/desaturation of precursors also controls levels of long chain PUFAs. These precursors, linoleic acid (LA, 18:2, n-6), and α-linolenic acid (ALNA, 18:3, n-3), for n-6 and n-3 pathways, respectively, are sequentially elongated and desaturated by mitochondrial enzymes to form AA and other 20–22 carbon PUFAs [14, 16–21]. Once PUFAs are obtained from the diet or synthesized, they are rapidly incorporated into glycerolipids and transported between organs in plasma associated with fatty acid binding proteins and lipoproteins. Cellular activation or pathophysiologic conditions result in the release of PUFAs by lipases [22–24]. Enzyme-mediated oxygenation or auto-oxidative processes generate potent inflammatory mediators from AA and other PUFAs [12, 25–39]. Thus, PUFAs are implicated in human diseases such as rheumatoid arthritis, cardiovascular dysfunction, tumor growth and metastasis, diabetic neuropathy, cirrhosis of the liver, as well as Alzheimer's disease and other neurological disorders [33, 40–48].

Arachidonate Remodeling and Inflammation, edited by Alfred N. Fonteh and Robert L. Wykle

Dietary sources of fatty acids and mechanism of action

Human breast milk is a rich source of EFAs such as gamma linolenic acid (GLA, 18:3, n-6) and low amounts are found in meats [5, 49–51]. High levels are found in plants, marine and fungal oils, mainly as components of triglycerides [52–54]. Major sources of GLA include primrose oil, where GLA is mainly acylated at the *sn*-3 position of triglycerides, blackcurrant seed oil with acylation at the *sn*-3 position, borage oil (*sn*-2 position) and fungal oils (*sn*-2 and *sn*-3 positions). Because of the importance of GLA in physiology and pathophysiology, many biotechnological approaches have been undertaken to engineer plants or unicellular organisms to produce GLA [54, 55]. For example, transgenic tobacco plants expressing Δ-6 desaturase or bacteria or yeast mutants have been designed for enhanced GLA production [54, 56–58]. GLA may act to ameliorate the symptoms of a number of diseases due to its capacity to modify lipid composition of cells and its role in the synthesis of anti-inflammatory lipid mediators [59–65]. Following GLA supplementation, there is an increase in the levels of the elongase product, dihomo gamma linolenic acid (DGLA) within inflammatory cells such as neutrophils, without a corresponding increase in the desaturase product, AA. However, there is a significant increase in AA content of serum, which may be detrimental in chronic disease conditions [62, 64]. A combination of elongase inhibitors or substrate competitors or feed back inhibitors may prevent such a build-up of AA within serum, while maintaining the positive attributes of GLA supplements. Elongated GLA is rapidly incorporated into cellular membranes and these reside in the same cellular glycerophospholipid pools, as does AA. Thus, they are equally utilized for the biosynthesis of eicosanoids by lipoxygenases (LO) and cyclo-oxygenase (COX) activities. The incorporation of elongase products into phospholipids is mediated by ligases and CoA-dependent acyl transferases, while remodeling within lipid pools is mediated by CoA-dependent and independent transacylases. Upon cell activation due to disease processes, phospholipases A_2 (PLA_2) may release DGLA, which is a substrate of COX and LO enzymes. Similar to GLA metabolism, fatty acids from the n-3 pathways can also be elongated to form eicosapentaenoic acid (EPA, Fig. 1) and subsequently incorporated/remodeled within glycerophospholipids using similar enzymatic activities. These metabolic pathways compete with AA metabolism because similar enzyme activities are involved. Thus, supplementation of human diets with GLA or EPA may result in a decrease in the formation of pro-inflammatory mediators such as LTB_4 and platelet activating factor formed during the rapid remodeling and reacylation of lipids. In addition to competing with AA, DGLA and EPA are converted to eicosanoids (1- or 3-series of prostanoids and 3- or 5 series of leukotrienes, respectively) that are either anti-inflammatory in nature or thousands of folds less active than AA-derived eicosanoids (2-series prostanoids and 4 series leukotrienes). Supplementation of human diets with GLA has been shown to reduce the signs and symptoms of chronic inflammatory diseases [42, 43, 46, 52, 66–74]. Being a precursor of AA, it is counter intuitive that GLA supplementation results in a decrease in pro-inflammatory lipid mediator biosynthesis; however, studies using neutrophils from subjects fed with GLA supplements have shown decreased LTB_4 and PAF biosynthesis [64]. Recent *in vitro* and *in vivo* studies have shown that human neutrophils do not express Δ-5-desaturase activity; thus, dietary GLA supplementation leads to the accumulation of DGLA, rather than AA, in cellular glycerolipids [62]. Levels of GLA, other n-6 fatty acids, and n-3 PUFAs available to cells after ingestion of nutraceuticals will likely be determined by pharmacokinetic parameters, stereo-specificity of lipases that release free fatty acids from triglycerides and the levels of free fatty acids that are controlled by rapid acylation and remodeling processes. Together, various studies reveal that the conversion of AA precursors by endogenous elongase activity to AA structural analogues such as DGLA or EPA can be an effective strategy for controlling the biosynthesis of mediators of inflammation.

Enzyme activities that control AA and PUFA levels

The synthesis of 20 and 22 carbon PUFAs

Long chain PUFAs, which serve as eicosanoid precursors, may be obtained either through the diet or by elongation and desaturation of 18 carbon fatty acids (Fig. 1). Sprecher has also proposed that 22 carbon fatty acids are formed in mammalian systems by 2 elongation of 20 carbon fatty acids followed by a Δ-6 desaturation reaction and retro conversion [75, 76]. Deficient retroconversion of 22:6 (n-3) to 20:5 (n-3) has been demonstrated in fibroblasts from subjects with Zellweger Syndrome, suggesting a defect in β-oxidation and peroxisomal function [77, 78].

AA (20:4, n-6) and DHA (22:6, n-3) are major acyl components of membrane phospholipids [79]. The brain, especially non-myelin membranes, is very rich in AA and DHA. Dietary deficiency in precursors of AA and DHA has been shown to reduce their levels in the central nervous system, resulting in altered learning behavior and impaired visual function. In addition to dietary levels, free levels of these fatty acids may also be influenced by oxidation for energy, and their distribution and transport determined by acylation into phospholipids, triglycerides and cholesterol esters. Thus key enzymes responsible for controlling levels of these PUFA include elongase, desaturase, acyl CoA-synthetase and acyl CoA transferases.

Elongase enzymes

Elongations of 18 carbon fatty acids, combined with desaturation (discussed below), generate 20 and 22 carbon fatty acids. Elongase activity converts GLA (18:3 to n-6) to DGLA (20:3, n-6) and also AA (20:4, n-6) to docosatetraenoic acid (DTA,

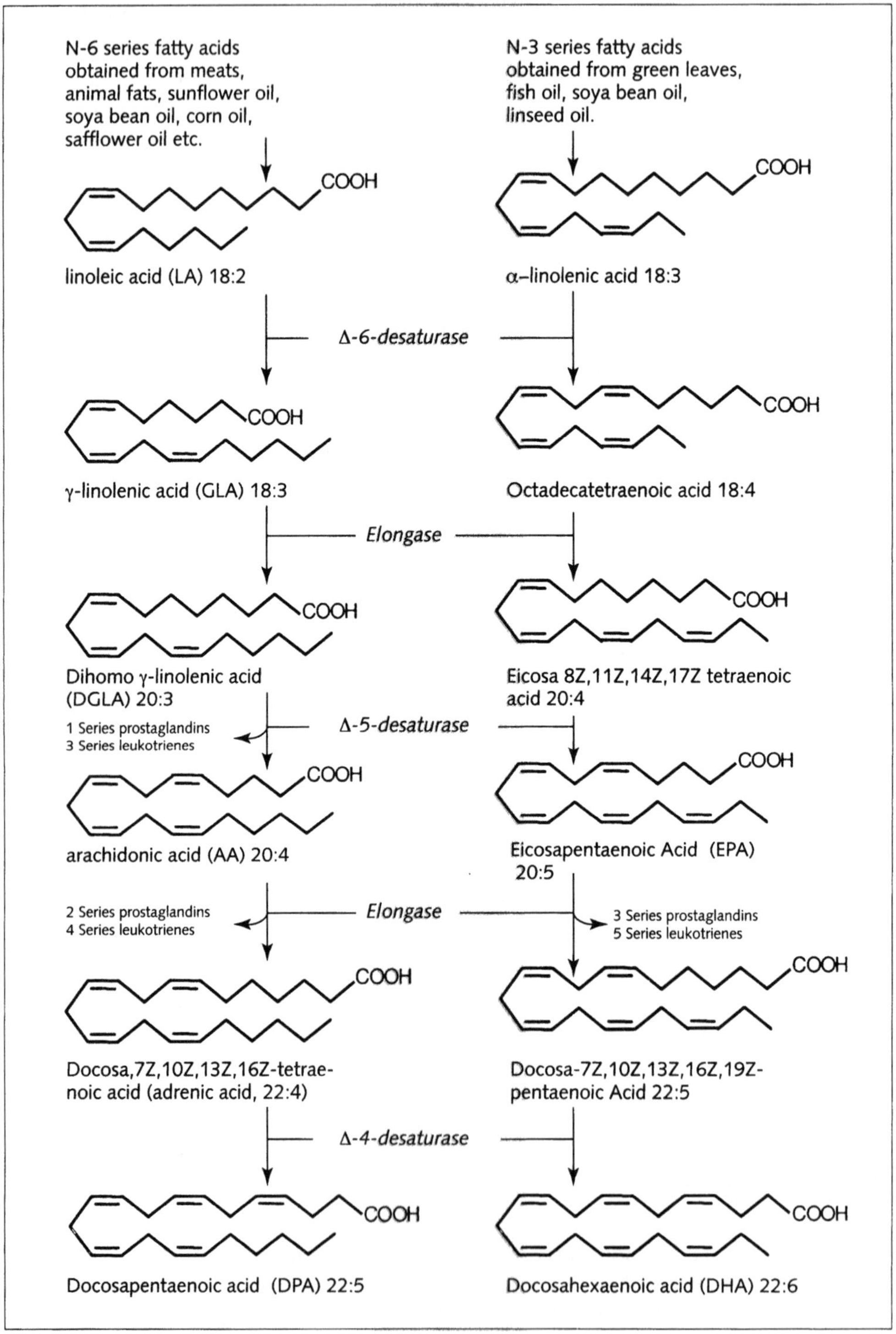
N-6 series fatty acids obtained from meats, animal fats, sunflower oil, soya bean oil, corn oil, safflower oil etc.
N-3 series fatty acids obtained from green leaves, fish oil, soya bean oil, linseed oil.
COOH
linoleic acid (LA) 18:2
α–linolenic acid 18:3
Δ-6-desaturase
γ-linolenic acid (GLA) 18:3
Octadecatetraenoic acid 18:4
Elongase
Dihomo γ-linolenic acid (DGLA) 20:3
Eicosa 8Z,11Z,14Z,17Z tetraenoic acid 20:4
1 Series prostaglandins
3 Series leukotrienes
Δ-5-desaturase
arachidonic acid (AA) 20:4
Eicosapentaenoic Acid (EPA) 20:5
2 Series prostaglandins
4 Series leukotrienes
Elongase
3 Series prostaglandins
5 Series leukotrienes
Docosa,7Z,10Z,13Z,16Z-tetraenoic acid (adrenic acid, 22:4)
Docosa-7Z,10Z,13Z,16Z,19Z-pentaenoic Acid 22:5
Δ-4-desaturase
Docosapentaenoic acid (DPA) 22:5
Docosahexaenoic acid (DHA) 22:6

22:4, n-6). Similarly, elongase converts stearidonic acid (SDA, 18:4 n-3) to eicosatetraenoic acid (20:4, n-3) and EPA (20:5 n-3) to docosapentaenoic acid (DPA, 22:5 n-3) (Fig. 1). While our understanding of the importance of elongation of fatty acids is advanced in unicellular organisms and plants, knowledge of the biochemical properties of elongase enzymes in animals is limited because these enzymes have only recently been cloned from animal sources. Human elongase (ELOVL5) uses monounsaturated and PUFA as substrates [80]. Two cDNA clones with 56.4% and 58% amino acid identity to ELOVL5 encoding a 296 amino acid peptide have recently been expressed in yeast and shown to elongate 20 and 22 carbon fatty acids [80, 81]. Studies using various other cloning and expression strategies have been applied to bacteria, yeast and plants to provide the following mechanistic details about elongases [82–85]. The elongation process is thought to be a four-step process involving condensation of malonyl-CoA with acyl-CoA to form carbon dioxide and an elongated β-ketoacyl-CoA. This first reaction catalyzed by β-ketoacyl CoA synthase is substrate specific and rate limiting. This condensation reaction is then followed by reduction to β-hydroxyacyl-CoA, dehydration to form enoyl-CoA and a final reduction to form acyl CoA.

Desaturase enzymes

Since 20 and 22 carbon PUFAs are synthesized from LA (18:3, n-6) and ALNA (18:3, n-3), the synthesis or ingestion of these precursors dictates the availability of these 20 and 22 carbon PUFAs. Sequential elongation and desaturation of precursors by enzymes are key events in the synthetic pathways. Three desaturation enzymes are involved in this process (Fig. 1). These include a Δ-6 desaturase that converts 18:2 (n-6) or 18:3 (n-3) to 18:3 (n-6) and 18:4 (n-3), respectively, a Δ-5 desaturase that converts 20:3 (n-6) or 20:4 (n-3) to 20:4 (n-6) and 20:5 (n-3), respectively and a Δ-4 desaturase that converts 22:4 (n-6) or 22:5 (n-3) to 22:4 (n-6) and 22:6 (n-3), respectively. Desaturases are complexes consisting of NADPH-cytochrome b5 reductase, cytochrome b5 and Δ-5 or Δ-6 enzymes [57, 86–92]. The liver contains high levels of desaturases and is the site for the synthesis of most

Figure 1
Elongation and desaturation of n-6 and n-3 fatty acids.
LA (18:2, n-6) and ALNA (18:3, n-3) are precursors for long chain PUFA synthesis by elongase and desaturase activities. PUFAs are incorporated into phospholipids and subsequently utilized for the synthesis of prostanoids designated series 1, series 2 or series 3 from DGLA, AA and EPA, respectively. Leukotrienes generated from these same fatty acids are designated series 3, 4 and 5, respectively.

PUFAs [87, 93, 94]. Human lung, brain and heart have recently been shown to contain comparable levels of Δ-5 desaturase, while very low but detectable enzyme expression is described in other tissues. Δ-6 is comparatively more abundant in most tissues than Δ-5 desaturase, although age-related or nutritional-regulated processes may alter this balance. Human Δ-5 and Δ-6 enzymes have recently been cloned [87, 88, 95, 96]. These genes are located on chromosome 11, have 75% nucleotide homology and 61% amino acid homology. Δ-5 is a single transcript of ~ 4.4 kb (444 amino acid peptide) while Δ-6 is 3.4 kb transcript. Δ-5 desaturase has two membrane-spanning domains characteristic of membrane anchored proteins, three histidine-rich regions suspected for heme iron binding and contain cytochrome b5 domain at the N terminus. In PUFA biosynthesis, Δ-6 is considered the rate-limiting step in the pathway (Fig. 1). It is still controversial whether there are specific desaturases for n-3 and n-6 pathways. Studies using the enzyme inhibitor, N-ethyl maleimide showed differential inhibition of n-6 product formation, suggesting that there may be different desaturases specific for these pathways [97]. However, expression of cloned enzymes has shown that the same activity is responsible for both pathways [87]. Several studies have implicated regulation of desaturase activities on dietary intake of various fatty acids and on hormonal levels. For example, when animals are fed an EFA deficient diet, Δ-6 and Δ-5 activities increase. Similarly, supplementation of animal diets with n-6 or n-3 fatty acids reduces Δ-6 and Δ-5 desaturase activities [20, 98, 99]. In contrast, supplementation of diets with oleic acid (18:1, n-9) does not affect desaturase activity. While the biochemical mechanisms that account for these dietary regulations have not been determined, it is possible that these enzymes share a regulatory sequence due to their localization and reverse sequence orientation on the same chromosome. These desaturases are also located on a chromosome that has been implicated in obesity, raising the possibility that these activities have pathologic importance. Further evidence for the involvement of desaturases in diseases is provided by studies showing that Δ-6 activity is induced by peroxisome proliferators and by insulin administration to diabetic animals [100–102]. As described above, 20 and 22 carbon PUFAs are implicated in many biological processes such as brain development, cognition, reproduction, inflammation, homeostasis, and pain [95, 101, 103, 104]. Δ-6 desaturase abnormalities have been described in brain autopsy samples from subjects with Alzheimer's disease, where there is elevated LA and a corresponding reduction in 20:4 (n-6), 22:4 (n-6) and 22:6 (n-3) [103]. Desaturases may influence these events and diseases by generating signaling molecules that control gene expression and influence lipid metabolism, thermogenesis and cell differentiation.

Supplementation studies show that in brain astrocytes, the rate of DHA synthesis is limited by elongation [105]. Addition of ALNA or EPA to these cells resulted in a decrease in 22:5 (n-3) and 22:6 (n-3) synthesis, whereas supplementation with LA (18:2, n-6) did not affect 22:5 (n-3) synthesis. Astrocytes also secret cholesterol esters and phosphatidylethanolamine (PE) containing AA and DHA. Similar studies by Laposata and colleagues used a fibrosarcoma cell line deficient in EFA to show the importance of elongation and desaturation in PUFA synthesis [20]. During n-6 fatty acid supplementation, these cells synthesize EPA *via* elongation while n-3 fatty acid supplementation results in a decrease in the desaturation of GLNA to AA and decrease in AA elongation to 22:4 (n-6). Interestingly, these supplementation strategies also shifted the majority of EPA from phospholipids to triglycerides. These supplementation strategies also resulted in the formation of series 1, 2 or 3 prostanoids. Together, these studies show that there is co-ordinate regulation of n-3 and n-6 pathways by modulation of elongation and desaturation activities. These pathways are also important in directing the distribution of PUFAs into glycerolipids and in directing the types of eicosanoids that are formed.

Incorporation of PUFAs into phospholipids

20 and 22 carbon PUFAs obtained from the diet or through elongase and desaturase reactions are preferentially incorporated into phospholipids. The incorporation process involves PUFA activation by acyl CoA synthetase also known as ligase, followed by acyl CoA transfer into glycerophospholipids.

Acyl CoA synthetase (ACS)

ACS catalyzes the reaction of a long chain carboxylic acid with ATP and CoA to form acyl-CoA, AMP and diphosphate. The enzyme acts on a variety of saturated and unsaturated fatty acids but enzymes from different tissues show differences in specificity. The enzyme from the liver acts on C6–C20 fatty acids while the enzyme from the brain acts on fatty acids with up to 24 carbons [106, 107]. ACS has been purified from many sources and Majerus and colleagues have described isoforms that are specific for AA [108–110]. Five ACSs have been cloned in mammalian tissues and their expression has helped elucidate substrate requirement and cellular distribution of these enzymes [111–116]. ACS2 and ACS3 mRNA are expressed in brain but not in liver, while ACS4 and ACS5 mRNA are expressed in steroid producing cells and intestines. ACS4 is a 75 kDa protein that is modified by translation and shows various changes in its expression during brain development [117]. ACS1 and ACS5 show very broad substrate specificity while ACS4 has marked preference for AA and EPA. Likewise, ACS3 has been shown to be AA specific and to be expressed in brain and heart tissues. ACS belongs to a super family of enzymes with two luciferase-like regions, an AMP and fatty acid binding sites located in both luciferase-like regions. ACS1, 2 and 3 have 60% amino acid identity but may be independently regulated by diet and hormones [118, 119]. Once formed by ACS, fatty acyl CoA may be metabolized *via* β-oxidation, used for glycerolipid synthesis, elongated or desaturated, used for synthesis of cholesterol esters, protein acylation or may serve as signaling molecules [120, 121]. Long chain ACS is linked to several pathologic conditions including X-linked adrenoleukodystrophy and polymorphism of the ACS gene is linked to risk factors of hypertriglyceridemia, visceral obesity and hypertension [122–124] and is deleted in a family with Alport syndrome, alliptocytosis and mental retardation [115, 116, 118, 122, 125–129].

CoA-dependent acyl transferase

Once acyl CoAs are formed, they are rapidly incorporated into glycerolipids by CoA-dependent acyl transferase (Fig. 2). Two major forms of acyl transferases synthesize phosphatitic acid by sequentially adding acyl CoA to glycerol-3-phosphate (Kennedy pathway) [130, 131]. In addition, acylation of lysophospholipids described by Lands and colleagues utilizes acyl CoAs [132–134]. In these acylation reactions, selectivity depends on whether the acyl CoA is saturated or unsaturated, the concentration of the acceptor molecules and the types of bond at the *sn*-1 position of lysophospholipid acceptors. This selectivity distinguishes *de novo* synthesis from the incorporation of low AA amounts into glycerolipids. Studies using human cells have shown that very low free levels of 20 and 22 carbon PUFAs are present under resting conditions [135]. ACS and CoA acyl transferase activities play the crucial role of controlling these levels by activating and rapidly incorporating fatty acids into glycerophospholipids.

Transacylase

Both CoA-DT and CoA-IT move acyl groups between phospholipid pools. Due to the selectivity of these activities, they play major roles in asymmetrically distributing fatty acids in phospholipid subclasses. CoA-DT transfers fatty acids from one phospholipid to another utilizing an acyl CoA and lysophospholipid intermediate. Waku and colleagues have demonstrated that the preferred substrates for CoA-DT are AA, LA and SA [136–140]. The major activity that transfers 20- and 22-carbon fatty acids between phospholipid subclasses is known as CoA-independent transacylase (CoA-IT). CoA-IT is specific in transferring these fatty acids from mainly diacyl phospholipid donors to 1-alkyl- or 1-alk-1-enyl-linked lysophospholipid acceptors and does not require a cofactor (CoA) or ATP for its activity [141–143]. Thus, CoA-IT is responsible for the fact that the majority of 20- and 22-carbon PUFAs are found in ether-linked phospholipids and not in the early ester-linked phospholipids in which they are initially synthesized. The main acceptors are 1-alkyl-2-lyso glycerophosphocholine or 1-alk-1-enyl-2-lyso glycerophosphoethanolamine (Fig. 2). CoA-IT thus generates the major precursor of platelet activating factor. CoA-IT is a

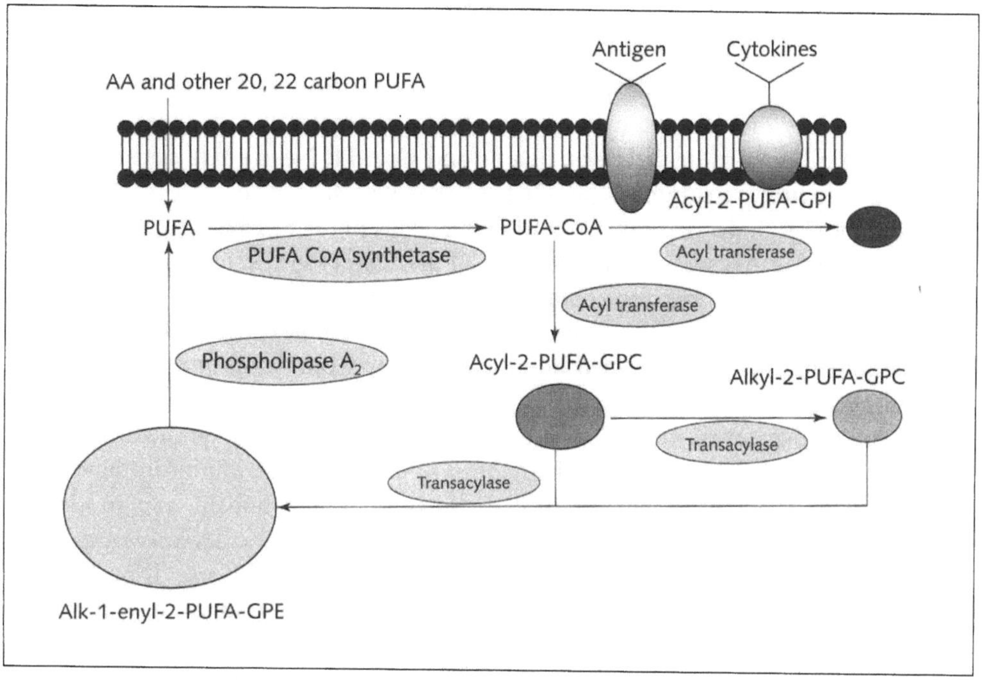

Figure 2
Remodeling of AA in inflammatory cells
AA and other PUFAs presented to mast cells or human neutrophils are rapidly incorporated into 1-acyl-linked glycerophospholipids, (1-acyl-2-PUFA-GPC and 1-acyl-2-PUFA-GPI). Transacylase activity transfers AA to ether-linked phospholipid subclasses (1-alk-1-enyl-2-PUFA-GPE and 1-alkyl-2-PUFA-GPC). This remodeling process is completed by the slow release of AA from these major ether pools and is enhanced when these cells are activated by antigen, cytokines or bacterial peptides such as FMLP or endotoxins.

membrane bound enzyme that has not been purified or cloned. However, gel filtration studies and native gel electrophoresis data suggest that CoA-IT is a ~60 kDa protein that is found in great quantities in microsomal fractions [144]. Various biochemical and pharmacological approaches have been utilized to gain a comprehensive insight into CoA-IT activity and function. Using radiolabeled assays or mass spectroscopic experiments, we have identified specific phospholipid subclasses and molecular species that are donors or acceptors of PUFAs during the remodeling process in neutrophils or mast cells [142]. Importantly, these studies have also shown how PUFAs get incorporated into inflammatory cells and get remodeled between phospholipid subclasses [145, 146]. Further evidence lending credence to these biochemical studies comes from experiments in which CoA-IT inhibitors block the movement of AA between phospholipid subclasses (discussed below).

Role of inhibitors in modulating PUFA levels

PUFA levels may be manipulated through dietary control or through inhibition of specific enzymes that maintain PUFA homeostasis. Currently, advances have been limited to inhibitors of desaturases, ligases and CoA-IT. Needleman and colleagues used a Δ-6 desaturase inhibitor (SC-26196) that was shown to increase LA levels, concomitant with a decrease in AA and DHA levels in mice [147, 148]. Interestingly, the decrease in AA was specific for PC, PI and cholesterol esters and not for PS, PI or triglycerides, and the molecular species formed were determined by the fatty acids at the *sn*-1 position of the phospholipid. These studies suggest that phospholipid remodeling mediated by Δ-6 desaturase is important in *de novo* synthesis of glycerolipids and cholesterol esters.

Similar to desaturase and elongase, inhibitors of acyl CoA-synthetase activity have the potential of influencing PUFA biosynthesis. Various inhibitors have been used to show the importance of ACS in fatty acid metabolism and in disease processes. Of these inhibitors, products of the Streptomycin species known as triacsin (Fig. 3A) have been shown to be very potent ACS inhibitors (Tab. 1) [113, 149–152]. In addition to triacsin, structurally related compounds having an N-hydroxytriazone moiety (WS-1228) that are known hypotensive vasodilators inhibit ACS activity. These inhibitors show that there are different ACSs in cells that show different selectively by inhibiting the activation of some fatty acids. Additionally, inhibitor studies show that there are likely different acyl CoA pools in different cells. Importantly, triacsin interferes with PAF formation, suggesting that it blocks reacylation by enhancing the supply of lysoPAF [136]. Triacsin blocks *de novo* synthesis of triglycerides and phospholipids but not the reacylation of lysophospholipids or the synthesis of cholesterol [151]. In some cells, triacsin inhibits OA incorporation without affecting AA and controls fatty acid distribution into glycerolipids during neuronal outgrowth [127]. In addition to triacsin, the potent antidiabetic agents, thiazolidinedione drugs (triglitazones) achieve their pharmacologic effects in part by direct interaction with ACS4, independent of PPAR-γ [113]. Similar to triacsin, triglitazones inhibit long chain fatty acid incorporation into cellular lipids. Medium chain ACS is also inhibited by quinolone antimicrobial agents and by nonsteroidal anti-inflammatory agents [153].

Inhibitors of AA-remodeling

Ether lipids play a prominent role in cell biology especially in inflammation, cancer biology and in CNS pathophysiology. As described above, AA is transferred from1-acyl-linked phospholipids to 1-ether-linked phospholipids, leading to the latter class containing most of the mass of cellular AA. Inflammatory cells such as mast cells, macrophages or neutrophils contain high amounts of ether-linked phospholipids

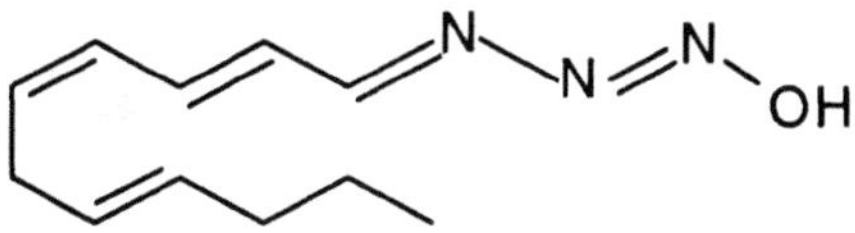

A) 2E, 4E, 7E-Undecatriene-1-triazene (Triacsin)

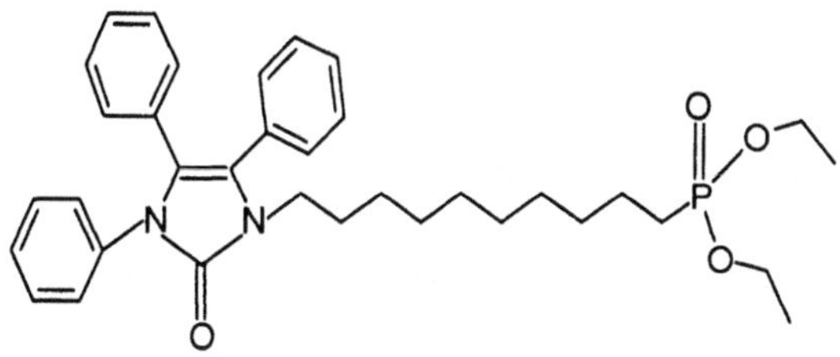

B) Diethyl 7-(3,4,5-triphenyl-2-oxo-2,3-dihydro-imidazole-1-yl) hepatane phosphonate (SK&F 98625)

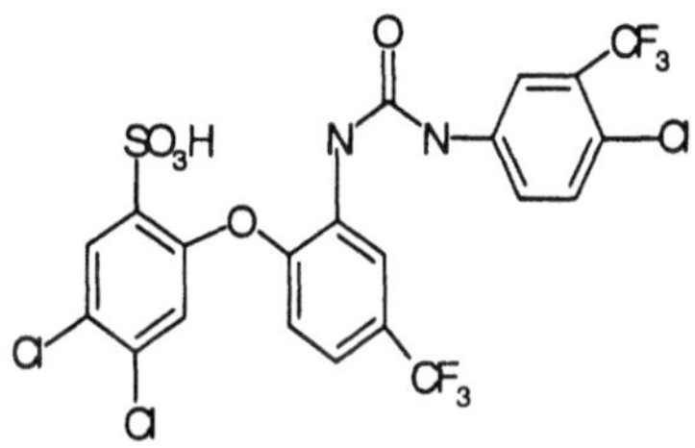

C) 2-[2-(3-4chloro-3-(trifloromethyl)phenyl)ureido]-4 (trifluoromethylphenoxy]-4,5-dichlorobenzenesulfonic acid (SK&F 45905)

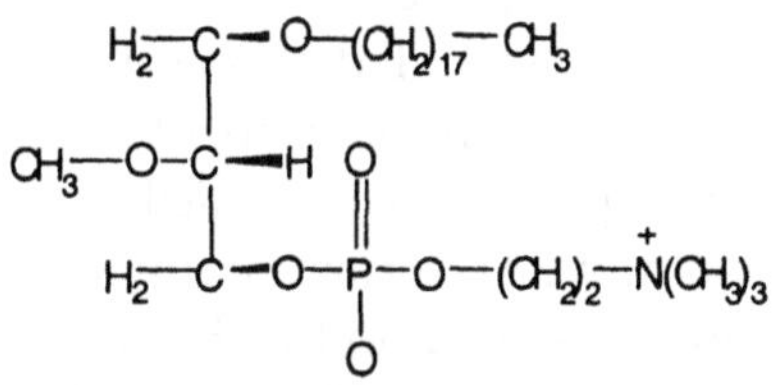

D) 1-*O*-Octadecyl-2-*O*-methyl-*sn*-glycero-3-phosphorylcholine (ET-18-OCH3)

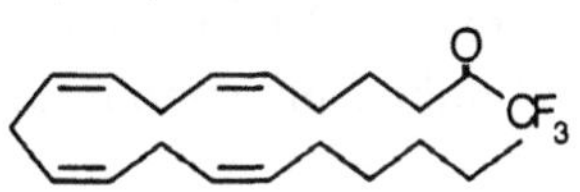

E) Arachidonoyltrifluoromethyl ketone

Figure 3
Inhibitors of AA or PUFA metabolism.

Table 1 - Inhibitors of AA incorporation and remodeling

Inhibitor	**Enzyme**	**IC_{50}**	**Refs.**
Triacsin A	ACS	17–18 μM	[152]
Triacsin C		4–10 μM	
SK&F 98625	CoA-IT	< 11 μM	[142]
	CoA-D transferase	50 μM	
	PLA_2	> 50 μM	
SK&F 45906	CoA-IT	< 20 μM	[142]
	CoA-D transferase	35 μM	
	PLA_2	ND	
ET-18-O-OCH_3	CoA-IT	0.5 μM	[159]
$AACOCF_3$	ACS	~ 5 μM	[160]
	CoA-IT	3 μM	
	$cPLA_2$	> 10 μM	
	$sPLA_2$	ND	
	$iPLA_2$	ND	

ND denotes not determined.

that are rich in arachidonate. These ether lipids provide most of the AA that is released by stimulated cells. As well as providing lyso PAF that is acetylated to form PAF, radiolabel studies show that ether lipids provide the majority of AA that is used for eicosanoid biosynthesis [135, 146, 154]. Laposata has provided additional evidence for a role of ether lipids in eicosanoid and PAF formation by demonstrating that depletion of AA in ether lipids reduces eicosanoid and PAF formation [155]. Because 1-alkyl-2-AA-GPC is the common precursor for LTB_4 and PAF, inhibitors of the formation of this precursor are likely to have significant anti-inflammatory properties. One approach that has been investigated by Winkler and colleagues is the inhibition of the transfer of AA from 1-acyl-linked phospholipids to 1-ether-linked phospholipids [142, 156–159]. CoA-IT has been the major focus on the development of various inhibitors described below.

Diethyl 7-(3,4,5-triphenyl-2-oxo-2,3-dihydro-imidazole-1-yl) hepatane phosphonate (SK&F 98625, Figs 3B and C) 2-[2-(3-4chloro-3 (trifloromethyl) phenyl) ureido]-4-(trifluoromethylphenoxy]-4,5-dichlorobenzenesulfonic acid (SK&F 45905, Fig. 3C)-SK&F 98625 and SK&F 45905 are two inhibitors of CoA-IT which were developed with the assumption that the mechanism of action of CoA-IT resembled that of acyl transferases such as LCAT or ACAT and that the inhibitor should have the capacity to recognize similar substrates as PLA_2s [142]. Thus, analogues of ACAT inhibitors (SK&F 98625) and a PLA_2 inhibitor (SK&F 45905) have shown great promise as inhibitors of CoA-IT (Figs. 3B and 3C). These inhibitors block the movement of AA from 1-acyl-linked phospholipids to 1-ether-linked phospholipids in cell-free and whole cell assays with $IC_{50} < 20\ \mu M$ (Tab. 1), 3–5-fold lower than their IC_{50}s for inhibition of PLA_2 or other transferases. CoA-IT inhibition produces a decrease in PAF and LTB_4 biosynthesis, concomitant with an increase in intracellular and extracellular free AA levels [157, 161]. Inhibition of lipid mediator biosynthesis likely accounts for the anti-inflammatory effects of these compounds in animal models of inflammation, while the build-up of free AA may account for induction of cancer cell apoptosis by these same compounds [143, 145, 159, 162, 163]. While none of the above inhibitors have been clinically developed, they have revealed several critical aspects about levels of control of lipid mediator biosynthesis and the potential of targeting specific enzymes for therapeutic benefits.

1-*O*-Octadecyl-2-*O*-methyl-*sn*-glycer-3-phosphorylcholine (ET-18-OCH_3)

ET-18-OCH_3 (Fig. 3D) is an ether lipid that has been shown to induce apoptosis of cancer cells and to display a range of other biological properties. In addition to its inhibitory effects on kinases, ET-18-OCH_3 has been shown by Winkler and colleagues to be a potent inhibitor of CoA-IT ($IC_{50} = 0.5\ \mu M$) [159].

Arachidonoyl trifluoromethyl ketone ($AACOCF_3$)

$AACOCH_3$ (Fig. 3E) has been extensively used as a specific slow binding inhibitor of $cPLA_2$ or $iPLA_2$ in mixed micelle assays [164]. However, other studies have shown that $AACOCF_3$ is a very non-specific inhibitor of many enzyme activities including CoA-IT ($IC_{50} = 3\ \mu M$), 5-LO (90% inhibition by 10 μM) and anandamide amidase, which is completely inhibited by 7.5 μM [160]. Higher concentrations of $AACOCF_3$ inhibited all AA-metabolic processes, including the incorporation of AA into cellular phospholipid pools. The implications of $AAOCF_3$ inhibition of CoA-IT have been many fold. First, treatment of human neutrophils with < 10 μM resulted in the inhibition of LTB_4 and PAF formation without significant effects on AA release, suggesting that there are distinct arachidonate pools within inflammatory cells; one of which is destined for PAF and LTB_4 formation and is linked to the CoA-IT-mediated formation of the common precursor 1-alkly-2-AA-GPC. A second revealing aspect of these studies is that AA release alone is not sufficient for lipid mediator biosynthesis. It requires the incorporation of AA into specific AA pools, accompanied by the remodeling of AA into ether-linked phospholipid pools for lipid mediators to be formed. Thus, remodeling may not only be needed to maintain membrane structure, but may be an essential process that transfers AA to specific membrane regions such the endoplasmic reticulum where it is further metabolized by 5-LO.

Conclusion

AA is an EFA derived from the diet or synthesized by the liver or brain cells using LA as the major precursor. LA is sequentially elongated and desaturated to form AA. AA is incorporated into cellular lipids by CoA-dependent enzymes and then remodeled within various phospholipid subclasses and molecular species by transacylases. Specific inhibitors of the aforementioned enzyme activities can be used to control AA levels. These inhibitors have been shown to alter the distribution of AA within cellular pools in inflammatory cells and to influence the biosynthesis of lipid mediators. Thus, use of combinatorial chemistry and structure-activity relationships of existing inhibitors, will likely result in the discovery of new and potent classes of anti-inflammatory agents. These inhibitors will be beneficial in controlling inflammatory diseases such as arthritis and asthma. In addition to specific inhibitors, nutraceuticals are becoming an essential component of the health management regimen of many people in industrialized countries. Examination of fatty acid supplements shows that some diets may have dangerous side effects due to the build up of AA levels in serum. Therefore, it is important that biochemical modifications and all key enzymes responsible for controlling AA be studied in order to avoid any detrimental side effects. Biochemical side effects may be avoided by using combinations of fatty acids in dietary supplements or specific inhibitors of key enzymes. Dissecting how AA and other PUFA pools are altered in various diseases provides pathophysiologic mechanisms that will be the basis of future therapeutic design or the design of efficacious dietary supplements.

References

1 Das UN (1999) Essential fatty acids in health and disease. *J Assoc Physicians India* 47: 906–911

2 Faergeman NJ, Knudsen J (1997) Role of long-chain fatty acyl-CoA esters in the regulation of metabolism and in cell signaling. *Biochem J* 323 (Pt 1): 1–12

3 Farooqui AA, Horrocks LA, Farooqui T (2000) Deacylation and reacylation of neural membrane glycerophospholipids. *J Mol Neurosci* 14: 123–135

4 Gibson RA, Makrides M (2000) n-3 Polyunsaturated fatty acid requirements of term infants. *Am J Clin Nutr* 71: 251S–255S

5 Innis SM (1993) Essential fatty acid requirements in human nutrition. *Can J Physiol Pharmacol* 71: 699–706
6 Kalmijn S (2000) Fatty acid intake and the risk of dementia and cognitive decline: A review of clinical and epidemiological studies. *J Nutr Health Aging* 4: 202–207
7 Zurier RB (1993) Fatty acids, inflammation and immune responses. *Prostaglandins Leukot Essent Fatty Acids* 48: 57–62
8 Bourre JM, Bonneil M, Clement M, Dumont O, Durand G, Lafont H, Nalbone G, Piciotti M (1993) Function of dietary polyunsaturated fatty acids in the nervous system. *Prostaglandins Leukot Essent Fatty Acids* 48: 5–15
9 Mead JF, Dhopeshwarkar GA (1971) Types of fatty acids in brain lipids, their derivation and function. Lipids, malnutrition & the developing brain. *Ciba Found Symp* 59–72
10 Nakamura MT, Cho HP, Xu J, Tang Z, Clarke SD (2001) Metabolism and functions of highly unsaturated fatty acids: An update. *Lipids* 36: 961–964
11 Roelcke U, Heil J (1993) Fatty acids as markers of the blood-cerebrospinal fluid barrier function in man. *Funct Neurol* 8: 189–192
12 Serhan CN, Clish CB, Brannon J, Colgan SP, Chiang N, Gronert K (2000) Novel functional sets of lipid-derived mediators with anti-inflammatory actions generated from omega-3 fatty acids *via* cyclooxygenase 2-nonsteroidal anti-inflammatory drugs and transcellular processing. *J Exp Med* 192: 1197–1204
13 Uauy R, Peirano P, Hoffman D, Mena P, Birch D, Birch E (1996) Role of essential fatty acids in the function of the developing nervous system. *Lipids* 31 (Suppl): S167–S176
14 Bezard J, Blond JP, Bernard A, Clouet P (1994) The metabolism and availability of essential fatty acids in animal and human tissues. *Reprod Nutr Dev* 34: 539–568
15 Sprecher H (1981) Biochemistry of essential fatty acids. *Prog Lipid Res* 20: 13–22
16 Moore SA, Yoder E, Spector AA (1990) Role of the blood-brain barrier in the formation of long-chain omega-3 and omega-6 fatty acids from essential fatty acid precursors. *J Neurochem* 55: 391–402
17 Bazan HE, Careaga MM, SprecherH, Bazan NG (1982) Chain elongation and desaturation of eicosapentaenoate to docosahexaenoate and phospholipid labeling in the rat retina *in vivo*. *Biochim Biophys Acta* 712: 123–128
18 Buzzi M, Henderson RJ, Sargent JR (1996) The desaturation and elongation of linolenic acid and eicosapentaenoic acid by hepatocytes and liver microsomes from rainbow trout, Oncorhynchus mykiss, fed diets containing fish oil or olive oil. *Biochim Biophys Acta* 1299: 235–244
19 Cook HW, Byers DM, Palmer FB, Spence MW, Rakoff H, Duval SM, Emken EA (1991) Alternate pathways in the desaturation and chain elongation of linolenic acid, 18: 3(n-3), in cultured glioma cells. *J Lipid Res* 32: 1265–1273
20 Rubin D, Laposata M (1992) Cellular interactions between n-6 and n-3 fatty acids: A mass analysis of fatty acid elongation/desaturation, distribution among complex lipids, and conversion to eicosanoids. *J Lipid Res* 33: 1431–1440
21 Singer P, Berger I, Wirth M, Godicke W, Jaeger W, Voigt S (1986) Slow desaturation and

elongation of linoleic and alpha-linolenic acids as a rationale of eicosapentaenoic acid-rich diet to lower blood pressure and serum lipids in normal, hypertensive and hyperlipemic subjects. *Prostaglandins Leukot Med* 24: 173–193
22 Fonteh AN, Bass DA, Marshall LA, Seeds M, Samet JM, Chilton FH (1994) Evidence that secretory phospholipase A_2 plays a role in arachidonic acid release and eicosanoid biosynthesis by mast cells. *J Immunol* 152: 5438–5446
23 Hurt-Camejo E, Camejo G, Peilot H, Oorni K, Kovanen P (2001) Phospholipase A_2 in vascular disease. *Circ Res* 89: 298–304
24 Balsinde J, Balboa MA, Insel PA, Dennis EA (1999) Regulation and inhibition of phospholipase A_2. *Annu Rev Pharmacol Toxicol* 39: 175–189
25 Samuelsson B, Funk CD (1989) Enzymes involved in the biosynthesis of leukotriene B_4. *J Biol Chem* 264: 19469–19472
26 Samuelsson B (2000) The discovery of the leukotrienes. *Am J Respir Crit Care Med* 161: S2–S6
27 Serhan CN, Nicolaou KC, Webber SE, Veale CA, Haeggstrom J, Puustinen TJ, Samuelsson B (1987) Stereochemistry and biosynthesis of lipoxins. *Adv Prostaglandin Thromboxane Leukot Res* 17A: 90–93
28 DeWitt DL, el Harith EA, SmithWL (1989) Molecular cloning of prostaglandin G/H synthase. *Adv Prostaglandin Thromboxane Leukot Res* 19: 454–457
29 Smith WL, DeWitt DL, Garavito RM (2000) Cyclo-oxygenases: Structural, cellular, and molecular biology. *Annu Rev Biochem* 69: 145–182
30 Smith WL, Meade EA, DeWitt DL (1994) Interactions of PGH synthase isozymes-1 and -2 with NSAIDs. *Ann NY Acad Sci* 744: 50–57
31 Smith WL, Lands WE (1972) Oxygenation of unsaturated fatty acids by soybean lipoxygenase. *J Biol Chem* 247: 1038–1047
32 Chilton FH, Patel M, Fonteh AN, Hubbard WC, Triggiani M (1993) Dietary n-3 fatty acid effects on neutrophil lipid composition and mediator production. Influence of duration and dosage. *J Clin Invest* 91: 115–122
33 Heller A, Koch T, Schmeck J, van Ackern K (1998) Lipid mediators in inflammatory disorders. *Drugs* 55: 487–496
34 Serhan CN, Clish CB, Brannon J, Colgan SP, Gronert K, Chiang N (2000) Anti-microinflammatory lipid signals generated from dietary N-3 fatty acids *via* cyclooxygenase-2 and transcellular processing: A novel mechanism for NSAID and N-3 PUFA therapeutic actions. *J Physiol Pharmacol* 51: 643–654
35 Sperling RI (1991) Dietary omega-3 fatty acids: Effects on lipid mediators of inflammation and rheumatoid arthritis. *Rheum Dis Clin North Am* 17: 373–389
36 Sperling RI, Robin JL, Kylander KA, Lee TH, Lewis RA, Austen KF (1987) The effects of N-3 polyunsaturated fatty acids on the generation of platelet-activating factor-acether by human monocytes. *J Immunol* 139: 4186–4191
37 Montine TJ, Beal MF, Cudkowicz ME, O'Donnell H, Margolin RA, McFarland L, Bachrach AF, Zackert WE, Roberts LJ, Morrow JD (1999) Increased CSF F2-isoprostane concentration in probable AD. *Neurology* 52: 562–565

38 Morrow JD, Awad JA, Boss HJ, Blair IA, Roberts LJ (1992) Non-cyclo-oxygenase-derived prostanoids (F2-isoprostanes) are formed *in situ* on phospholipids. *Proc Natl Acad Sci USA* 89: 10721–10725

39 Roberts LJ, Brame CJ, Chen Y, Morrow JD (1999) Novel eicosanoids. Isoprostanes and related compounds. *Methods Mol Biol* 120: 257–285

40 Blackburn GL (1992) Nutrition and inflammatory events: Highly unsaturated fatty acids (omega-3 vs omega-6) in surgical injury. *Proc Soc Exp Biol Med* 200: 183–188

41 Calder PC (2001) Polyunsaturated fatty acids, inflammation, and immunity. *Lipids* 36: 1007–1024

42 Duchen K, Bjorksten B (2001) Polyunsaturated n-3 fatty acids and the development of atopic disease. *Lipids* 36: 1033–1042

43 Fan YY, Chapkin RS (1998) Importance of dietary gamma-linolenic acid in human health and nutrition. *J Nutr* 128: 1411–1414

44 Geerling BJ, Houwelingen AC, Badart-Smook A, Stockbrugger RW, Brummer RJ (1999) Fat intake and fatty acid profile in plasma phospholipids and adipose tissue in patients with Crohn's disease, compared with controls. *Am J Gastroenterol* 94: 410–417

45 Gibson RA (1988) The effect of diets containing fish and fish oils on disease risk factors in humans. *Aust N Z J Med* 18: 713–722

46 Horrocks LA, Yeo YK (1999) Health benefits of docosahexaenoic acid (DHA). *Pharmacol Res* 40: 211–225

47 Kankaanpaa P, Sutas Y, Salminen S, Lichtenstein A, Isolauri E (1999) Dietary fatty acids and allergy. *Ann Med* 31: 282–287

48 Petroni A, Bertagnolio B, La Spada P, Blasevich M, Papini N, Govoni S, Rimoldi M, Galli C (1998) The beta-oxidation of arachidonic acid and the synthesis of docosahexaenoic acid are selectively and consistently altered in skin fibroblasts from three Zellweger patients *versus* X-adrenoleukodystrophy, Alzheimer and control subjects. *Neurosci Lett* 250: 145–148

49 Gibson RA, Makrides M (2001) Long-chain polyunsaturated fatty acids in breast milk: Are they essential? *Adv Exp Med Biol* 501: 375–383

50 Hamosh M (1995) Lipid metabolism in pediatric nutrition. *Pediatr Clin North Am* 42: 839–859

51 Uauy R, Birch E, Birch D, Peirano P (1992) Visual and brain function measurements in studies of n-3 fatty acid requirements of infants. *J Pediatr* 120: S168–S180

52 Sinclair AJ, Murphy KJ, Li D (2000) Marine lipids: Overview "news insights and lipid composition of Lyprinol". *Allerg Immunol (Paris)* 32: 261–271

53 Galle AM, Joseph M, Demandre C, Guerche P, Dubacq JP, Oursel A, Mazliak P, Pelletier G, Kader JC (1993) Biosynthesis of gamma-linolenic acid in developing seeds of borage, *Borago officinalis* L . *Biochim Biophys Acta* 1158: 52–58

54 Garcia-Maroto F, Garrido-Cardenas JA, Rodriguez-Ruiz J, Vilches-Ferron M, Adam AC, Polaina J, Alonso DL (2002) Cloning and molecular characterization of the delta 6-desaturase from two echium plant species: production of GLA by heterologous expression in yeast and tobacco. *Lipids* 37: 417–426

55 Alsted AL, Hoy CE (1992) Fatty acid profiles of brain phospholipid subclasses of rats fed n-3 polyunsaturated fatty acids of marine or vegetable origin. A two generation study. *Biochim Biophys Acta* 1125: 237–244
56 Sayanova O, Smith MA, Lapinskas P, Stobart AK, Dobson G, Christie WW, Shewry PR, Napier JA (1997) Expression of a borage desaturase cDNA containing an N-terminal cytochrome b5 domain results in the accumulation of high levels of delta6-desaturated fatty acids in transgenic tobacco. *Proc Natl Acad Sci USA* 94: 4211–4216
57 Aki T, Shimada Y, Inagaki K, Higashimoto H, Kawamoto S, Shigeta S, Ono K, Suzuki O (1999) Molecular cloning and functional characterization of rat delta-6 fatty acid desaturase. *Biochem Biophys Res Commun* 255: 575–579
58 de Antueno RJ, Knickle LC, Smith H, Elliot ML, Allen SJ, Nwaka S, Winther MD (2001) Activity of human Delta 5 and Delta 6 desaturases on multiple n-3 and n-6 polyunsaturated fatty acids. *FEBS Lett* 509: 77–80
59 Miller CC, Ziboh VA (1988) Gammalinolenic acid-enriched diet alters cutaneous eicosanoids. *Biochem Biophys Res Commun* 154: 967–974
60 Singer P, Jaeger W, Berger I, Barleben H, Wirth M, Richter-Heinrich E, Voigt S, Godicke W (1990) Effects of dietary oleic, linoleic and alpha-linolenic acids on blood pressure, serum lipids, lipoproteins and the formation of eicosanoid precursors in patients with mild essential hypertension. *J Hum Hypertens* 4: 227–233
61 Ziboh VA, Fletcher MP (1992) Dose-response effects of dietary gamma-linolenic acid-enriched oils on human polymorphonuclear-neutrophil biosynthesis of leukotriene B_4. *Am J Clin Nutr* 55: 39–45
62 Barham JB, Edens MB, Fonteh AN, Johnson MM, Easter L, Chilton FH (2000) Addition of eicosapentaenoic acid to gamma-linolenic acid-supplemented diets prevents serum arachidonic acid accumulation in humans. *J Nutr* 130: 1925–1931
63 Chilton L, Surette ME, Swan DD, Fonteh AN, Johnson MM, Chilton FH (1996) Metabolism of gammalinolenic acid in human neutrophils. *J Immunol* 156: 2941–2947
64 Johnson MM, Swan DD, Surette ME, Stegner J, Chilton T, Fonteh AN, Chilton FH (1997) Dietary supplementation with gamma-linolenic acid alters fatty acid content and eicosanoid production in healthy humans. *J Nutr* 127: 1435–1444
65 Johnson MM, Vaughn B, Triggiani M, Swan DD, Fonteh AN, Chilton FH (1999) Role of arachidonyl triglycerides within lipid bodies in eicosanoid formation by human polymorphonuclear cells. *Am J Respir Cell Mol Biol* 21: 253–258
66 Frenoux JM, Prost ED, Belleville JL, Prost JL (2001) A polyunsaturated fatty acid diet lowers blood pressure and improves antioxidant status in spontaneously hypertensive rats. *J Nutr* 131: 39–45
67 Gerster H (1995) The use of n-3 PUFAs (fish oil) in enteral nutrition. *Int J Vitam Nutr Res* 65: 3–20
68 Guichardant M, Traitler H, Spielmann D, Sprecher H, Finot PA (1993) Stearidonic acid, an inhibitor of the 5-lipoxygenase pathway. A comparison with timnodonic and dihomogammalinolenic acid. *Lipids* 28: 321–324

69 Harel Z, Riggs S, Vaz R, White L, Menzies G (2001) Omega-3 polyunsaturated fatty acids in adolescents: Knowledge and consumption. *J Adolesc Health* 28: 10–15

70 Lewis NM, Seburg S, Flanagan NL (2000) Enriched eggs as a source of N-3 polyunsaturated fatty acids for humans. *Poult Sci* 79: 971–974

71 Mantzioris E, Cleland LG, Gibson RA, Neumann MA, Demasi M, James MJ (2000) Biochemical effects of a diet containing foods enriched with n-3 fatty acids. *Am J Clin Nutr* 72: 42–48

72 Shoda R, Matsueda K, Yamato S, Umeda N (1995) Therapeutic efficacy of N-3 polyunsaturated fatty acid in experimental Crohn's disease. *J Gastroenterol* 30 (Suppl) 8: 98–101

73 Sperling RI (1995) Eicosanoids in rheumatoid arthritis. *Rheum Dis Clin North Am* 21: 741–758

74 Zurier RB, Rossetti RG, Jacobson EW, DeMarco DM, Liu NY, Temming JE, White BM, Laposata M (1996) gamma-Linolenic acid treatment of rheumatoid arthritis. A randomized, placebo-controlled trial. *Arthritis Rheum* 39: 1808–1817

75 Sprecher H (1996) New advances in fatty-acid biosynthesis. *Nutrition* 12: S5–S7

76 Sprecher H, Chen Q (1999) Polyunsaturated fatty acid biosynthesis: A microsomal-peroxisomal process. *Prostaglandins Leukot Essent Fatty Acids* 60: 317–321

77 Gronn M, Christensen E, Hagve TA, Christophersen BO (1990) The Zellweger syndrome: Deficient conversion of docosahexaenoic acid (22: 6(n-3)) to eicosapentaenoic acid (20: 5(n-3)) and normal delta 4-desaturase activity in cultured skin fibroblasts. *Biochim Biophys Acta* 1044: 249–254

78 Wanders RJ, van Roermund CW, van Wijland MJ, Schutgens RB, Heikoop J, Van den BH, Schram AW, Tager JM (1987) Peroxisomal fatty acid beta-oxidation in relation to the accumulation of very long chain fatty acids in cultured skin fibroblasts from patients with Zellweger syndrome and other peroxisomal disorders. *J Clin Invest* 80: 1778–1783

79 Rapoport SI, Chang MC, Spector AA (2001) Delivery and turnover of plasma-derived essential PUFAs in mammalian brain. *J Lipid Res* 42: 678–685

80 Leonard AE, Kelder B, Bobik EG, Chuang LT, Lewis CJ, Kopchick JJ, Mukerji P, Huang YS (2002) Identification and expression of mammalian long-chain PUFA elongation enzymes. *Lipids* 37: 733–740

81 Parker-Barnes JM, Das T, Bobik E, Leonard AE, Thurmond JM, Chaung LT, Huang YS, Mukerji P (2000) Identification and characterization of an enzyme involved in the elongation of n-6 and n-3 polyunsaturated fatty acids. *Proc Natl Acad Sci USA* 97: 8284–8289

82 Chuang LT, Leonard AE, Liu JW, Mukerji P, Bray TM, Huang YS (2001) Inhibitory effect of conjugated linoleic acid on linoleic acid elongation in transformed yeast with human elongase. *Lipids* 36: 1099–1103

83 Matsuzaka T, Shimano H, Yahagi N, Yoshikawa T, Amemiya-Kudo M, Hasty AH, Okazaki H, Tamura Y, IizukaY, Ohashi K et al (2002) Cloning and characterization of a mammalian fatty acyl-CoA elongase as a lipogenic enzyme regulated by SREBPs. *J Lipid Res* 43: 911–920

84 Moon YA, Shah NA, Mohapatra S, Warrington JA, Horton JD (2001) Identification of a mammalian long chain fatty acyl elongase regulated by sterol regulatory element-binding proteins. *J Biol Chem* 276: 45358–45366
85 Zank TK, Zahringer U, Lerchl J, Heinz E (2000) Cloning and functional expression of the first plant fatty acid elongase specific for Delta(6)-polyunsaturated fatty acids. *Biochem Soc Trans* 28: 654–658
86 Biagi PL, Hrelia S, Stefanini GF, Zunarelli P, Bordoni A (1990) Delta-6-desaturase activity of human liver microsomes from patients with different types of liver injury. *Prostaglandins Leukot Essent Fatty Acids* 39: 39–42
87 Cho HP, Nakamura M, Clarke SD (1999) Cloning, expression, and fatty acid regulation of the human delta-5 desaturase. *J Biol Chem* 274: 37335–37339
88 de Antueno RJ, Allen SJ, Ponton A, Winther MD (2001) Activity and mRNA abundance of Delta-5 and Delta-6 fatty acid desaturases in two human cell lines. *FEBS Lett* 491: 247–251
89 Dias VC, Parsons HG (1995) Modulation in delta 9, delta 6, and delta 5 fatty acid desaturase activity in the human intestinal CaCo-2 cell line. *J Lipid Res* 36: 552–563
90 Horrobin DF (1993) Fatty acid metabolism in health and disease: The role of delta-6-desaturase. *Am J Clin Nutr* 57: 732S–736S
91 Rodriguez A, Sarda P, Nessmann C, Boulot P, Leger CL, Descomps B (1998) Delta 6- and delta 5-desaturase activities in the human fetal liver: Kinetic aspects. *J Lipid Res* 39: 1825–1832
92 Sato M, Adan Y, Shibata K, Shoji Y, Sato H, Imaizumi K (2001) Cloning of rat delta 6-desaturase and its regulation by dietary eicosapentaenoic or docosahexaenoic acid. *World Rev Nutr Diet* 88: 196–199
93 Chambaz J, Ravel D, Manier MC, Pepin D, Mulliez N, Bereziat G (1985) Essential fatty acids interconversion in the human fetal liver. *Biol Neonate* 47: 136–140
94 Duffin KL, Obukowicz MG, Salsgiver WJ, Welsch DJ, Shieh C, Raz A, Needleman P (2001) Lipid remodeling in mouse liver and plasma resulting from delta 6 fatty acid desaturase inhibition. *Lipids* 36: 1203–1208
95 Leonard AE, Kelder B, Bobik EG, Chuang LT, Parker-Barnes JM, Thurmond JM, Kroeger PE, Kopchick JJ, Huang YS, Mukerji P (2000) cDNA cloning and characterization of human Delta 5-desaturase involved in the biosynthesis of arachidonic acid. *Biochem J* 347 Pt 3: 719–724
96 Melin T, Nilsson A (1997) Delta-6-desaturase and delta-5-desaturase in human Hep G2 cells are both fatty acid interconversion rate limiting and are up-regulated under essential fatty acid deficient conditions. *Prostaglandins Leukot Essent Fatty Acids* 56: 437–442
97 Rodriguez A, Sarda P, Boulot P, Leger CL, Descomps B (1999) Differential effect of N-ethyl maleimide on delta 6-desaturase activity in human fetal liver toward fatty acids of the n-6 and n-3 series. *Lipids* 34: 23–30
98 Hrelia S, Lopez Jimenez JA, Bordoni A, Nvarro SZ, Horrobin DF, Rossi CA, Biagi PL (1995) Essential fatty acid metabolism in cultured rat cardiomyocytes in response to

either N-6 or N-3 fatty acid supplementation. *Biochem Biophys Res Commun* 216: 11–19

99 Ramanadham S, Zhang S, Ma Z, Wohltmann M, Bohrer A, Hsu FF, Turk J (2002) Delta 6-, stearoyl CoA-, and Delta 5-desaturase enzymes are expressed in beta-cells and are altered by increases in exogenous PUFA concentrations. *Biochim Biophys Acta* 1580: 40–56

100 de Alaniz MJ, Marra CA (1992) Glucocorticoid and mineralocorticoid hormones depress liver delta 5 desaturase activity through different mechanisms. *Lipids* 27: 599–604

101 Horrobin DF (1981) Loss of delta-6-desaturase activity as a key factor in aging. *Med Hypotheses* 7: 1211–1220

102 Igal RA, Mandon EC, deGDI (1991) Abnormal metabolism of polyunsaturated fatty acids in adrenal glands of diabetic rats. *Mol Cell Endocrinol* 77: 217–227

103 Nakada T, Kwee IL, Ellis WG (1990) Membrane fatty acid composition shows delta-6-desaturase abnormalities in Alzheimer's disease. *Neuroreport* 1: 153–155

104 Tollesson A, Frithz A, Berg A, Karlman G (1993) Essential fatty acids in infantile seborrheic dermatitis. *J Am Acad Dermatol* 28: 957–961

105 Innis SM, Dyer RA (2002) Brain astrocyte synthesis of docosahexaenoic acid from n-3 fatty acids is limited at the elongation of docosapentaenoic acid. *J Lipid Res* 43: 1529–1536

106 Reddy TS, Bazan NG (1985) Long-chain acyl CoA synthetase in microsomes from rat brain gray matter and white matter. *Neurochem Res* 10: 377–386

107 Uberti MA, Pierce J, Weis MT (2003) Molecular characterization of a rabbit long-chain fatty acyl CoA synthetase that is highly expressed in the vascular endothelium. *Biochim Biophys Acta* 1645: 193–204

108 Laposata M, Reich EL, Majerus PW (1985) Arachidonoyl-CoA synthetase. Separation from non-specific acyl-CoA synthetase and distribution in various cells and tissues. *J Biol Chem* 260: 11016–11020

109 Neufeld EJ, Sprecher H, Evans RW, Majerus PW (1984) Fatty acid structural requirements for activity of arachidonoyl-CoA synthetase. *J Lipid Res* 25: 288–293

110 Wilson DB, Prescott SM, Majerus PW (1982) Discovery of an arachidonoyl coenzyme A synthetase in human platelets. *J Biol Chem* 257: 3510–3515

111 Cao Y, Traer E, Zimmerman GA, McIntyre TM, Prescott SM (1998) Cloning, expression, and chromosomal localization of human long-chain fatty acid-CoA ligase 4 (FACL4). *Genomics* 49: 327–330

112 Vessey DA, Lau E, Kelley M, Warren RS (2003) Isolation, sequencing, and expression of a cDNA for the HXM-A form of xenobiotic/medium-chain fatty acid: CoA ligase from human liver mitochondria. *J Biochem Mol Toxicol* 17: 1–6

113 Kim JH, Lewin TM, Coleman RA (2001) Expression and characterization of recombinant rat Acyl-CoA synthetases 1, 4, and 5. Selective inhibition by triacsin C and thiazolidinediones. *J Biol Chem* 276: 24667–24673

114 Minekura H, Kang MJ, Inagaki Y, Suzuki H, Sato H, Fujino T, Yamamoto TT (2001)

Genomic organization and transcription units of the human acyl-CoA synthetase 3 gene. *Gene* 278: 185–192
115 Oikawa E, Iijima H, Suzuki T, Sasano H, Sato H, Kamataki A, Nagura H, Kang MJ, Fujino T, Suzuki H et al (1998) A novel acyl-CoA synthetase, ACS5, expressed in intestinal epithelial cells and proliferating preadipocytes. *J Biochem (Tokyo)* 124: 679–685
116 Tang PZ, Tsai-Morris CH, Dufau ML (2001) Cloning and characterization of a hormonally regulated rat long chain acyl-CoA synthetase. *Proc Natl Acad Sci USA* 98: 6581–6586
117 Cao Y, Murphy KJ, McIntyre TM, Zimmerman GA, Prescott SM (2000) Expression of fatty acid-CoA ligase 4 during development and in brain. *FEBS Lett* 467: 263–267
118 Cho YY, Kang MJ, Ogawa S, Yamashita Y, Fujino T, Yamamoto TT (2000) Regulation by adrenocorticotropic hormone and arachidonate of the expression of acyl-CoA synthetase 4, an arachidonate-preferring enzyme expressed in steroidogenic tissues. *Biochem Biophys Res Commun* 274: 741–745
119 Mandon EC, de GDI, Brenner RR (1988) Long-chain fatty acyl-CoA synthetase of rat adrenal microsomes. Effect of ACTH and epinephrine. *Mol Cell Endocrinol* 56: 123–131
120 Singh H, Derwas N, Poulos A (1987) Beta-oxidation of very-long-chain fatty acids and their coenzyme A derivatives by human skin fibroblasts. *Arch Biochem Biophys* 254: 526–533
121 Tol VA (1975) Aspects of long-chain acyl-COA metabolism. Mol Cell Biochem 7: 19–31
122 Iwai N, Mannami T, Tomoike H, Ono K, Iwanaga Y (2003) An Acyl-CoA Synthetase Gene Family in Chromosome 16p12 May Contribute to Multiple Risk Factors. *Hypertension* 41: 1041–1046
123 Shimabukuro M, Zhou YT, Levi M, Unger RH (1998) Fatty acid-induced beta cell apoptosis: A link between obesity and diabetes. *Proc Natl Acad Sci USA* 95: 2498–2502
124 Shimomura I, Tokunaga K, Jiao S, Funahashi T, Keno Y, Kobatake T, Kotani K, Suzuki H, Yamamoto T, Tarui S et al (1992) Marked enhancement of acyl-CoA synthetase activity and mRNA, paralleled to lipoprotein lipase mRNA, in adipose tissues of Zucker obese rats (fa/fa). *Biochim Biophys Acta* 1124: 112–118
125 Antinozzi PA, Segall L, Prentki M, McGarry JD, Newgard CB (1998) Molecular or pharmacologic perturbation of the link between glucose and lipid metabolism is without effect on glucose-stimulated insulin secretion. A re-evaluation of the long-chain acyl-CoA hypothesis. *J Biol Chem* 273: 16146–16154
126 Fulgencio JP, Kohl C, Girard J, Pegorier JP (1996) Troglitazone inhibits fatty acid oxidation and esterification, and gluconeogenesis in isolated hepatocytes from starved rats. *Diabetes* 45: 1556–1562
127 Li J, Wurtman RJ (1999) Heterogeneous long chain acyl-CoA synthetases control distribution of individual fatty acids in newly-formed glycerolipids of neuronal cells undergoing neurite outgrowth. *Neurochem Res* 24: 739–750
128 Piccini M, Vitelli F, Bruttini M, Pober BR, Jonsson JJ, Villanova M, Zollo M, Borsani G, Ballabio A, Renieri A (1998) FACL4, a new gene encoding long-chain acyl-CoA syn-

thetase 4, is deleted in a family with Alport syndrome, elliptocytosis, and mental retardation. *Genomics* 47: 350–358

129 Yamada T, Shinnoh N, Kondo A, Uchiyama A, Shimozawa N, Kira J, Kobayashi T (2000) Very-long-chain fatty acid metabolism in adrenoleukodystrophy protein-deficient mice. *Cell Biochem Biophys* 32: 239–246

130 Carter JR, Kennedy EP (1966) Enzymatic synthesis of cytidine diphosphate diglyceride. *J Lipid Res* 7: 678–683

131 Colodzin M, Kennedy EP (1965) Biosynthesis of diphosphoinositide in brain. *J Biol Chem* 240: 3771–3780

132 Husbands DR, Lands WE (1968) Properties of glycerol phosphate acyl transferase from pigeon liver particles. *Biochem J* 110: 50P–51P

133 Hill EE, Lands WE (1968) Incorporation of long-chain and polyunsaturated acids into phosphatidate and phosphatidylcholine. *Biochim Biophys Acta* 152: 645–648

134 Lands WE (2000) Stories about acyl chains. *Biochim Biophys Acta* 1483: 1–14

135 Fonteh AN, Chilton FH (1993) Mobilization of different arachidonate pools and their roles in the generation of leukotrienes and free arachidonic acid during immunologic activation of mast cells. *J Immunol* 150: 563–570

136 Masuzawa Y, Sugiura T, Sprecher H, Waku K (1989) Selective acyl transfer in the reacylation of brain glycerophospholipids. Comparison of three acylation systems for 1-alk-1'-enylglycero-3-phosphoethanolamine, 1-acylglycero-3-phosphoethanolamine and 1-acylglycero-3-phosphocholine in rat brain microsomes *Biochim Biophys Acta* 1005: 1–12

137 Ojima A, Nakagawa Y, Sugiura T, Masuzawa Y, Waku K (1987) Selective transacylation of 1-O-alkylglycerophosphoethanolamine by docosahexaenoate and arachidonate in rat brain microsomes. *J Neurochem* 48: 1403–1410

138 Sugiura T, Kudo N, Ojima T, Mabuchi-Itoh K, Yamashita A, Waku K (1995) Coenzyme A-dependent cleavage of membrane phospholipids in several rat tissues: ATP-independent acyl-CoA synthesis and the generation of lysophospholipids. *Biochim Biophys Acta* 1255: 167–176

139 Sugiura T, Masuzawa Y, Waku K (1985) Transacylation of 1-O-alkyl-SN-glycero-3-phosphocholine (lyso platelet-activating factor) and 1-O-alkenyl-SN-glycero-3-phosphoethanolamine with docosahexaenoic acid (22: 6 omega 3). *Biochem Biophys Res Commun* 133: 574–580

140 Sugiura T, Masuzawa Y, Nakagawa Y, Waku K (1987) Transacylation of lyso platelet-activating factor and other lysophospholipids by macrophage microsomes. Distinct donor and acceptor selectivities. *J Biol Chem* 262: 1199–1205

141 Chilton FH, Fonteh AN, Surette ME, Triggiani M, Winkler JD (1996) Control of arachidonate levels within inflammatory cells. *Biochim Biophys Acta* 1299: 1–15

142 Chilton FH, Fonteh AN, Sung CM, Hickey DM, Torphy TJ, Mayer RJ, Marshall LA, Heravi JD, Winkler JD (1995) Inhibitors of CoA-independent transacylase block the movement of arachidonate into 1-ether-linked phospholipids of human neutrophils. *Biochemistry* 34: 5403–5410

143 Surette ME, Winkler JD, Fonteh AN, Chilton FH (1996) Relationship between arachidonate--phospholipid remodeling and apoptosis. *Biochemistry* 35: 9187–9196
144 Winkler JD, Sung CM, Bennett CF, Chilton FH (1991) Characterization of CoA-independent transacylase activity in U937 cells. *Biochim Biophys Acta* 1081: 339–346
145 Fonteh AN, LaPorte T, Swan D, McAlexander MA (2001) A decrease in remodeling accounts for the accumulation of arachidonic acid in murine mast cells undergoing apoptosis. *J Biol Chem* 276: 1439–1449
146 Fonteh AN, Chilton FH (1992) Rapid remodeling of arachidonate from phosphatidylcholine to phosphatidylethanolamine pools during mast cell activation. *J Immunol* 148: 1784–1791
147 Obukowicz MG, Raz A, Pyla PD, Rico JG, Wendling JM, Needleman P (1998) Identification and characterization of a novel delta 6/delta 5 fatty acid desaturase inhibitor as a potential anti-inflammatory agent. *Biochem Pharmacol* 55: 1045–1058
148 Obukowicz MG, Welsch DJ, Salsgiver WJ, Martin-Berger CL, Chinn KS, Duffin KL, Raz A, Needleman P (1998) Novel, selective delta 6 or delta 5 fatty acid desaturase inhibitors as anti-inflammatory agents in mice. *J Pharmacol Exp Ther* 287: 157–166
149 Omura S, Tomoda H, Xu QM, Takahashi Y, Iwai Y (1986) Triacsins, new inhibitors of acyl-CoA synthetase produced by *Streptomyces* sp. *J Antibiot (Tokyo)* 39: 1211–1218
150 Hartman EJ, Omura S, Laposata M (1989) Triacsin C: A differential inhibitor of arachidonoyl-CoA synthetase and non-specific long chain acyl-CoA synthetase. *Prostaglandins* 37: 655–671
151 Igal RA, Wang P, Coleman RA (1997) Triacsin C blocks *de novo* synthesis of glycerolipids and cholesterol esters but not recycling of fatty acid into phospholipid: Evidence for functionally separate pools of acyl-CoA. *Biochem J* 324 (Pt 2): 529–534
152 Tomoda H, Igarashi K, Omura S (1987) Inhibition of acyl-CoA synthetase by triacsins. *Biochim Biophys Acta* 921: 595–598
153 Kasuya F, Hiasa M, KawaiY, Igarashi K, Fukui M (2001) Inhibitory effect of quinolone antimicrobial and non-steroidal anti-inflammatory drugs on a medium chain acyl-CoA synthetase. *Biochem Pharmacol* 62: 363–367
154 Fonteh AN, Samet JM, Chilton FH (1995) Regulation of arachidonic acid, eicosanoid, and phospholipase A_2 levels in murine mast cells by recombinant stem cell factor. *J Clin Invest* 96: 1432–1439
155 Laposata M, Kaiser SL, Capriotti AM (1988) Icosanoid production can be decreased without alterations in cellular arachidonate content or enzyme activities required for arachidonate release and icosanoid synthesis. *J Biol Chem* 263: 3266–3273
156 Winkler JD, Sung CM, Chabot-Flecher M, Griswold DE, Marshall LA, Chilton FH, Bondinell W, Mayer RJ (1998) Beta-lactams SB 212047 and SB 216754 are irreversible, time-dependent inhibitors of coenzyme A-independent transacylase. *Mol Pharmacol* 53: 322–329
157 Winkler JD, Fonteh AN, Sung CM, Heravi JD, Nixon AB, Chabot-Fletcher M, Griswold D, Marshall LA, Chilton FH (1995) Effects of CoA-independent transacylase inhibitors

on the production of lipid inflammatory mediators. *J Pharmacol Exp Ther* 274: 1338–1347

158 Winkler JD, Fonteh AN, Sung CM, Huang L, Chabot-Fletcher M, Marshall LA, Chilton FH (1995) Inhibition of CoA-independent transacylase reduces inflammatory lipid mediators. *Adv Prostaglandin Thromboxane Leukot Res* 23: 89–91

159 Winkler JD, Eris T, Sung CM, Chabot-Fletcher M, Mayer RJ, Surette ME, Chilton FH (1996) Inhibitors of coenzyme A-independent transacylase induce apoptosis in human HL-60 cells. *J Pharmacol Exp Ther* 279: 956–966

160 Fonteh AN (2002) Differential effects of arachidonoyl trifluoromethyl ketone on arachidonic acid release and lipid mediator biosynthesis by human neutrophils. Evidence for different arachidonate pools. *Eur J Biochem* 269: 3760–3770

161 Winkler JD, Sung CM, Hubbard WC, Chilton FH (1992) Evidence for different mechanisms involved in the formation of lyso platelet-activating factor and the calcium-dependent release of arachidonic acid from human neutrophils. *Biochem Pharmacol* 44: 2055–2066

162 Surette ME, Fonteh AN, Bernatchez C, Chilton FH (1999) Perturbations in the control of cellular arachidonic acid levels block cell growth and induce apoptosis in HL-60 cells. *Carcinogenesis* 20: 757–763

163 Trimboli AJ, Waite BM, Atsumi G, Fonteh AN, Namen AM, Clay CE, Kute TE, High KP, Willingham MC, Chilton FH (1999) Influence of coenzyme A-independent transacylase and cyclo-oxygenase inhibitors on the proliferation of breast cancer cells. *Cancer Res* 59: 6171–6177

164 Ghomashchi F, Loo R, Balsinde J, Bartoli F, Apitz-Castro R, Clark JD, Dennis EA, Gelb MH (1999) Trifluoromethyl ketones and methyl fluorophosphonates as inhibitors of group IV and VI phospholipases A_2: Structure-function studies with vesicle, micelle, and membrane assays. *Biochim Biophys Acta* 1420: 45–56

Remodeling of arachidonic acid in inflammatory cells of the human lung

Massimo Triggiani[1], Giorgio Giannattasio[1], Francescopaolo Granata[1], Stefania Loffredo[1], Francesca W. Rossi[1], Salvatore Salzano[2], Gianni Marone[1]

[1]Division of Clinical Immunology and Allergy, University of Naples Federico II, Via Pansini 5, 80131 Naples, Italy; [2]Experimental Endocrinology and Oncology Institute "G. Salvatore", Consiglio Nazionale delle Ricerche (CNR), Naples, Italy

Introduction

Arachidonic acid (AA) is a key molecule in the modulation of several pathophysiological events in mammalian cells, including gene expression, membrane signal transduction, cell differentiation and apoptosis, exocytosis and generation of eicosanoids [1–4]. It has been now convincingly demonstrated that the intracellular levels of free AA within mammalian cells are crucial for most of these events and, therefore, they are carefully regulated by complex biochemical reactions [5, 6]. These reactions are catalyzed by enzymes involved in both AA mobilization and re-esterification into the storage sites and transfer from one intracellular pool to another [7, 8].

Under normal circumstances, AA is stored within different phospholipid and neutral lipid pools. The distribution of AA in the lipid pools of the cell is a highly dynamic phenomenon that may follow routes different from cell to cell and in the same cell exposed to various conditions [9, 10]. The changes in the distribution of AA within the glycerolipid pools have been referred to as remodeling. In the last ten years a large body of evidence has accumulated indicating that remodeling of AA between lipid pools is an important mechanism to regulate the intracellular levels of free and esterified AA and to determine the quantities of AA that can be mobilized at any time from the storage pools [11–13].

The rate and extent of AA remodeling is strongly influenced by the stage of maturation and differentiation of the cell and by its state of activation [14–17]. Increasing evidence also suggests that AA remodeling may be biochemically or pharmacologically modulated and that this type of intervention has relevant and complex effects on a variety of intracellular processes [18–21].

AA metabolism is a crucial biochemical event in cells involved in inflammatory and immune responses. In both humans and experimental animals, inflammatory cells contain relatively large quantities of esterified AA that is utilized primarily for the production of prostaglandins, thromboxanes, leukotrienes and other metabo-

Arachidonate Remodeling and Inflammation, edited by Alfred N. Fonteh and Robert L. Wykle

lites involved in inflammation [22, 23]. In addition, AA in inflammatory cells has been implicated in the degranulation process leading to the release of preformed mediators [4, 24, 25], in the production of cytokines and chemokines [26–28], activation of respiratory burst [1, 29], chemotaxis and phagocytosis [2, 30] as well as in the survival of cells recruited at sites of inflammation [13, 31].

Arachidonic acid pools in human inflammatory cells

The bulk of AA in human inflammatory cells under resting conditions is esterified into various phospholipid pools [6]. Data obtained with cells isolated from peripheral blood of healthy donors including neutrophils [32–34] and eosinophils [35] show that the majority of AA in unstimulated cells is esterified in phosphatidylethanolamine (PE), followed by phosphatidylcholine (PC), phosphatidylinositol (PI) and phosphatidylserine (PS). In these cells a small percentage of the total cellular AA (1–5%) is usually associated with neutral lipid pools, primarily triglycerides (TG), diglycerides (DG) and cholesterol esters [36, 37]. Most of the AA in resting inflammatory cells isolated from blood is esterified into the 1-alkyl and the 1-alk-1'-enyl (plasmalogen) subclasses of PE and the 1-alkyl subclasses of PC [32, 33].

The distribution of AA in inflammatory cells resident within tissues, i.e., the lung or the gastrointestinal tract, is to some extent different from that of inflammatory cells circulating in the blood. For example, mast cells purified from the human lung contain approximately 50% of the total cellular AA esterified into TG [38]. Similarly, macrophages isolated from the lung or the peritoneal cavity contain 10–25% of AA associated with TG [39, 40].

Major differences also exist in the distribution of AA within phospholipid pools between tissue and blood inflammatory cells. For example, the majority of AA in mast cells purified from the human lung or skin is esterified into 1-acyl-PC and 1-acyl-PE [41, 42], making these, and not the 1-ether-linked, the major AA-containing phospholipid subclasses. As previously mentioned, it is believed that the distribution of AA into glycerolipid pools may to some extent dictate the quality and quantity of eicosanoid produced upon cell activation [32, 43, 44]. Several studies have shown that some classes and subclasses of membrane phospholipids represent preferential substrates for PLA_2s [31, 45]. In other words, certain intracellular pools of AA are predominantly involved in the production of eicosanoids whereas others may represent primarily storage pools [39, 43]. A good correlation between the amount of AA in certain phospholipid pools (1-ether-linked) and products (eicosanoids) has been demonstrated in several inflammatory cells such as neutrophils, eosinophils, macrophages and mast cells [43, 44, 46, 47]. In particular, only AA derived from the 1-alkyl subclasses of phosphatidylcholine and phosphatidylethanolamine has been closely linked to eicosanoid synthesis. In contrast, AA derived from other classes of phospholipids (phosphatidylinositol and phosphatidylserine) or from the 1-acyl subclass of phosphatidylcholine and phosphatidylethanolamine may be preferentially secreted as free AA or may remain inside the cell functioning as a second messenger.

Remodeling of arachidonic acid in inflammatory cells migrating to the lung

The possibility of comparing cells in a relatively resting state (isolated from blood) with those retrieved by bronchoalveolar lavage (BAL) in patients with inflammatory diseases of the lung offered a great opportunity to explore AA remodeling and changes in lipid mediator synthesis induced by the cell recruitment in an inflamed tissue.

In initial studies we compared the total cellular content and the distribution of endogenous AA in the glycerolipid classes of neutrophils isolated from peripheral blood or from the BAL of patients with adult respiratory distress syndrome (ARDS), an acute and severe inflammation of the lung with extensive neutrophilic infiltration [36]. Neutrophils from the BAL of ARDS patients contain an amount of total AA almost four times higher than that found in neutrophils from peripheral blood of the same patients (4.3 ± 1.2 *versus* 1.3 ± 0.4 nmol/10^6 cells). The major difference between the distribution of AA in the glycerolipid classes between neutrophils from peripheral blood and those from the BAL of ARDS patients is the accumulation of large quantities of AA in TG [36]. BAL neutrophils contain a pool of AA associated with TG that is almost ten-fold larger than that in resting blood neutrophils. AA in TG accounted for $28.7 \pm 7.9\%$ *versus* $3.1 \pm 1.2\%$ of total cellular AA in BAL and peripheral blood neutrophils, respectively. In addition, a smaller but significant increase in the percentage of AA esterified in PC can be detected in BAL neutrophils. The percentage of AA in PE, which is the major pool in blood neutrophils, is reduced by 30% in BAL neutrophils. These data indicate that there is a significant increase in the total cellular content of AA and a major remodeling of AA within glycerolipid pools in human neutrophils migrating from the blood to the lung and the alveolar space. In particular a large pool of AA in TG is constituted in neutrophils entering the inflamed lung *in vivo* [48].

A similar observation has been made in human eosinophils. Blood eosinophils from normal donors contain 3–5% of AA into the TG pool [38]. However, when the eosinophils are isolated from the blood of patients with hypereosinophilic syndromes, they are found to contain up to 30% of AA esterified into TG [36]. This observation suggests that in inflammatory cells, such as eosinophils and neutrophils activated in the blood or recruited in the lung during an inflammatory response, there is a remodeling of AA pools, with an increase of AA associated with the TG pool.

Subsequent studies suggested that the pool of AA associated with TG may be located in the lipid bodies, non-membrane-bound cytoplasmic organelles whose function is unknown [49]. Lipid bodies are abundant in human lung macrophages and mast cells, but they are very few in resting blood neutrophils and eosinophils [49, 50]. However, there is a dramatic increase in the number of lipid bodies in neutrophils and eosinophils isolated from the BAL of ARDS and asthmatic patients, respectively [36]. In these cells the increase in the number of lipid bodies correlates significantly with the accumulation of AA in the TG pool [36].

The remodeling of AA pools observed in the cells isolated *ex vivo* can be reproduced *in vitro*. For example, peripheral blood neutrophils from normal donors significantly increase their TG pool of AA and the number of lipid bodies when incubated with high concentrations of exogenous AA [37]. Similarly, eosinophils from normal donors acquire the same characteristics of the eosinophils recruited in the airways of asthmatic patients when they are cultured with GM-CSF for 24 hrs in the presence of exogenous AA [38, 46]. In these experimental conditions, both neutrophils and eosinophils acquire the morphologic features of "hypodense" cells, similar to those found in BAL eosinophils and macrophages of asthmatic patients. Taken together, these observations suggest that a large increase in the amount of AA occurs in cells recruited in an inflammatory area, such as the eosinophils and the neutrophils in the lungs of asthmatics and of patients with ARDS, respectively. This increase is mostly due to the expansion of the TG pool, which presumably functions as a large reservoir for AA.

Remodeling of arachidonic acid into TG pool induced by cell differentiation and maturation in the lung

The aforementioned studies have shown that TG may be the largest arachidonate-containing pool in certain inflammatory cells. For example, mast cells isolated from the human lung contain up to 50% of endogenous AA esterified in TG [38]. Other cells such as macrophages, eosinophils, monocytes, endothelial cells and muscular cells contain 5–30% of the total cellular AA associated with TG [38, 39, 51, 52]. The large variability in the AA content of the TG pool from cell to cell is in contrast with the relatively constant distribution of this fatty acid within the phospholipid pools.

The reasons for the wide variation in the TG pool of AA are not known. Like other pools of AA, that associated with TG is a dynamic pool and its size can greatly change under various conditions. For example, accumulation of AA in TG occurs in neutrophils when they are recruited into an inflammatory area *in vivo* [36]. This phenomenon can be reproduced *in vitro* by incubating neutrophils or HL-60 cells with high concentration of exogenous AA [36, 53, 54]. Antigen-induced activation of murine mast cells or ionophore-induced activation of human macrophages results in a transient increase in the AA content of the TG pool [14, 39]. These data suggest that the AA pool in TG is not constitutive but it may be induced by different stimuli. Understanding the cellular events that are able to promote the accumulation of AA into the TG pool may provide important information on the mechanisms regulating AA processing in inflammatory cells.

Previous works have consistently shown that little or no store of AA in TG is found in immature and/or actively proliferating cell lines. For example, in contrast with mature tissue mast cells, bone marrow-derived mast cells (BMMC) have little or no AA esterified in TG [14]. This observation led to the hypothesis that cellular differentiation and/or inhibition of cell proliferation may induce AA accumulation into the TG pool. We have tested this hypothesis in the human promonocytic cell line U937, an experimental model used to explore the biochemical modifications associated with cell differentiation. U937 can be differentiated into macrophage-like cells upon exposure to such diverse agents as IFN-γ, phorbol esters, retinoic acid (RetAc) and dimethylsulfoxide [55–57]. Differentiation of U937 is associated with inhibition of cell proliferation and with the acquisition of morphological and surface markers typical of mature macrophages [58].

In our experiments, differentiation of U937 into macrophages was achieved by incubation with RetAc, PMA, or both, either in the presence or in the absence of exogenous AA supplementation.

Incubation of AA with PMA or with PMA plus RetAc, but not with the AA supplement alone, blocked U937 proliferation and induced the expression of CD23, a marker of mature macrophages [59].

In addition, differentiation of U937 with PMA induced a significant increase in the total cellular content of AA from 0.9 ± 0.2 to 1.4 ± 0.4 nmoles/10^6 cells. This accumulation of AA was further enhanced in U937 incubated with PMA plus RetAc (to 1.7 ± 0.6 nmoles/10^6 cells). When differentiation of U937 with PMA plus RetAc occurred in the presence of the AA supplement, the total cellular content of AA reached 2.4 ± 0.9 nmoles/10^6 cells. Figure 1 shows the effect of U937 differentiation on the amount of AA into the major glycerolipid classes. Undifferentiated U937 contain AA esterified mostly in PE and to a lesser extent in PC and PI/PS. An average of $3.3 \pm 1.8\%$ of the total AA is associated with neutral lipids, 95% of which is esterified into TG. Exposure of U937 to PMA results in a significant increase in the percentage of AA associated with TG ($8.1 \pm 2.8\%$ of total cellular AA). When U937 are incubated with RetAc plus PMA, accumulation of AA into TG is higher than in U937 exposed to PMA alone ($11.0 \pm 4.6\%$). Finally, differentiation of U937 with RetAc plus PMA in the presence of AA supplement results in the esterification of $18.8 \pm 6.9\%$ of cellular AA into TG.

While the changes in the percentage of AA esterified into TG are the most striking, a significant increase in the percentage of AA esterified in PC is also observed in cells differentiated in the presence of exogenous AA (Fig. 1)

The increase in the TG pool of AA in differentiated U937 is due to an increase in the mass size of TG rather then to the substitution of other fatty acids stored in TG with AA. In fact, the mass amount of TG, expressed as μg/10^6 cells, increased

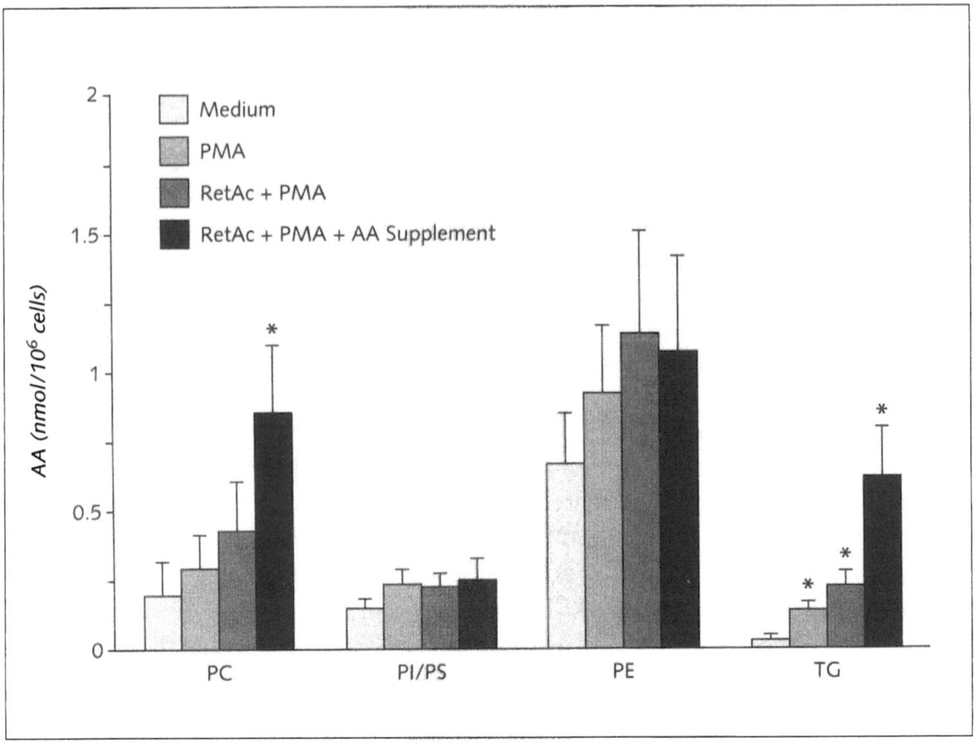

Figure 1
Distribution of AA in the glycerolipid classes of undifferentiated and differentiated U937
U937 were incubated for 96 hrs in medium alone, in the presence of PMA, RetAc and PMA or RetAc and PMA, with an AA supplement (10 μM). At the end of incubation, the lipids were extracted and AA was determined in each pool by NICI-GC/MS. The data are expressed as nmoles of AA/10^6 cells and are the mean ± SE of five experiments.
**$p < 0.05$ versus medium*

from 0.78 ± 0.26 to 1.82 ± 0.56 and to 2.35 ± 0.68, respectively, in undifferentiated, PMA and RetAc plus PMA differentiated cells. These findings suggest that the accumulation of AA into TG during differentiation of U937 is mostly due to the mass expansion of the TG pool.

Previous findings in the human neutrophil suggested that an increase in the total cellular content of AA is a major factor leading to AA accumulation in the TG pool [36]. The observation that AA supplement enhances the remodeling into the TG pool induced by cell differentiation further supports this hypothesis. To confirm the relationship between the total cellular content of AA and the amount of AA esterified in TG, we used mass spectrometry to determine the amount of endogenous AA (total and TG-associated) in U937 at various stages of differentiation induced by

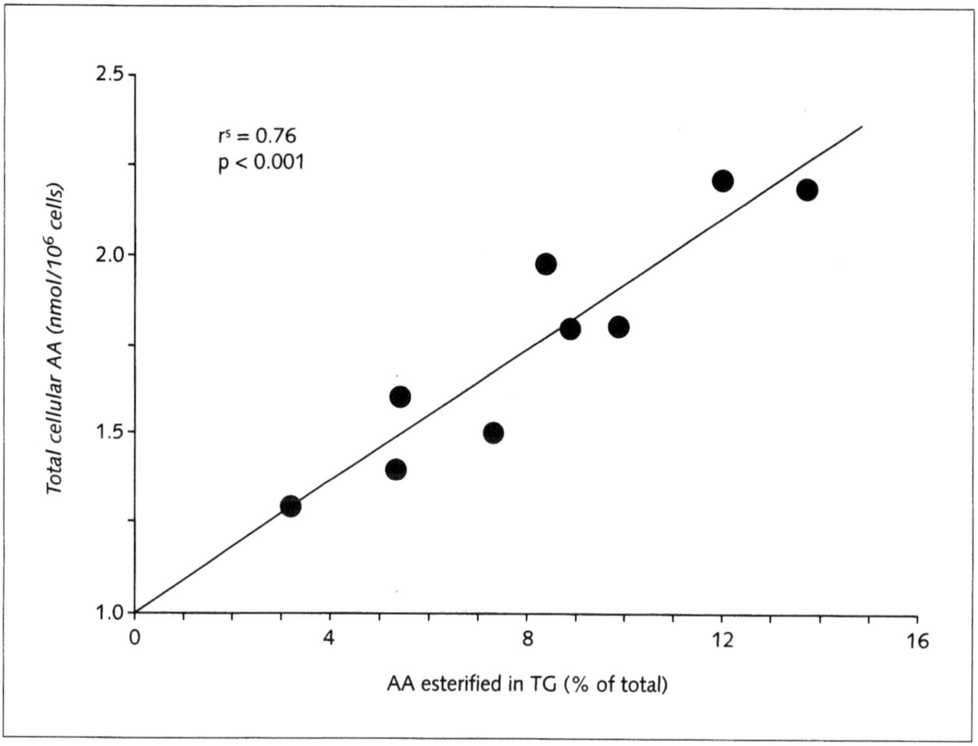

Figure 2
Correlation between the total cellular AA content and the percentage of AA esterified in TG in U937 at different stages of differentiation
U937 were incubated for 48–96 hrs in the presence of RetAc + PMA. At different time points the cells were harvested and the lipids were extracted. TG were isolated by TLC and endogenous AA was liberated by alkaline hydrolysis. Total cellular AA and AA associated with TG were determined by NICI-GC/MS. The data are obtained with three different preparations of U937.

RetAc plus PMA. A significant correlation can be found between the total cellular content of AA and the percentage of AA associated with TG in U937 at different stages of differentiation (Fig. 2).

This result suggested that accumulation of AA in the TG pool could be related to an increase in the capacity of differentiated U937 cells to incorporate exogenous AA. To test this, control and U937 exposed to RetAc, PMA or RetAc plus PMA for 96 hrs were incubated with radiolabeled AA (10^{-9} M) for another 12 hrs. Differentiation with PMA significantly increased the capacity of U937 to take up exogenous AA and this effect was increased up to 2.5-fold in U937 differentiated in the presence of the AA supplement.

Lipid bodies are non-membrane-bound cytoplasmic organelles detectable in most mammalian cells [49] by fluorescent staining with Nile red [60]. Indirect evidence suggests that these organelles may be the subcellular location of the TG pool of AA in human inflammatory cells [61]. Figure 3 shows the flow cytometry scan of Nile red-stained U937 after 96 hrs incubation with PMA or RetAc plus PMA. Incubation with PMA, but not with RetAc alone, increases the fluorescence intensity of U937 compared to control cells. Incubation with RetAc plus PMA further increases cell fluorescence compared to cells incubated with PMA alone. Direct microscope counting [62] and flow cytometry gave comparable results both indicating that differentiation of U937 with PMA or with RetAc plus PMA increased the number of cytoplasmic lipid bodies. This observation further supports the hypothesis that lipid bodies may be the cellular repository for the *de novo* formed TG pool.

Arachidonic acid pools in different populations of human lung macrophages

The above mentioned observations indicated that activation of inflammatory cells, e.g., by recruitment into an inflammatory area, or their differentiation and maturation may lead to significant changes in AA distribution within intracellular pools. The recent identification of different populations of macrophages in the human lung has been a timely occasion to confirm this hypothesis.

Centrifugation over density gradients of cells obtained by mechanical or enzymatic dispersion of the human lung tissue has been largely used to enrich the macrophage population. This procedure consistently yields two fractions of macrophages clearly distinct by their cell density. High density macrophages (HDM) have density comprised between 1.065 and 1.078 and account for two-thirds of the total macrophages in the human lung. Low density macrophages (LDM) have density between 1.039 and 1.052 and constitute the remaining one-third of the total lung macrophages [63]. LDM also have a larger area and diameter and appear highly vacuolated as compared to HDM. The two macrophage populations differ significantly in their expression of surface markers and in the production of cytokines

Figure 3
FACS scan of lipid bodies-associated fluorescence in undifferentiated and differentiated U937
U937 were incubated for 96 hrs in medium alone or in the presence of RetAc (10 nM), PMA (10 ng/ml) or both. At the end of the incubation, the cells were washed and incubated with Nile red. The cells were then washed and fluorescence was measured by flow cytometry. Fluorescence intensity units are indicated as a log scale on the abscissa. Each panel reports the peak fluorescence of undifferentiated cells for comparison.

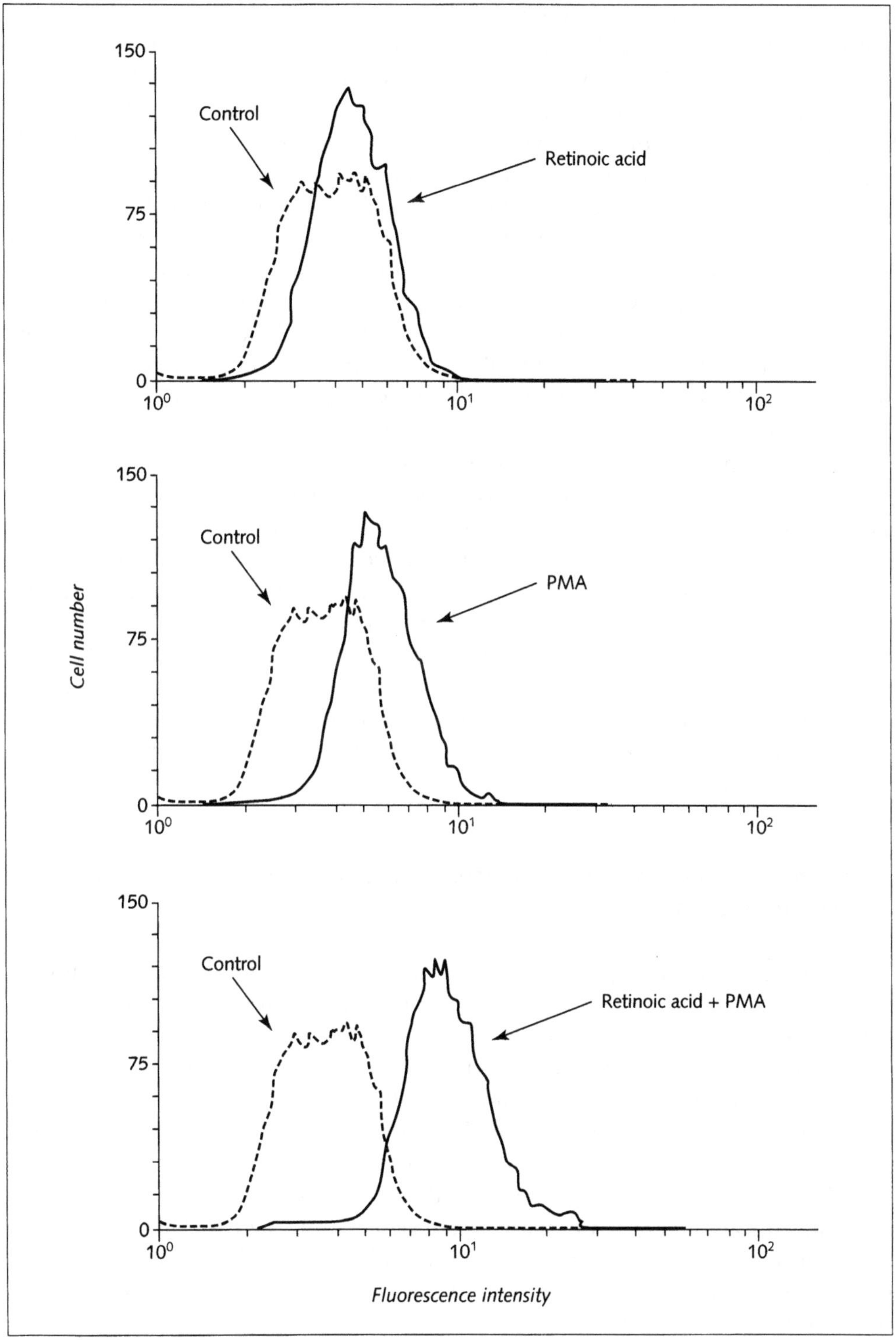
150
75
0
Control
Retinoic acid
10^0
10^1
10^2
Cell number
Control
PMA
Control
Retinoic acid + PMA
Fluorescence intensity

in response to physiologic stimuli. HDM predominantly express co-stimulatory molecules (CD40 and CD86) whereas LDM express predominantly activation markers (CD63 and CD64). In addition, HDM are highly responsive to lipopolysaccharide and histamine, producing large quantities of IL-10 and IL-12. In contrast, LDM are stimulated more effectively than HDM by secretory PLA_2 (group IA, IB, II and X) [64] and by opsonized zymosan and they release mostly IL-6 and TNF-α.

We have recently initiated to explore the AA distribution into the lipid pools of HDM and LDM. HDM and LDM do not differ significantly in their total AA cellular content. However, LDM contain significantly higher percentage of AA esterified into TG (22% *versus* 10%) and a larger proportion of AA esterified into 1-acyl-PC and 1-acyl-PE as compared to HDM. The HDM population contains the majority of AA into 1-alk-1'enyl-PE and 1-alkyl-PC. LDM also show a significantly higher number of cytoplasmatic lipid bodies per cell as compared to HDM (12 ± 5 *versus* 5 ± 3) (Triggiani et al., unpublished observations).

The mechanisms underlying the development of two subpopulations of macrophages in the human lung are presently unclear. Preliminary evidence suggests that the hypodense (LDM) cells may represent more mature macrophages in a pre-activated state, whereas HDM may be less mature cells more recently entered into the lung. This hypothesis is supported by the observation that HDM display several morphological, functional and biochemical features, including the AA distribution, similar to those of blood monocytes, the circulating precursor of tissue macrophages. These observations lend further support to the concept that maturation of the cells and/or their activation induced in the lung by local micro-environmental factors induces a profound remodeling of intracellular AA pools and the accumulation of AA into TG.

Biochemical functions of the triglyceride pool of arachidonic acid in inflammatory cells

While the role of the phospholipid pools as a source of AA for eicosanoid synthesis has been extensively studied, very little is known on the function of the TG pool, either constitutively present or induced by cell activation or differentiation. It has been clearly shown that TG are not substrates for PLA_2s [37, 39]. To address the role of TG as a source of AA in stimulated cells we have performed experiments in lung macrophages stimulated with the PMA or with the Ca^{2+} ionophore A23187. AA is released primarily from PC and PI in macrophages stimulated with PMA and from PC and PE in macrophages stimulated with A23187 [40]. In contrast to phospholipids, the amount of AA in the TG pool significantly increased at the early time points during cell activation induced by either PMA or A23187. These results suggest that TG do not act as a source of AA but rather as a pool to reincorporate AA released from phospholipids.

The rapid reacylation of free AA into TG in stimulated macrophages suggests at least two hypotheses. First, TG may act as a reacylation pool during the early phase of cell activation and thus contribute to maintain low intracellular concentrations of free AA. As mentioned before, free AA is a second messenger involved in the regulation of several cell functions and its intracellular concentrations must be closely regulated at any time during cell activation. A second hypothesis is that enzymes responsible for the transfer of AA into TG [54] are activated in stimulated cells as a mechanism to prevent excessive AA mobilization and depletion of this fatty acid. Should this be the case, the rapid reacylation of free AA into TG may be an important factor to limit the amount of AA available for the biosynthesis of eicosanoids. The hypothesis that TG are not an immediate source of AA for eicosanoid synthesis is confirmed by the data of Johnson et al. in the human neutrophils [37]. These cells supplemented in culture with exogenous AA to induce the accumulation of AA into TG produce the same amount of leukotrienes as the un-supplemented cells [37]. In addition, differential labeling of the endogenous pools of AA indicate that, even in neutrophils with a large TG pool of AA, the vast majority of AA converted to eicosanoids derives from the phospholipid pools.

Even though the TG may not be a direct source of AA during cell activation, this pool may still be important in the long-term regulation of cellular levels of AA. Experiments performed with human macrophages in which the cells were allowed to recover after stimulation, clearly demonstrate that, when the stimulus is removed, AA is progressively transferred from TG to phospholipids [39]. Therefore, among several possibilities, TG may be a late source of AA required to support eicosanoid biosynthesis once the phospholipid pools have been partially depleted. Alternatively, cell activation may in the delayed phase accelerate the transfer of AA from TG to phospholipids, a process normally occurring in resting cells [11]. The latter may be an important mechanism by which the cells may refill the early releasable phospholipid pools with AA.

Although these observations do not conclusively define the function of the TG pool, they indicate that phospholipid and TG pools have distinct biochemical roles in the release of AA in stimulated inflammatory cells. While phospholipid pools appear to be the major source of AA for immediate eicosanoid biosynthesis, the TG pool may play a crucial role to regulate intracellular levels of free AA and to maintain an adequate supply of AA to the phospholipid pools during the delayed phase of cell activation.

Closing remarks

Remodeling of AA within intracellular pools of inflammatory cells is gaining increasing attention as an important regulatory mechanism of several cell functions. The dynamic distribution of this fatty acid into long-term storage or rapidly-releasing pools has been shown to modulate profoundly not only the pro-inflammatory potential of the cell but also its life cycle and differentiation programs.

An interesting observation that is emerged in the last decade is that human inflammatory cells may constitutively contain or build-up, under certain circumstances, a large pool of AA associated with TG. This is an expandable, high-capacity pool presumably located into cytoplasmic lipid bodies. Exposure of the cells to high concentrations of exogenous AA or to certain cytokines and growth factors as well as their activation by pro-inflammatory stimuli can induce the mass expansion of and the accumulation of AA into the TG pool. Formation of the TG pool may be also associated with the acquisition of a differentiated or hypodense cell phenotype.

The biochemical functions of the TG pool of AA in inflammatory cells are not yet defined. This pool is not an immediate source of AA in stimulated cells and it may rather function as a recapture pool for AA mobilized from phospholipids. Understanding the mechanisms involved in the TG pool formation and the definition of its role in the AA remodeling may open new perspectives in the modulation of eicosanoid synthesis and in the biochemical regulation of inflammatory responses in human diseases.

References

1 Abramson SB, Leszczynska-Piziak J, Weissmann G (1991) Arachidonic acid as a second messenger. Interactions with a GTP-binding protein of human neutrophils. *J Immunol* 147: 231–236

2 Lennartz MR, Brown EJ (1991) Arachidonic acid is essential for IgG Fc receptor-mediated phagocytosis by human monocytes. *J Immunol* 147: 621–626

3 Rotman EI, Brostrom MA, Brostrom CO (1992) Inhibition of protein synthesis in intact mammalian cells by arachidonic acid. *Biochem J* 282: 487–494

4 Cockcroft S (1992) G-protein-regulated phospholipases C, D and A_2-mediated signalling in neutrophils. *Biochim Biophys Acta* 1113: 135–160

5 Irvine RF (1982) How is the level of free arachidonic acid controlled in mammalian cells? *Biochem J* 204: 3–16

6 Chilton FH, Fonteh AN, Surette ME, Triggiani M, Winkler JD (1996) Control of arachidonate levels within inflammatory cells. *Biochim Biophys Acta* 1299: 1–15

7 MacDonald JI, Sprecher H (1991) Phospholipid fatty acid remodeling in mammalian cells. *Biochim Biophys Acta* 1084: 105–121

8 Yamashita A, Sugiura T, Waku K (1997) Acyltransferases and transacylases involved in fatty acid remodeling of phospholipids and metabolism of bioactive lipids in mammalian cells. *J Biochem (Tokyo)* 122: 1–16

9 Reinhold SL, Zimmerman GA, Prescott SM, McIntyre TM (1989) Phospholipid remodeling in human neutrophils. Parallel activation of a deacylation/reacylation cycle and platelet-activating factor synthesis. *J Biol Chem* 264: 21652–21659

10 Waku K (1992) Origins and fates of fatty acyl-CoA esters. *Biochim Biophys Acta* 1124: 101–111

11 Chilton FH, Murphy RC (1986) Remodeling of arachidonate-containing phosphoglycerides within the human neutrophil. *J Biol Chem* 261: 7771–7777

12 Tessner TG, Greene DG, Wykle RL (1990) Selective deacylation of arachidonate-containing ethanolamine-linked phosphoglycerides in stimulated human neutrophils. *J Biol Chem* 265: 21032–21038

13 Fonteh AN, Chilton FH (1992) Rapid remodeling of arachidonate from phosphatidylcholine to phosphatidylethanolamine pools during mast cell activation. *J Immunol* 148: 1784–1791

14 Nakamura T, Fonteh AN, Hubbard WC, Triggiani M, Inagaki N, Ishizaka T, Chilton FH (1991) Arachidonic acid metabolism during antigen and ionophore activation of the mouse bone marrow derived mast cell. *Biochim Biophys Acta* 1085: 191–200

15 Winkler JD, Eris T, Sung CM, Chabot-Fletcher M, Mayer RJ, Surette ME, Chilton FH (1996) Inhibitors of coenzyme A-independent transacylase induce apoptosis in human HL-60 cells. *J Pharmacol Exp Ther* 279: 956–966

16 Samet JM, Fonteh AN, Galli SJ, Tsai M, Fasano MB, Chilton FH (1996) Alterations in arachidonic acid metabolism in mouse mast cells induced to undergo maturation *in vitro* in response to stem cell factor. *J Allergy Clin Immunol* 97: 1329–1341

17 Fonteh AN, LaPorte T, Swan D, McAlexander MA (2001) A decrease in remodeling accounts for the accumulation of arachidonic acid in murine mast cells undergoing apoptosis. *J Biol Chem* 276: 1439–1449

18 Laposata M, Kaiser SL, Capriotti AM (1988) Icosanoid production can be decreased without alterations in cellular arachidonate content or enzyme activities required for arachidonate release and icosanoid synthesis. *J Biol Chem* 263: 3266–3273

19 Triggiani M, Connell TR, Chilton FH (1990) Evidence that increasing the cellular content of eicosapentaenoic acid does not reduce the biosynthesis of platelet-activating factor. *J Immunol* 145: 2241–2248

20 Chilton FH, Patel M, Fonteh AN, Hubbard WC, Triggiani M (1993) Dietary n-3 fatty acid effects on neutrophil lipid composition and mediator production. Influence of duration and dosage. *J Clin Invest* 91: 115–122

21 Winkler JD, Fonteh AN, Sung CM, Heravi JD, Nixon AB, Chabot-Fletcher M, Griswold D, Marshall LA, Chilton FH (1995) Effects of CoA-independent transacylase inhibitors on the production of lipid inflammatory mediators. *J Pharmacol Exp Ther* 274: 1338–1347

22 Chilton FH, Lichtenstein LM (1990) Lipid mediators of the allergic reaction. *Chem Immunol* 49: 173–205

23 Holtzman MJ (1991) Arachidonic acid metabolism. Implications of biological chemistry for lung function and disease. *Am Rev Respir Dis* 143: 188–203

24 Murakami M, Hara N, Kudo I, Inoue K (1993) Triggering of degranulation in mast cells by exogenous type II phospholipase A_2. *J Immunol* 151: 5675–5684

25 Hii CS, Huang ZH, Bilney A, Costabile M, Murray AW, Rathjen DA, Der CJ, Ferrante

A (1998) Stimulation of p38 phosphorylation and activity by arachidonic acid in HeLa cells, HL60 promyelocytic leukemic cells, and human neutrophils. Evidence for cell type-specific activation of mitogen-activated protein kinases. *J Biol Chem* 273: 19277–19282

26 Camandola S, Leonarduzzi G, Musso T, Varesio L, Carini R, Scavazza A, Chiarpotto E, Baeuerle PA, Poli G (1996) Nuclear factor κB is activated by arachidonic acid but not by eicosapentaenoic acid. *Biochem Biophys Res Commun* 229: 643–647

27 Peplow PV (1996) Actions of cytokines in relation to arachidonic acid metabolism and eicosanoid production. *Prostaglandins Leukot Essent Fatty Acids* 54: 303–317

28 Serhan CN, Haeggstrom JZ, Leslie CC (1996) Lipid mediator networks in cell signaling: update and impact of cytokines. *Faseb J* 10: 1147–1158

29 Nielson CP, Bayer C, Hodson S, Hadjokas N (1992) Regulation of the respiratory burst by cyclic 3',5'-AMP, an association with inhibition of arachidonic acid release. *J Immunol* 149: 4036–4040

30 Locati M, Riboldi E, Bonecchi R, Transidico P, Bernasconi S, Haribabu B, Morris AJ, Mantovani A, Sozzani S (2001) Selective induction of phospholipase D_1 in pathogen-activated human monocytes. *Biochem J* 358: 119–125

31 Dennis EA, Rhee SG, Billah MM, Hannun YA (1991) Role of phospholipase in generating lipid second messengers in signal transduction. *Faseb J* 5: 2068–2077

32 Chilton FH, Ellis JM, Olson SC, Wykle RL (1984) 1-O-alkyl-2-arachidonoyl-*sn*-glycero-3-phosphocholine. A common source of platelet-activating factor and arachidonate in human polymorphonuclear leukocytes. *J Biol Chem* 259: 12014–12019

33 Mueller HW, O'Flaherty JT, Greene DG, Samuel MP, Wykle RL (1984) 1-O-alkyl-linked glycerophospholipids of human neutrophils: distribution of arachidonate and other acyl residues in the ether-linked and diacyl species. *J Lipid Res* 25: 383–388

34 Strum JC, Emilsson A, Wykle RL, Daniel LW (1992) Conversion of 1-O-alkyl-2-acyl-sn-glycero-3-phosphocholine to 1-O-alk-1'-enyl-2-acyl-*sn*-glycero-3-phosphoethanolamine. A novel pathway for the metabolism of ether-linked phosphoglycerides. *J Biol Chem* 267: 1576–1583

35 Ojima-Uchiyama A, Masuzawa Y, Sugiura T, Waku K, Saito H, Yui Y, Tomioka H (1988) Phospholipid analysis of human eosinophils: High levels of alkylacylglycerophosphocholine (PAF precursor). *Lipids* 23: 815–817

36 Triggiani M, Oriente A, Seeds MC, Bass DA, Marone G, Chilton FH (1995) Migration of human inflammatory cells into the lung results in the remodeling of arachidonic acid into a triglyceride pool. *J Exp Med* 182: 1181–1190

37 Johnson MM, Vaughn B, Triggiani M, Swan DD, Fonteh AN, Chilton FH (1999) Role of arachidonyl triglycerides within lipid bodies in eicosanoid formation by human polymorphonuclear cells. *Am J Respir Cell Mol Biol* 21: 253–258

38 Calabrese C, Triggiani M, Marone G, Mazzarella G (2000) Arachidonic acid metabolism in inflammatory cells of patients with bronchial asthma. *Allergy* 55 (Suppl) 61: 27–30

39 Triggiani M, Oriente A, Marone G (1994) Differential roles for triglyceride and phos-

pholipid pools of arachidonic acid in human lung macrophages. *J Immunol* 152: 1394–1403

40 Triggiani M, Oriente A, de Crescenzo G, Rossi G, Marone G (1995) Biochemical functions of a pool of arachidonic acid associated with triglycerides in human inflammatory cells. *Int Arch Allergy Immunol* 107: 261–263

41 Triggiani M, Schleimer RP, Warner JA, Chilton FH (1991) Differential synthesis of 1-acyl-2-acetyl-*sn*-glycero-3-phosphocholine and platelet-activating factor by human inflammatory cells. *J Immunol* 147: 660–666

42 Triggiani M, Fonteh AN, Chilton FH (1992) Factors that influence the proportions of platelet-activating factor and 1-acyl-2-acetyl-*sn*-glycero-3-phosphocholine synthesized by the mast cell. *Biochem J* 286 (Pt 2): 497–503

43 Chilton FH, Connell TR (1988) 1-ether-linked phosphoglycerides. Major endogenous sources of arachidonate in the human neutrophil. *J Biol Chem* 263: 5260–5265

44 Fonteh AN, Chilton FH (1993) Mobilization of different arachidonate pools and their roles in the generation of leukotrienes and free arachidonic acid during immunologic activation of mast cells. *J Immunol* 150: 563–570

45 Mayer RJ, Marshall LA (1993) New insights on mammalian phospholipase A_2s; comparison of arachidonoyl-selective and -nonselective enzymes. *Faseb J* 7: 339–348

46 Triggiani M, Schleimer RP, Tomioka K, Hubbard WC, Chilton FH (1992) Characterization of platelet-activating factor synthesized by normal and granulocyte-macrophage colony-stimulating factor-primed human eosinophils. *Immunology* 77: 500–504

47 Marone G, Casolaro V, Patella V, Florio G, Triggiani M (1997) Molecular and cellular biology of mast cells and basophils. *Int Arch Allergy Immunol* 114: 207–217

48 Triggiani M, De Marino V, de Crescenzo G, Marone G (1997) Arachidonic acid remodeling in human inflammatory cells migrating to the lung *in vivo*. *Int Arch Allergy Immunol* 113: 190–192

49 Weller PF, Monahan-Earley RA, Dvorak HF, Dvorak AM (1991) Cytoplasmic lipid bodies of human eosinophils. Subcellular isolation and analysis of arachidonate incorporation. *Am J Pathol* 138: 141–148

50 Dvorak AM, Dvorak HF, Peters SP, Shulman ES, MacGlashan DW Jr, Pyne K, Harvey VS, Galli SJ, Lichtenstein LM (1983) Lipid bodies: cytoplasmic organelles important to arachidonate metabolism in macrophages and mast cells. *J Immunol* 131: 2965–2976

51 Sivarajan M, Hall ER, Wu KK, Rafelson ME, Manner C (1984) Regulation of intracellular arachidonate in normal and stressed endothelial cells. *Biochim Biophys Acta* 795: 271–276

52 Lewis GS, Wood DL, Caldwell DW (1992) Incorporation of [^{14}C] arachidonic acid into ovine conceptus and endometrial lipids. *Prostaglandins* 44: 135–144

53 Blank ML, Smith ZL, Snyder F (1992) Contributing factors in the trafficking of ^{3}H-arachidonate between phospholipids. *Biochim Biophys Acta* 1124: 262–272

54 Blank ML, Smith ZL, Snyder F (1993) Arachidonate-containing triacylglycerols: biosynthesis and a lipolytic mechanism for the release and transfer of arachidonate to phospholipids in HL-60 cells. *Biochim Biophys Acta* 1170: 275–282

55 Ralph P, Harris PE, Punjabi CJ, Welte K, Litcofsky PB, Ho MK, Rubin BY, Moore MA, Springer TA (1983) Lymphokine inducing "terminal differentiation" of the human monoblast leukemia line U937: A role for gamma interferon. *Blood* 62: 1169–1175

56 Roux-Lombard P, Cruchaud A, Dayer JM (1986) Effect of interferon-gamma and 1a,25-dihydroxyvitamin D_3 on superoxide anion, prostaglandins E_2, and mononuclear cell factor production by U937 cells. *Cell Immunol* 97: 286–296

57 Ways DK, Dodd RC, Bennett TE, Hooker JL, Earp HS (1988) Effect of retinoic acid on phorbol ester-stimulated differentiation and protein kinase C-dependent phosphorylation in the U937 human monoblastoid cell. *Cancer Res* 48: 5779–5787

58 Rovera G, O'Brien TG, Diamond L (1979) Induction of differentiation in human promyelocytic leukemia cells by tumor promoters. *Science* 204: 868–870

59 Stone RM, Imamura K, Datta R, Sherman ML, Kufe DW (1990) Inhibition of phorbol ester-induced monocytic differentiation and c-fms gene expression by dexamethasone: potential involvement of arachidonic acid metabolites. *Blood* 76: 1225–1232

60 Greenspan P, Fowler SD (1985) Spectrofluorometric studies of the lipid probe, Nile red. *J Lipid Res* 26: 781–789

61 Weller PF, Dvorak AM (1985) Arachidonic acid incorporation by cytoplasmic lipid bodies of human eosinophils. *Blood* 65: 1269–1274

62 Caulfield JP, Hein A, Rothenberg ME, Owen WF, Soberman RJ, Stevens RL, Austen KF (1990) A morphometric study of normodense and hypodense human eosinophils that are derived *in vivo* and *in vitro*. *Am J Pathol* 137: 27–41

63 Petraroli A, Balestrieri B, Giannattasio G, Frattini A, Triggiani M, Marone G Phenotypical and functional heterogeneity of human lung macrophages. *Allergy; in press*

64 Triggiani M, Granata F, Oriente A, De Marino V, Gentile M, Calabrese C, Palumbo C, Marone G (2000) Secretory phospholipases A_2 induce beta-glucuronidase release and IL-6 production from human lung macrophages. *J Immunol* 164: 4908–4915

Arachidonate remodeling, platelet-activating factor signaling, and the inflammatory response in the central nervous system

Royal D. Saunders[1] and Nicolas G. Bazan[2]

[1]Department of Biology, Xavier University of Louisiana, New Orleans, LA 70125, USA; [2]LSU Neuroscience Center, Louisiana State University Health Sciences Center, New Orleans, LA 70112, USA

Introduction

Arachidonate remodeling is a very prevalent process in cells of the central nervous system. As in other organs, there are a wide variety of phospholipid molecular species that display active arachidonate remodeling. In neurons, these events contribute to the maintenance of excitable membrane properties by providing a fluid environment for proteins to perform their functions. In addition, arachidonate remodeling allows the controlled release of free arachidonic acid. This fatty acid is a messenger in itself, and also serves as a precursor to a wide variety of biologically-active messengers, the eicosanoids. Recently, several novel messengers derived from arachidonic acid have been identified in addition to the well known prostaglandins. These new messengers include the endocannabinoids, which elicit potent modulatory actions in the nervous system. Overall, arachidonate remodeling in the nervous system is clearly engaged in cell function.

In the nervous system there is a very close relationship between arachidonate remodeling and inflammation. Ischemia promotes the rapid and selective release of brain free arachidonic acid [1]. This effect results in a major imbalance in arachidonate-remodeling pathways by enhancing the free arachidonic acid pool. This tells us that injury is critical in setting into motion this imbalance. Seizures and neurodegenerative diseases including Alzheimer's disease involve profound alterations in arachidonate remodeling [2, 3].

Platelet-activating factor

Although platelet-activating factor (PAF, 1-O-alkyl-2-acetyl-*sn*-glycero-3-phosphocholine) was originally described as an inducer of platelet aggregation, it is now recognized as a physiologic mediator in synaptic plasticity. In addition, PAF is accumulated in neuronal degeneration that results from the activation of the injury

Arachidonate Remodeling and Inflammation, edited by Alfred N. Fonteh and Robert L. Wykle

/inflammatory response. The synthesis and turnover of PAF are components of the "remodeling" of arachidonic acid-containing choline plasmalogens [4] (Fig. 1).

There are three routes for the synthesis of PAF. The nervous system contains alkyl-arachidonoyl-GPC, the PAF precursor, a very minor component of membrane phospholipids. This pool of choline-containing phospholipids is characterized by a high concentration of arachidonate (or docosahexaenoate) in the *sn*-2 position. Most of our current knowledge is on the arachidonate-containing PAF precursor. Arachidonate is hydrolyzed from the *sn*-2 position of alkyl-acyl GPC by phospholipase A_2, and then replaced with an acetyl group by an acetyltransferase. Other major routes of PAF synthesis in the brain may be a *de novo* pathway by the transfer of choline to alkylacetylglycerol *via* cholinephosphotransferase or by a coenzyme-A-independent route [5]. There is an additional pathway for PAF formation that is particularly important during central nervous system (CNS) injury. After oxidative injury, fragmentation of *sn*-2 PUFA (mainly arachidonate and docosahexaenoate) of some species of phosphatidylcholine phospholipids can result in the formation of molecules with PAF-like activity. The presence of a shorter-chain peroxidated fatty acid at the C-2 position is recognized by the PAF receptor. The release of PUFA from the PAF precursor also generates bioactive molecules, such as leukotrienes, which may act as neuronal second messengers or as pro-inflammatory mediators.

PAF is an important modulator of neural function, but its over-production plays a role in neural dysfunction. Although PAF at relatively high concentrations may disrupt neural function by perturbing membrane function [6, 7], it is also a highly bioactive molecule.

Cellular responses to PAF

Phospholipase A_2 (PLA_2) releases arachidonic acid or docosahexaenoic acid [8]. These PUFA may be metabolized by one of several cyclooxygenase (COX) enzymes into other bioactive lipids (see below). The remaining lyso-lipid (1-O-alkyl-2-lyso-*sn*-glycero-3-phosphocholine or lyso-PAF) is acetylated, producing the bioactive PAF.

PAF binds to high- and low-affinity receptors in several cell types, including platelets, basophils, and lung tissue [9]. PAF attains high concentrations in the hippocampus in response to injury [10]. In addition to seven transmembrane domain receptor on the cell surface (including the synapse), high-affinity intracellular binding sites are present [11]. Moreover, a diversity of K_m parameters for PAF-receptor binding suggests heterogeneity of PAF-binding sites in the nervous system.

The synthesis of PAF results in physiologic responses in cells, including Ca^{2+} mobilization [12], activation of protein kinase [13], accumulation of arachidonic acid [14], and modulation of gene expression [15, 16]. This multiplicity of cellular responses to PAF is not surprising, considering that the PAF receptor belongs to the

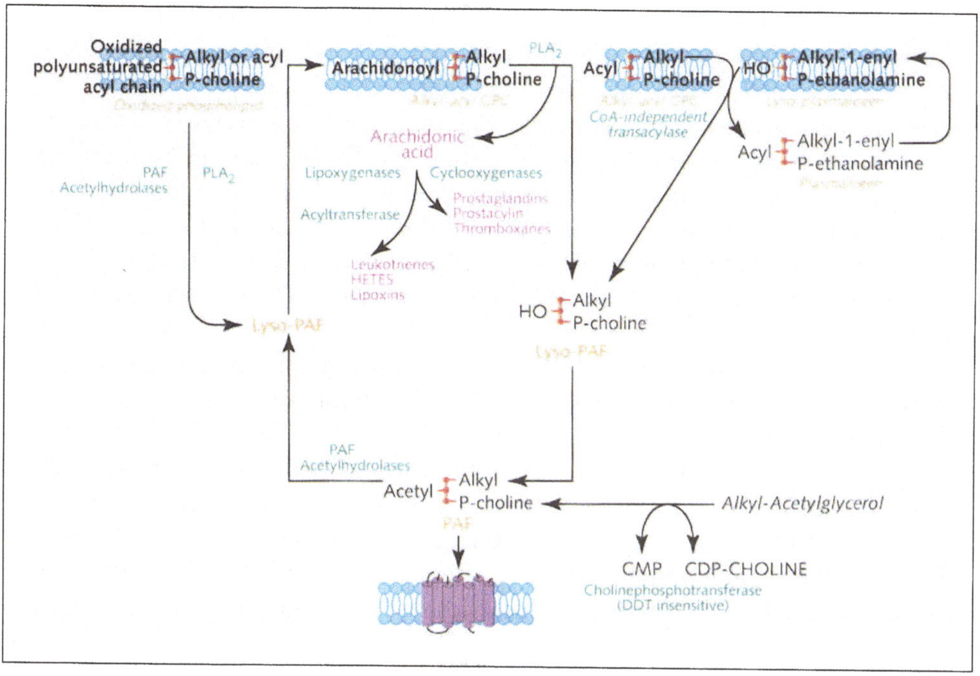

Figure 1
Platelet-activating factor (PAF) synthesis and degradation
The remodeling pathway or "PAF cycle" leads from the membrane phospholipid PAF precursor, alkyl-acyl (usually arachidonoyl)-glycerophosphorylcholine (GPC, on the left) to the biologically active PAF (at right), and includes the production of lyso-PAF, which is generated from the PAF precursor alkyl-acyl-GPC either directly by the action of phospholipase A_2, or by the transfer of the sn-2 *acyl moiety to a lyso-plasmalogen (top), which is itself mobilized from membrane plasmalogen by phospholipase action. The* de novo *route of PAF synthesis (upper right) involves the direct transfer of a choline moiety to alkyl-acetylglycerol. Note that PAF acetylhydrolase (bottom) inactivates all PAF molecules, regardless of their biosynthetic route, and additionally inactivates oxidatively damaged phospholipids (shorter peroxidated acyl group at C_2) that possess biological activity at the PAF receptor.*

family of receptors linked to the G-protein [17]. PAF may elicit cellular responses by activating phospholipase C (PLC), with the resultant hydrolyzing of PIP_2. In an *in vitro* model using [^{3}H]-inositol-labeled platelets, a four- to five-fold increase in IP_3 and IP_2 was observed within five seconds of treatment with PAF [18]. The response was independent of extracellular [Ca^{2+}] and was inhibited by the PAF antagonist CV 3988, indicating direct activation of PLC by the binding of PAF to its receptor.

As a consequence of PLC activation, the increase in intracellular [Ca^{2+}] may activate PLA_2. PLA_2 is also activated during the same time course that PAF synthesis occurs *via* the remodeling pathway. Several studies indicate that PAF antagonists effectively block the rise in intracellular [Ca^{2+}]. For example, in platelets a PAF antagonist decreases the PAF-induced rise in intracellular Ca^{2+} [9].

When the plasma membrane-bound PLC is activated, a series of intracellular events occur, which are involved in the pathologic changes during brain ischemia and seizures [19]. Triazolobenzodiazepines (alprazolam and trizolam) and some benzodiazepines antagonize PAF effects [20]. Because PLC and PLA_2 activities are elevated in brain undergoing ischemia [4, 19], investigators have targeted the PAF receptor as a potential site for the reduction of the progression of damage after CNS injury.

The PAF antagonist BN 50739 prevents postischemic hyporeperfusion and edema formation in rabbit spinal cord [21]. PAF antagonists restore cerebral blood flow and also enhance survival during the post-ischemic phase [22, 23]. The mechanisms of protection by PAF antagonists have been studied in several models. Post-treatment with BN 52021 reduced free fatty acid (FFA) and diacylglycerol accumulation after ischemic injury, and pre-treatment with BN 52021 reduced FFA release following ischemic injury in mouse brain, although minuscule effect on PIP_2 hydrolysis was observed [24]. However, in this study the investigators reported no significant reduction in FFA accumulation and PLC activation after electroconvulsive shock (ECS).

PAF receptor antagonists also have proved beneficial in neuronal recovery as measured by electrophysiologic parameters [25]. In this study, a PAF receptor antagonist of the neolignan family, kadsurenone, enhanced neuronal recovery as measured by cortical somatosensory evoked potential. The protective effect appeared to be independent of effects on platelet aggregation, as evidenced by ^{111}I-labeled platelet distribution, which was the same in control and kadsurenone-treated brain. Thus, PAF antagonists may improve post-ischemic recovery by mechanisms separate from prevention of platelet aggregation. Histologic examination showed that CA1 hippocampal neurons in rats are protected from ischemic damage by the PAF antagonist ginkgolide B [26].

PAF receptor antagonists elicit neuroprotection in a rodent model of brain ischemia-reperfusion [27], suggesting that PAF may work at the synapse. In this study, the accumulation of free polyunsaturated fatty acids following ischemia was decreased by the PAF antagonists, which also restored cerebral blood flow [27]. Polyunsaturated fatty acid release accompanying ischemia probably results from synaptic phospholipase(s) A_2 activity [28]. The same PAF antagonist that was neuroprotective in ischemia-reperfusion also selectively displaced radiolabeled PAF binding in synaptic membranes [11]. For this reason, the synaptic membrane binding site was proposed to modulate glutamate neurotransmitter release [29]. Moreover, intracellular PAF-binding sites isolated from a microsomal fraction were distinct from synaptic membrane sites with regard to their responses to several antagonists [11]. At the time of those studies, PAF was already known to activate early-response genes [15, 30, 31]; therefore the intracellular receptor was proposed to be the signaling linkage to gene expression. Cloning of the seven transmembrane domain PAF receptor [17, 32, 33] and the discovery of specific PAF receptor-mediated Ca^{2+} influx into neurons [34] has shed more light on neural PAF signaling. In studies of long-term potentiation in hippocampal neurons isolated from PAF-receptor-deficient mice, both incidence and size of LTP, defined as increased excitatory postsynaptic potentials, were decreased in the knockout mice, as compared to wild-type mice, and PAF-receptor antagonists reduced LTP in wild-type mice, but not in PAF-receptor knockout mice. These results lend additional support to the hypothesis that PAF is involved in hippocampal synaptic plasticity [35]. The intracellular PAF receptors have been more extensively characterized in subsequent studies that have identified one form in endosomes [36] and another in the nuclear membrane [37]. Both intracellular forms of the receptor may actually be components of the intracellular microsomal form that was previously described [11]. We now know that the PAF seven transmembrane domain receptor is expressed in neurons, astrocytes, and microglia, as well as endothelial cells. It is still unclear whether the intracellular PAF receptor(s) differ in molecular structure from, or instead are an intracellular state of, the cell-surface PAF seven transmembrane domain receptor. We do not yet know whether the cell-surface PAF receptor internalizes, or whether a cell-surface PAF receptor destined to insertion in the membrane is already active. If that is the case, does PAF itself internalize to access either of these intracellular PAF-receptor forms? Two possible mechanisms might explain the role of PAF in gene transcription:

1. The cell-surface receptor triggers the signaling cascade, or;
2. The intracellular form establishes interactions with specific kinases/phosphatases or transcription factors with or without specific scaffolding proteins.

In pathologic conditions involving oxidative stress such as ischemia or seizures, the rates of PAF synthesis and degradation that maintain a modulated PAF pool size become mismatched and PAF concentration increases to where it becomes a pro-inflammatory messenger and mediator of neurotoxicity. In this capacity PAF activates COX-2 expression [38], expression of several early-response genes that encode transcription factors [15, 30], apoptosis, and polymorphonuclear leukocyte (PMN) adhesion to microvessels [39], which has critical consequences for cell survival. Leukocyte infiltration mediates neural injury in head trauma, stroke, spinal cord injury, and other diseases and enhances the synthesis and release of IL-6, IL-8, IL-10, TNF-α, and of other mediators of the inflammatory response. PAF binding to its receptor activates phospholipases and results in additional PAF synthesis, along with that of prostaglandins and leukotrienes. While overall, PAF is a potent neuronal injury messenger, it also plays a prominent role in astrocytes and in microglial cells. Although many of these actions have been studied in non-neural cells they are assumed to occur in the nervous system as well.

In summary, excessive PAF promotes neuronal damage, and PAF-receptor antagonists elicit neuroprotection in various models of neural injury [21, 23, 25–27, 40].

Phospholipase A_2, arachidonate metabolism, and gene expression

Synaptic membrane-bound phospholipases play a key role in neuronal plasticity in long-term potentiation, which is a model of learning and memory [41, 42]. However, during seizures, ischemia, trauma, and neurodegeneration, these signaling mechanisms are overstimulated and lead to excitotoxic brain damage [1, 43]. Thus, during acute neuronal stimulation by excitatory glutamatergic neurotransmission, the balance of levels of mediators of neuronal plasticity and injury determines whether the neurons recover or become damaged.

PLA_2 enzymes are classified into three types: calcium-dependent PLA_2 ($cPLA_2$ or type IV), calcium-independent ($iPLA_2$ or type VI), and low-molecular-weight secretory PLA_2 ($sPLA_2$) [44]. $cPLA_2$ displays a high selectivity for phospholipids containing PUFA in the *sn*-2 position. The regulation of $cPLA_2$ activity by agonists that increase cytosolic Ca^{2+} involves translocation of the enzyme from the cytosolic compartment to the nuclear membrane or the endoplasmic reticulum, where phospholipids are hydrolyzed, thus releasing PUFA, mainly arachidonic and docosahexaenoic acids [45]. Two important enzymes, the inducible cyclooxygenase-2 (COX-2) and lipoxygenase (LOX), are located in the nuclear membrane and metabolize the PUFA into prostaglandins (PGs) and leukotrienes (LTs).

Several observations indicate that PGs, synthesized through the COX-2 pathway, play a pivotal role in neuropathology. In the brain, COX-2 is constitutively expressed, mainly in the cortex and hippocampus [46, 47]. Not only has the COX-2 pathway been implicated in neuronal plasticity, but expression of the enzyme is also induced during cerebral ischemia and seizures [19, 48]. Inhibition of COX-2 activity prevents ischemia and NMDA-induced cell death [49, 50].

Arachidonate and COX-2 gene expression

Because COX-2 activity is involved in neuronal death after CNS injury, the regulation of expression of the COX-2 gene is an important event in the utilization of free arachidonic acid during the activation of the remodeling pathway. Nuclear factor (NF)-κB is a DNA-binding protein that controls the transcription of the COX-2 gene. NF-κB binds to at least two sites in the mouse, rat, and human COX-2 gene proximal promoter [51, 52]. The human gene encoding for expression of COX-2 may be involved in the pathology of slower-onset neurodegeneration, such as in Alzheimer's disease (AD) [53]. Examination of nuclear protein extracts taken from aging control and AD-affected brain neocortical nuclei revealed a significant correlation between the levels of NF-κB-DNA binding and the production of COX-2 RNA. This study provided evidence linking NF-κB-DNA binding to up-regulation of transcription of the gene coding for COX-2. In addition, oxidative stress due to exposure of reactive oxygen species (ROS) contributes to neuronal death during aging, disease, and injury [54], and there is a link between exposure of neurons to ROS and NF-κB activation [55, 56].

Glutamate receptor signaling, $sPLA_2$, arachidonate, and COX-2 expression

Brain ischemia and seizures promote PAF accumulation that contributes to further glutamate release and COX-2 transcription, both of which increase neuronal cell damage and death. The interplay between PAF accumulation, EAA receptor activation, and a secretory PLA_2 are an additional factor. For example in hippocampal neurons, stimulation of the NMDA receptor by PAF activates several mitogen-activated protein (MAP) kinases, including c-JUN NH_2-terminal kinase (JNK), p38, and extracellular signal-regulated kinase (ERK). The hetrazepine BN50730 (an intracellular PAF receptor antagonist) inhibits both NMDA-stimulated MAP kinases and neuronal cell death [57].

When $sPLA_2$ or glutamate was injected into the right striatum of male Wistar rats, there was a dose-dependent response in terms of neurologic abnormalities and tissue damage. When low levels (10–20 pmol) of $sPLA_2$ or glutamate (2.5 μmol) were injected into the striatum, no neurologic abnormalities or tissue damage occurred. However, when 20 pmol $sPLA_2$ and 2.5 μmol glutamate were co-injected, the animals became apathetic for several hours after injection and then displayed circling toward the side of injection in the following days [58]. The animals that were co-injected displayed extensive histologic damage in the right hemisphere, and in several rats the damage extended into the contralateral hemisphere. Thus, the investigators demonstrated a synergistic excitotoxic action *in vivo* exerted by $sPLA_2$ and glutamate.

Although the synergism of neuronal damage between $sPLA_2$ and glutamate is dramatic, it is noteworthy that the two may act by different but overlapping mechanisms. In primary cortical neuron cultures, $sPLA_2$ and glutamate display the same synergy in eliciting neuronal cell death as demonstrated *in vivo* [59]. In this study, ^3H-arachidonic acid (AA)-labeled cultures were treated with mildly toxic doses of $sPLA_2$ and glutamate. Glutamate treatment resulted in the release of AA predominately from phosphatidylethanolamine (PE), whereas $sPLA_2$ treatment resulted in AA release from phosphatidylcholine (PC). Co-treatment with $sPLA_2$ and glutamate resulted in a greater degradation in both PE and PC, although the NMDA antagonist MK-801 blocked only the glutamate effects. Therefore, the neurotoxicity induced by glutamate and $sPLA_2$, although synergistic, involves the hydrolysis of different phospholipid pools. In addition, both $sPLA_2$ and glutamate induce COX-

2 expression, although during different time intervals [60]. In this study, the expression of COX-2 in rat striatum was examined using *in situ* hybridization and immunohistochemistry. Injection of sPLA2 induced COX-2 expression (4 hrs after injection), whereas glutamate induced COX-2 expression within 2 hrs after injection. This time differential suggests separate signaling mechanisms for $sPLA_2$ and glutamate. In addition, recent observations [61] indicate that $sPLA_2$ affects neuronal responses to glutamate through the opening of Ca^{2+} channels, which provides more evidence that $sPLA_2$, by releasing AA, is central in CNS pathology following injury.

Conclusions

PAF is a potent inducer of platelet aggregation, but the specific PAF receptors present on neurons may cause neuronal cell damage when over-activated by excessive PAF production. Because there are multiple pathways for PAF synthesis in the brain, the PAF receptor has been targeted for intervention in conditions of CNS degeneration, including traumatic injury and slower-onset degeneration such as Alzheimer's disease. In fact, the synthesis of PAF initiates several mechanisms leading to neuronal damage including:

1. The release of arachidonic acid that is metabolized *via* specific enzymes into the inflammatory compounds prostaglandins and leukotrienes. PGs and LTs also contribute to circulatory abnormalities and infiltration of blood cells that exacerbate CNS injury. Eicosanoids can also counteract the inflammatory response, such as lipoxins [62]. We have relatively scarce information regarding anti-inflammatory regulation by eicosanoids in brain.
2. The synthesis of PAF results in Ca^{2+} mobilization within neurons resulting in the activation of enzymes (particularly phospholipases A_2) leading to further action on arachidonate remodeling.
3. The release of PAF during CNS injury increases glutamate release and expression of the COX-2 gene. This in turn activates synthesis of eicosanoids.
4. PAF accumulation is related to excitatory amino acid receptor activation as well as the activation of $sPLA_2$. Glutamate and $sPLA_2$ are synergistic in causing neuronal cell damage [58]. These studies provide evidence that PAF can potentiate damage through more than one pathway, and that arachidonate remodeling is involved [63].

One of the conclusions from these studies is that therapeutic interventions may be possible to prevent PAF damage to neurons. In particular, novel PAF antagonists enhance neuronal survival after CNS injury. There is widespread signaling disruption during CNS injury that contributes to neuronal damage. PAF antagonism, however, provides an effective target to modulate multiple pro-inflammatory pathways, and as a consequence, neuroprotection.

References

1 Bazan NG, Allan G (1998) Platelet-activating factor and other bioactive lipids. In: MD Ginsberg, J Bogousslavsky (eds): *Cerebrovascular Disease, Pathophysiology, Diagnosis and Management*. Blackwell Science Publishers: Malden, Mass, 532–555

2 Bazan NG, Lukiw WJ (2002) Cyclooxygenase-2 and presenilin-1 gene expression induced by interleukin-1beta and amyloid beta 42 peptide is potentiated by hypoxia in primary human neural cells. *J Biol Chem* 277: 30359–30367

3 Bazan NG, Palacios-Pelaez R, Lukiw WJ (2002) Hypoxia signaling to genes: Significance in Alzheimer's disease. *Mol Neurobiol* 26: 283–298

4 Bazan NG (1990) Involvement of arachidonic acid and platelet-activating factor in the response of the nervous system to ischemia and convulsions. In: NG Bazan (ed): *Lipid Mediators in Experimental Brain Damage and Experimental Epilepsy*. Basel: Karger, 241–252

5 Snyder F (1995) Platelet-activating factor and its analogs: metabolic pathways and related intracellular pathways. *Biochim Biophys Acta* 1254: 231–249

6 Sawyer DB, Anderson OS (1989) Platelet-activating factor is a general membrane perturbant. *Biophys Acta* 987: 129–132

7 Bratton DL, Harris RA, Clay KL, Henson PM (1988) Effects of platelet-activating factor on calcium-lipid interactions and lateral phase separations in phospholipid vesicles. *Biochim Biophys Acta* 943: 211–219

8 Suga K, Kaurasaki T, Blank ML, Synder F (1990) An arachidonoyl (polyenoic)-specific phospholipase A_2 activity regulates the synthesis of platelet-activating factor in granulocytic HL-60 cells. *J Biol Chem* 265: 12363–12371

9 Hwang S-B (1990) Specific receptors of platelet-activating factor, receptor heterogeneity, and signal transduction mechanisms. *J Lipid Med* 2: 123–158

10 Kumar R., Harvey S, Kester N, Hanahan D, Olson M (1988) Production and effects of platelet-activating factor in the rat brain. *Biochim Biophys Acta* 963: 375–383

11 Marcheselli VL, Rossowska MJ, Domingo MT, Braquet P, Bazan NG (1990) Distinct platelet-activating factor binding sites in synaptic endings and in intracellular membranes of rat cerebral cortex. *J Biol Chem* 265: 9140–9145

12 Bussolini F, Anglietta M, Sanavio F, Stacchini A, Lauri D, Camussi G (1985) Alkylether phosphoglycerides influence calcium fluxes into human endothelial cells. *J Immunol* 135: 2748–2753

13 Gomez-Cambronero, Wang E, Johnson G, Huang C-K, Shaafi R (1990) Platelet-activating factor induces tyrosine phosphorylation in human neutrophils. *J Biol Chem* 266: 6240–6245

14 Nakashima S, Sugunuma A, Sato M, Tohmatsu T, Nozawa Y (1989) Mechanism of

arachidonic acid liberation in platelet-activating factor stimulated human polymorphonuclear neutrophils. *J Immunol* 143: 1295–1302

15 Squinto SP, Block AL, Braquet P, Bazan NG (1989) Platelet-activating factor stimulates Fos/Jun/AP-1 transcriptional signaling system in human neuroblastoma cells. *J Neurosci Res* 24: 558–566

16 Schulam P, Kuruvilla A, Putcha G, Mangus L, Franklin-Johnson J, Shearer WT (1991) Platelet-activating factor induces phospholipid turnover, calcium flux, arachidonic acid liberation, eicosanoids generation and oncogene expression in a human B cell line. *J Immunol* 146: 1842–1848

17 Honda ZI, Nakamura M, Miki I, Minami M, Watanabe T, Seyama Y, Okado H, Toh H, Ito K, Miyamoto T et al (1991) Cloning by functional expression of platelet-activating factor receptor from guinea pig lung. *Nature* 349: 342–346

18 Shukla S (1985) Platelet-activating factor-stimulated formation of inositol triphosphate in platelet and its regulation by various agents including calcium, indomethacin, CV-3988 and forskolin. *Arch Biochem Biophys* 240: 674–681

19 Bazan NG (1970) Effects of ischemia and electroconvulsive shock on free fatty acid accumulation in brain. *Biochim Biophys Acta* 218: 1–10

20 Kornecki E, Ehrlich YH, Lenox RM (1985) Platelet-activating factor-induced aggregation of is specifically inhibited by triazolobenzodiazepines. *Science* 226: 1454–1456

21 Lindsberg PJ, Yue T-L, Fredericks KU, Hallenback JM, Feuerstein G (1990) Evidence for platelet-activating factor as a novel mediator in experimental stroke in rabbits. *Stroke* 21: 1452–1457

22 Spinnewyn B, Blavet N, Clostre F, Bazan NG, Braquet P (1987) Involvement of platelet-activating factor (PAF) in cerebral post-ischemic phase in Mongolian gerbils. *Prostaglandins* 34: 337–349

23 Gilboe DD, Kinter D, Fitzpatrick JH, Emoto SE, Esanu A, Braquet PG, Bazan NG (1991) Recovery of post-ischemic brain metabolism and function following treatment with a free radical scavenger and platelet-activating factor antagonists. *J Neurochem* 56: 311–319

24 Birkle DL, Kurian P, Braquet P, Bazan NG (1988) The platelet-activating factor antagonist BN 52021 decreases accumulation of free polyunsaturated fatty acid in mouse brain during ischemia and electroconvulsive shock. *J Neurochem* 51: 1900–1905

25 Kochanek PM, Dutka AJ, Kumaroo KK, Hallenbeck JM (1987) Platelet-activating factor receptor blockage enhances recovery after multi-focal brain ischemia. *Life Sci* 41: 2639–2644

26 Oberpichler H, Sauer D, Robberg C, Mennel H-D, Krieglstein J (1990) PAF antagonist ginkgolide B reduces post-ischemic neuronal damage in rat brain hippocampus. *J Cereb Blood Flow Metab* 10: 133–145

27 Panetta T, Marcheselli VL, Braquet P, Spinnewyn B, Bazan NG (1987) Effects of a platelet-activating factor antagonist (BN 52021) on free fatty acids, diacylglycerols, polyphosphoinositides and blood now in the gerbil brain: Inhibition of ischemia reperfusion induced cerebral injury. *Biochem Biophys Res Commun* 149:580–587

28 Bazan NG, MG Packard, L Teather , G Allan G. (1997) Bioactive lipids in excitatory neurotransmission and neuronal plasticity. *Neurochem Int* 30: 225–231
29 Clark GD, LT Happel, CF Zorumski , NG Bazan NG (1992) Enhancement of hippocampal excitatory synaptic transmission by platelet-activating factor. *Neuron* 9: 1211–1216
30 Squinto SP, Braquet P, Block AL, Bazan NG (1990) Platelet-activating factor activates HIV promoter in transfected SH-SY5Y neuroblastoma cells and MOLT-4 T lymphocytes. *J Mol Neurosci* 2: 79–84
31 Bazan NG, SP Squinto P Braquet , T Panetta , VL Marcheselli VL (1991) Platelet-activating factor and polyunsaturated fatty acids in cerebral ischemia or convulsions: Intracellular PAF-binding sites and activation of a Fos/Jun/AP-1 transcriptional signaling system. *Lipids* 26: 1236–1242
32 Kunz D, Gerard NP, Gerard C (1992) The human leukocyte platelet-activating factor receptor. CDNA cloning, cell surface expression, and construction of a novel epitope-bearing analog. *J Biol Chem* 267: 9101–9106
33 Sugimoto T, Tsuchimochi H, McGregor CG, Mutoh H, Shimizu T, Kurachi Y (1992) Molecular cloning and characterization of the platelet-activating factor receptor gene expressed in the human heart. *Biochem Biophys Res Commun* 189: 617–624
34 Bito H, Nakamura M, Honda Z, Izumi T, Iwatsubo T, Seyama A, Ogura Y, Kudo, T Shimizu T (1992) Platelet-activating factor (PAF) receptor in rat brain: PAF mobilizes intracellular Ca^{2+} in hippocampal neurons. *Neuron* 9: 285–294
35 Chen C, Magee JC, Marcheselli VL, Hardy M, Bazan NG (2001) Attenuated long-term potentiation in hippocampal dentate gyrus neurons of mice deficient in the platelet-activating factor receptor. *J Neurophysiol* 85: 384–390
36 Ihida K, Predescu D, Czekay R-P, Palade GE (1999) Platelet activating factor receptor (PAF-R) is found in a large endosomal compartment in human umbilical vein endothelial cells. *J Cell Sci* 112: 285–295
37 Marrache AM, Gobeil Jr F, Bernier SG, Stankova J, Rola-Pleszczynski M, Choufani S, Bkaily G, Bourdeau A, Sirois MG, Vazquez-Tello A et al (2002) Pro-inflammatory gene induction by platelet-activating factor mediated *via* its cognate nuclear receptor. *J Immunol* 169: 6474–6481
38 Bazan NG, BS Fletcher, HR Herschman, PK Mukherjee (1994) Platelet-activating factor and retinoic acid synergistically activate the inducible prostaglandin synthase gene. *Proc Natl Acad Sci USA* 91: 5252–5256
39 Prescott SM, Zimmerman GA, Stafforini DM, McIntyre TM (2000) Platelet-activating factor and related lipid mediators. *Annu Rev Biochem* 69: 419–445
40 Prehn JHM, Krieglstein J (1993) Platelet-activating factor antagonists reduce excitotoxic damage in cultured neurons from embryonic chick telencephalon and protect the rat hippocampus and neocortex from ischemic injury *in vivo*. *J Neurosci Res* 34: 179–188
41 Bliss TVP, Collingridge GL (1993) A synaptic model of memory: Long-term potentiation in the hippocampus. *Nature* 361: 31–39

42 Nakanishi S (1994) Metabotropic glutamate receptors: Synaptic transmission, modulation and plasticity. *Neuron* 13: 1031–1037
43 Bazan NG, Rodriguez de Turco EB, Allan G (1995) Mediators of injury in neurotrauma: Intracellular signal transduction and gene expression. *J Neurotrauma* 12: 791–814
44 Balsinde J, Balboa MA, Insel PA, Dennis EA (1999) Regulation and inhibition of PLA2. *Annu Rev Pharmacol Toxicol* 39: 175–189
45 Hirabayashi T, Kume K, Hirose K, Yokomizo T, Iino M, Itoh H, Shimizu T (1999) Critical duration of intracellular Ca^{2+} response required for continuous translocation and activation of cytosolic phospholipase A_2. *J Biol Chem* 274: 5163–5169
46 Yamagata K, Andreasson KI, Kaufmann WE, Barnes CA, Worley PF (1993) Expression of a mitogen-inducible cyclo-oxygenase in brain neurons: Regulation by synaptic activity and glucocorticoids. *Neuron* 11: 371–386
47 Vane JR, Bakhle YS, Botting RM (1998) Cyclo-oxygenases 1 and 2. *Annu Rev Pharmacol Toxicol* 38: 97–120
48 Miettinen S, Fusco FR, Yrjanheikki J, Keinanen R, Hirvonen T, Roivainen R, Narhi M, Hokfelt T, Koistinaho J (1997) Spreading depression and focal brain ischemia induce cyclooxygenase-2 in cortical neurons through N-methyl-D-aspartic acid-receptors and phospholipase A_2. *Proc Natl Acad Sci USA* 94: 6500–6505
49 Nakayama N, Uchimura K, Zhu RL, Nagayama T, Rose ME, Stetler RA, Isakson PC, Chen J, Graham SH (1998) Cyclo-oxygenase-2 inhibition prevents delayed cell death of CA1 hippocampal neurons following global ischemia. *Proc Natl Acad Sci USA* 95: 10954–10959
50 Hewett SJ, Uliasz TF, Vidwans AS, Hewett JA (2000) Cyclo-oxygenase-2 contributes to N-methyl-D-aspartate-mediated neuronal cell death in primary cortical cell culture. *J Pharmacol Exper Ther* 293: 417–425
51 Kosaka T, Miyata A, Ihara H, Hara S, Sugimoto T, Takeda O, Takahashi E, Tanabe T (1994) Characterization of the human gene (PTGS2) encoding prostaglandin-endoperoxidase 2 synthase. *Eur J Biochem* 221: 889–897
52 Newton R, Kuitert LM, Bergmann M, Adcock IM, Barnes PJ (1997) Evidence for involvement of NF-kappa B in the transcriptional control of COX-2 gene expression by IL-1beta. *Biochem Biophys Res Commun* 237: 28–32
53 Lukiw WJ, Bazan NG (1998) Strong nuclear factor-B-DNA binding parallels cyclo-oxygenase-2 (COX-2) gene transcription in sporadic Alzheimer's disease (AD) superior temporal lobe neocortex. *J Neurosci Res* 53: 583–592
54 Hensley K, Butterfield DA, Hall N, Cole P, Subramaniam R, Mark R, Mattson MP, Markesbery WR, Harris ME, Aksenov M et al (1996) Reactive oxygen species as causal agents in the neurotoxicity of the Alzheimer's disease-associated amyloid beta peptide. *Ann NY Acad Sci* 786: 120–134
55 Schmidt-Ullrich R, Memet S, Lillenbaum A, Feuillard J, Raphael M, Israel A (1996) NF-kappa B activity in transgenic mice: Developmental regulation and tissue specificity. *Development* 122: 2117–2128

56 O'Neill LA, Kaltschmidt C (1997) NF-kappa B: A crucial transcription factor for glial and neuronal cell function. *Trends Neurosci* 20: 252–258
57 Mukherjee PK, DeCoster MA, Campbell FZ, Davis RJ, Bazan NG (1999) Glutamate receptor signaling interplay modulates stress-sensitive mitogen-activated protein kinases and neuronal cell death. *J Biol Chem* 274: 6493–6498
58 Kolko M, Bruhn T, Christensen T, Lazdunski M, Lambeau G, Bazan NG, Diemer NH (1999) Secretory phospholipase A_2 potentiates glutamate-induced rat striatal neuronal death *in vivo*. *Neurosci Lett* 274: 167–170
59 Rodriguez de Turco EB, Jackson FR, DeCoster MA, Kolko M, Bazan NG (2002) Glutamate signaling and secretory phospholipase A_2 modulate the release of arachidonic acid from neuronal membranes. *J Neurosci Res* 86: 558–567
60 Kolko M, Nielsen M, Bazan NG, Diemer (2002) Secretory phospholipase A_2 induces delayed neuronal COX-2 expression compared with glutamate. *J Neurosci Res* 69: 169–177
61 DeCoster, MA, Lambeau G, Lazdunski M, Bazan NG (2002) Secreted phospholipase A_2 potentiates glutamate-induced calcium increase and cell death in primary neuronal cultures. *J Neurosci Res* 67: 634–645
62 Perretti M, Chiang N, La M, Fierro IM, Marullo S, Getting SJ, Solito E, Serhan CN (2002) Endogenous lipid- and peptide-derived anti-inflammatory pathways generated with glucocorticoid and aspirin treatment activate the lipoxin A_4 receptor. *Nat Med* 8: 1296–1302
63 Kolko M, DeCoster MA, de Turco EB, Bazan NG (1996) Synergy by secretory phospholipase A_2 and glutamate on inducing cell death and sustained arachidonic acid metabolic changes in primary cortical neuronal cultures. *J Biol Chem* 271: 32722–32728

Remodeling of arachidonate and other polyunsaturated fatty acids in Alzheimer's disease

Alfred N. Fonteh and Michael G. Harrington

Molecular Neurology Program, Huntington Medical Research Institutes, Pasadena CA 91101-1830, USA

Overview of AD

AD is the most common dementia, afflicting millions in developed countries [1]. AD is a neurodegenerative process characterized by irreversible destruction of the neocortex leading to impairment of memory and cognitive function, language deterioration, poor judgment, indifferent attitude, but preserved motor function. An estimated 7% of people above 65 years old have AD and its prevalence is expected to increase as better medical science increase life expectancy [1]. Although AD is the most characterized dementia, no effective treatments are available for controlling this debilitating disease. Definitive diagnosis of AD is made at autopsy. The pathogenic characteristics of AD include the cleavage of β-amyloid precursor protein (βAPP) to form neurotoxic β-amyloid (βA) peptides that are two amino acids longer than normal, and an aggregated insoluble polymer of βA that forms senile plaques [2, 3]. Enzymes known as secretases mediate processing of βAPP. In addition to plaques, the formation of intraneuronal tau pathology yielding widespread deposits of argyrophilic neurofibrillary tangles (NFT) is typical in AD. In addition to plaque formation, enhancement of inflammatory responses in the brain mediated by neurotoxic peptides is characteristic of AD. These peptides and inflammatory cytokines (IL-1, IL-6, and TNF-α) are implicated in the activation of glial cells and astrocytes [4].

The major risk factors associated with AD include genes, head trauma, environmental and demographic factors [5–8]. Four major genetic loci are associated with AD. These include APP, two presenilins and apolipoprotein E (apoE) [9, 10]. Mutations of APP are linked to early onset AD mainly as a result of increased formation of the plaque component of βA peptides. Various studies have shown that βA addition to cultures result in cell death [11–13]. Additionally, βA peptides can also induce activation of glial cells [14–16]. Induction of neuronal cell death and abnormal activation of cells are plausible mechanisms of action of βA. Mutations in the presenilin genes PS1 and PS2, implicated in early onset familial AD, cause abnormal

Arachidonate Remodeling and Inflammation, edited by Alfred N. Fonteh and Robert L. Wykle

βAPP processing with resultant overproduction of neurotoxic βA_{1-42} and related peptides [17, 18]. Various apoE alleles are found in humans, however, individuals homozygous for apoE4 are 3–4 times more likely to have AD [19, 20]. ApoE is linked to high-density lipoprotein (HDL) in plasma and cerebrospinal fluid (CSF) and functions mainly to provide lipids to cells. Unfortunately at this time, much less is known of the unique distribution, transport and biochemistry of CSF and brain apolipoproteins compared to the plasma. ApoE affects lipid transport, neurite growth *via* low density lipoprotein (LDL) receptors and processes critical to neurodegeneration [19]. ApoE forms are extensively O-glycosylated and glycation changes to apoE alter LDL and VLDL binding that can change specific receptor binding of these complexes [21, 22]. Furthermore, the increase in advanced glycation end products that occur in AD co-localize with ApoE4, linking this phenotype with the plaque deposition in AD [23–26].

In addition to genetic predisposition, repeated head trauma may contribute to early onset of AD, perhaps by creating a dysfunctional blood brain barrier (BBB) [6]. Changes in BBB may allow environmental agents (organic solvent, heavy cigarette smoking, industrial toxins, transition metals, Al, Fe) more access to the brain leading to higher incidences of AD [6, 7, 27–36]. Finally, demographic factors (age, gender, race, co-morbid conditions, especially cerebrovascular insults) may play important roles in AD [8, 35, 37]. While AD has been shown to be age-dependent, only small gender differences have been documented [38]. Overall, biochemical knowledge on the interplay of these contributing factors in AD pathogenesis is still limited.

Brain lipids

Lipids comprise approximately 50 percent of the dry matter of the brain and play important roles in brain function (signaling molecules, inducers of chemotaxis, pro- and anti-inflammatory mediator formation, ion channel, receptor and membrane functions and an energy source). Major cerebral lipids (also found in lesser quantities in other organs) include cholesterol and cholesterol esters, glycerolipids, sphingolipids and ceramide. Apart from water, cholesterol is greater than any other single constituent in the brain comprising 4–5% by mass. The most abundant form is free cholesterol that may be linked to other lipids or proteins. The fatty acid constituents of cholesterol esters are oleic, palmitic, palmitoleic, and arachidonic acids. Levels of these esters are altered in autopsied AD brain with a significant decrease in unsaturated fatty acids [39, 40]. Cholesterol levels are controlled by its synthesis, transport and metabolism. For example, higher activities of cholesterol ester hydrolase increase cholesterol levels, while lecithin cholesterol acyl transferase (LCAT) controls the reverse transport of cholesterol [41–43]. Conversion of cholesterol to esters or hydroxylation are critical events in AD pathology. Not surprisingly, levels of 24 hydroxy cholesterol in CSF or plasma are very good correlates of AD pathology [44, 45]. Only trace amounts of free fatty acids are found in the brain and CSF, while most fatty acids are acylated into glycerides and their levels and distribution within glycerolipids and brain regions are altered in diseased conditions [46–51]. The brain is rich in very long chain polyunsaturated fatty acids (PUFA) obtained from dietary sources or synthesized from precursors by brain cells. Our data show that CSF contains various enzymes involved in PUFA binding, transport and degradation [52]. Significant levels of the prostanoid metabolites, $PGF_{2\alpha}$ (100–500 pg/ml) and 6-keto $PGF_{1\alpha}$ (100 pg/ml) are found in CSF and these levels are increased manyfold in pAD and other head traumas [53–56]. Compared to plasma, the CSF contains less total lipid (400–800 mg/100 ml compared to 1–2 mg/100 ml) [57]. The distribution of lipids varies in brain regions or within cells. Certain membrane fractions are enriched with unsaturated fatty acids or with plasmalogen-linked phospholipids. Unsaturated fatty acids acylated onto phospholipids in myelin are mainly oleic acid while nerve endings contain mainly arachidonic acid (AA, 20:4, n-6) and docosahexaenoic acid (22:6, n-3) [57].

In the brain, turnover in lipids varies according to their precursor. The fastest phospholipid turnover occurs in phosphatidic acid (PA) and phosphatidylinositol (PI). In addition to the biosynthesis of phospholipids, there is rapid turnover of fatty acyl constituents. Changes in acyl constituents likely reflect function of fatty acids in the brain. Defects in lipid metabolism are enhanced in neurological disorders including AD. Various catabolic enzymes including phospholipases A_2 (PLA_2) and anabolic enzymes such as lysophospholipid acyltransferase that recycles lysophospholipids into intact glycerophospholipids and glycerophosphocholine phosphodiesterase are decreased in AD brain regions. Enzymes involved in *de novo* synthesis (ethanolamine kinase, choline kinase) are normal or just slightly altered in AD [57].

Biosynthesis and sources of AA and other PUFAs

Fatty acids are important in mammalian systems, being the structural elements of cell membranes, a source of energy and they provide signaling molecules that control cell growth, development, ion channel function and cell death. Most saturated and monounsaturated fatty acids are obtained from dietary sources or can be synthesized by fatty acid synthetic machinery. However, precursors of PUFAs that are needed for regular functioning of cells can only be derived from dietary sources. These fatty acids are known as essential fatty acids (EFA). The two major classes of essential fatty acids are the n-6, derived from linoleic acid (LA) and the n-3 derived from α-linolenic acid (ALNA). A series of desaturation (Δ6- and Δ5-desaturase) and elongase enzymes convert LA and ALNA to longer chain PUFAs (Fig. 1) [58]. The major PUFAs found in brain are arachidonic acid (AA, 20:4, n-6), docosahexaenoic acid (DHA, 22:6, n-3) and eicosapentaenoic acid (EPA, 20:5, n-3). DHA and EPA can also be derived from fish oils [59]. AA is the major substrate for the production of lipid mediators of inflammation. Evidence for a major role of PUFAs in brain physiology are provided by studies showing their enhanced elongation and desaturation during myelination. *In vitro* studies show that cerebral microvascular endothelium and astrocytes can produce DHA and AA *via* elongase and desaturase activities [60]. In contrast, neurons cannot produce PUFAs but get enriched with PUFAs if they are co-cultured with astrocytes and endothelial cells. In addition to synthetic pathways for PUFA, the brain may also obtain dietary PUFA *via* fatty acid binding proteins. Little is known of the plasma/CSF/brain differences in sources of PUFA. For example, does the brain obtain all PUFAs from the plasma or is there substantial *de novo* synthesis of PUFAs by the brain? Do PUFAs from the plasma diffuse into the brain or is there active transport through receptor-mediated processes? Is the PUFA transport mechanism similar to that of cholesterol utilizing lipoproteins (HDL and LDL) and their receptors? Is there a preference for n-6 or n-3 fatty acid biosynthesis by astrocytes? Are all apolipoprotein species found in the brain involved in PUFA transport? These important questions are being addressed as our knowledge about apolipoproteins is increased. For example, it is now known that various apolipoproteins present in CSF are composed of phospholipids that contain PUFA [19] and brain DHA is suggested to be important in brain development, plasma membrane composition, neuroplasticity, aging, neuronal ion channels and in signal transduction [61].

The major biochemical processes involved in AA metabolism and their potential to influence AD are summarized in Figure 1. Very low levels of free PUFA are maintained in mammalian cells. Free PUFAs in plasma are closely associated with lipoproteins, albumin or fatty acid binding proteins. Once presented to mammalian cells, PUFAs are rapidly converted to acyl CoAs by synthetases [62]. PUFA acyl CoAs are rapidly incorporated into glycerolipids by CoA dependent enzymes [63, 64]. This initial incorporation is accompanied by remodeling of PUFA into ether-linked plasmalogen lipid pools that comprise the major repository of PUFA in the brain. The remodeling and release of PUFA from glycerolipids is important in maintaining homeostasis and in regulating cell function [65–69].

Uptake and transport of PUFAs

PUFAs and other major brain lipids such as cholesterol are transported in plasma bound to lipoproteins (presumably also in CSF, but this is not confirmed). PUFAs are activated by synthetases and then incorporated into glycerolipids or cholesterol esters by CoA dependent acyl transferases. Once obtained from the diet, fatty acids are presented to cellular tissues as complexes associated with binding proteins, albumin or apolipoproteins. Upon contact with cells, proteins with high affinity for PUFAs (fatty acid translocase or fatty acid binding protein (FABP) mediate dissoci-

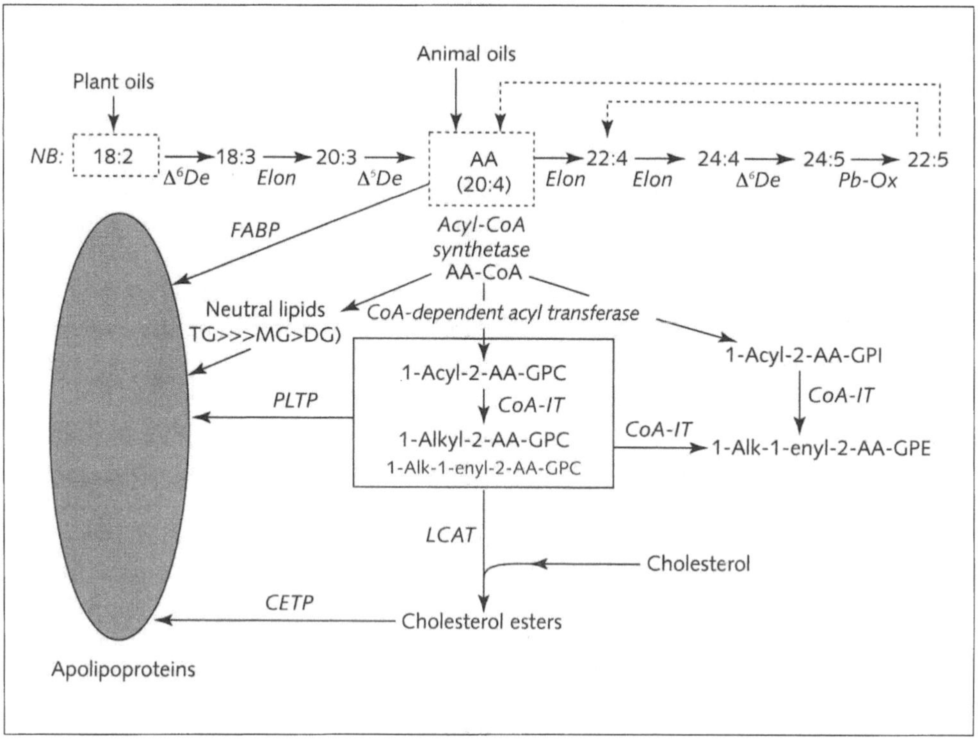

Figure 1
AA metabolism
AA is obtained from the diet or from elongation/desaturation of LA (18:2, n-6) or retro conversion of other longer chain PUFAs. AA is converted to AA-CoA by acyl CoA synthetase and is rapidly incorporated into cellular lipids by acyl transferase activity. AA in phospholipids is remodeled from 1-acyl-linked to 1-alkyl- or 1-alk-1-enyl-linked phospholipid classes by transacylases. Cholesterol esters are also formed by LCAT that is associated with apo-A1 from HDL. Various fatty acid binding proteins (FABP), phospholipid transport proteins (PLTP) and cholesterol ester transfer proteins are coupled with cells or lipoproteins and albumin to deliver lipids to tissues.

ation of these complexes. Flip-flop of fatty acids across the lipid bilayer is associated with cytosolic FABP or with caveolin-1 containing vesicles known as caveolae. Caveolae deliver lipids to subcellular organelles. The transport of fatty acids may be accompanied by activation by acyl CoA synthetase and targeting to specific membranes controlled by anchoring to phospholipid bilayers.

Edmond has suggested that there is selective uptake of essential PUFA by the brain based on studies showing that linoleic acid (18:2, n-6) enters the brain while oleic acid (18:1, n-9), cholesterol and other non-essential fatty acids do not enter the brain [70]. Capillary networks composed of endothelial cells have receptors for lipoproteins but do not use LDL-cholesterol. The recent discovery of a transmembrane monocaboxylic acid transporter (MCT) and fatty acid transport proteins (FATPs) that are positioned at the inner luminal membrane allow essential PUFAs to be transported into the brain [70].

In addition to esterification to glycerolipids, PUFAs are also acylated into cholesterol and are similarly transported bound to lipoproteins. Low-density lipoprotein (LDL) transports cholesterol from the liver to peripheral tissues while HDL transports cholesterol from the periphery to the liver for excretion as bile salts [71–74]. HDL is a macromolecule made up mainly of phospholipids, cholesterol and little triglycerides complexed to apolipoproteins. The major apolipoprotein in HDL is apoA-1 which is synthesized and secreted by the liver. Nascent HDL containing apoA-1 interacts with cells to acquire cholesterol and phospholipids. Properties and functions of HDL include promotion of cellular cholesterol efflux, reverse cholesterol transport, antioxidation (protects LDL from oxidation), anti-inflammation (induces nitric oxide) and anti-coagulation properties. HDL also transports cholesterol to steroidogenic tissues where it is used as the major precursor for the synthesis of steroid hormones. Cholesterol esters are also transferred to apoB-1 containing lipoproteins in exchange for triglycerides through the action of cholesterol ester transfer protein (CETP). The transport of cholesterol from tissues to the liver is commonly referred to as "reverse transport" and is linked to an enzyme known as lecithin cholesterol acyl transferase (LCAT) [42].

LCAT catalyzes the transfer of a fatty acid from lecithin (phosphatidylcholine) to cholesterol to form cholesterol esters and lysophosphatidylcholine (LPC) (Fig. 1). The reaction takes place on the surface of high-density lipoprotein (HDL) in several steps including interfacial binding to HDL and activation of LCAT by apo-A1, PC binding, acyl enzyme formation, release of LPC, cholesterol binding and is terminated by the release of cholesterol ester and LCAT. In addition to cholesterol, LCAT can also esterify steroid hormones such as pregnenolone and dehydroepiandrosterone and has been shown to convert 25-hydroxycholesterol, an AD marker, to diester by acylating carbon 3 and carbon 27 [75–77]. LCAT can also use phosphatidylethanolamine as a fatty acyl donor but has no reactivity towards phosphatidylserine or diacylglycerol. LCAT also transesterifies and hydrolyzes platelet-activating factor and oxidized PC molecules containing short chain fatty acids at the *sn*-2 position. LCAT is a 49–60 kDa glycoprotein consisting of β sheets and α-helical elements that was first identified in the late 1960s by Glomset [78]. The human LCAT gene located on chromosome 16 was first cloned by McLean and colleagues [79, 80] and subsequently chemically sequenced in 1987 [81, 82]. The 416 amino acids sequence has four heterogeneous N-glycosylation sites and six cysteine residues with two disulfide bonds that are implicated in lipoprotein binding but not in enzyme activation. An Asp-His-Ser catalytic triad catalyzes the transesterification reaction. Conversion of cholesterol to esters results in the removal of cholesterol from the surface of HDL concomitant with the accumulation cholesterol esters in the core of HDL. This process promotes the flux of more cholesterol from cell membranes into HDL and the HDL particle grows in size as more esters are transported to the HDL core. LCAT is synthesized mainly by liver cells and circulates in association with lipoproteins. The brain also produces LCAT and low levels of LCAT have also been found in CSF amounting to 2.2% of the activity in plasma. However, the function of LCAT in the brain has not been determined. In animals with LCAT deficiency, cholesterol and cholesterol esters accumulate in tissues and blood cells. This accumulation alters cell and lipoprotein structure and function resulting in familial LCAT deficiency or Fish eye disease [42, 83, 84].

Several studies have implicated cholesterol homeostasis with AD pathology. Cholesterol reducing drugs have no effect or very modest effect on LCAT expression. However, various substrates and fatty acids can influence LCAT expression or activity. For example, supplementation with LA increases LCAT activity while DHA reduces LCAT activity. Other evidence that LCAT is involved in cholesterol metabolism is derived mainly from transgenic animal studies. Transgenic mice over-expressing LCAT and fed with cholesterol increased apoE-containing cholesterol esters in HDL1 particles [85]. Over-expression of LCAT also delays the catabolism of apoA-1 while deficiency results in reduction in HDL-c and apoA-1. Positive correlation between LCAT activity or mass and HDL-c has been established [86]. Together, these studies show that LCAT is an important enzyme in regulating cholesterol and fatty acid levels as well as in controlling the biosynthesis of steroid hormones.

Degradation of brain lipids

Once AA and other PUFAs are acylated into glycerolipids of the brain, they can be degraded by enzyme-catalyzed processes or by auto-catalyzed oxidative processes. These processes generate potent mediators of inflammation and cytotoxic lipid radicals that are implicated in AD pathology.

Release of PUFA from lipids

Pro-inflammatory microglia, reactive astrocytes and their associated cytokines and chemokines are linked to the mobilization of PUFAs as an integral part of the inflammatory process [87]. The release of PUFAs is mediated by cellular and secretory lipases. Secretory phospholipase A_2 ($sPLA_2$) has the ability to hydrolyze HDL phospholipids. Inducible oxidoreductase enzymes and cytosolic phospholipase A_2 ($cPLA_2$) are strongly activated during brain trauma, epilepsy and AD [88]. Properties of secretory PLA_2 and intracellular PLA_2 are reviewed in the chapters by Barbour et al., Marion and Fonteh, and Balsinde et al., in this volume. Lipoprotein lipase (LPL) hydrolyzes triglycerides resulting in enhanced transfer of lipids from apolipoproteins to HDL. LPL is an extracellular enzyme that enables tissues to import fatty acids from triacylglyceride-rich lipoproteins. LPL occurs in all tissues including the brain where its function is not known. LPL activity is high in newborn brain and this activity increases during postnatal development. The hippocampus shows the highest LPL activity where levels are 5–11-fold higher than other sections of the brain.

Generation of lipid mediators

Cyclo-oxygenase (COX) also known as prostaglandin H synthase (PGHS) catalyzes the conversion of arachidonic acid (AA) to prostaglandin H_2 (PGH_2), an intermediate in the synthesis of prostaglandins (PGE_2, $PGF_{2\alpha}$, PGD_2, PGI_2), and thromboxane A_2 (TXB_2) [89]. mRNA and protein levels of one COX isoform (COX-2) have been shown to be induced by endotoxin and inflammatory cytokines in brain astrocytes and glial cells [90]. COX can also convert other PUFAs into oxidized lipids; however, AA is the best substrate. COX has two different enzymatic activities: the formation of PGG_2 from oxygen and AA by a cyclo-oxygenase; the reduction of PGG_2 to PGH_2 catalyzed by hydroperoxidase. The COX intermediate PGH_2 is the substrate for terminal synthases that convert it to PGD_2 (PGDS), PGE_2 (PGES), TXB_2 (TBXS) and PGI_2 (PGIS) [91] (Fig. 2). Induction of COX-2 plays an important role in inflammatory disease by increasing the formation of these lipid mediators. NSAIDs prevent the formation of lipid mediators of inflammation by inhibiting COX activity [92]. Our data show an increase of PGDS in CSF of AD subjects [93] and studies by Bazan's group show the induction of COX-2 mRNA in the brain of AD subjects [87]. Likewise Morrow and colleagues have shown an increase in PGE_2 levels in CSF from pAD subjects compared to controls [54]. Postmortem brain studies also show neuronal COX-2 content to be an indicator of early AD dementia [94]. Together with data showing the induction of $cPLA_2$ that releases AA from membrane lipids, these data suggest that pro-inflammatory lipid mediator pathways may play a role in the pathogenesis of AD. In addition to COX-derived products, PUFAs may also be converted to leukotrienes by the lipoxygenase (LO) pathways [95, 96]. Very little is known about LO pathways in the brain. However, various studies have shown that brain cells express LO [97, 98]. Any changes in levels of these products in AD compared with "controls" will suggest a role of LO-derived products in disease. Finally, PUFAs are also substrates of cytochrome P450 (CYP) enzymes [99]. These enzymes convert PUFAs such as AA to conjugated dienols, ω-terminal hydroxylated alcohols and cis-epoxyeicosatrienoic acids (EETs) [100]. These compounds have been shown to regulate cellular proliferation, inflammation and to be involved in many signaling pathways.

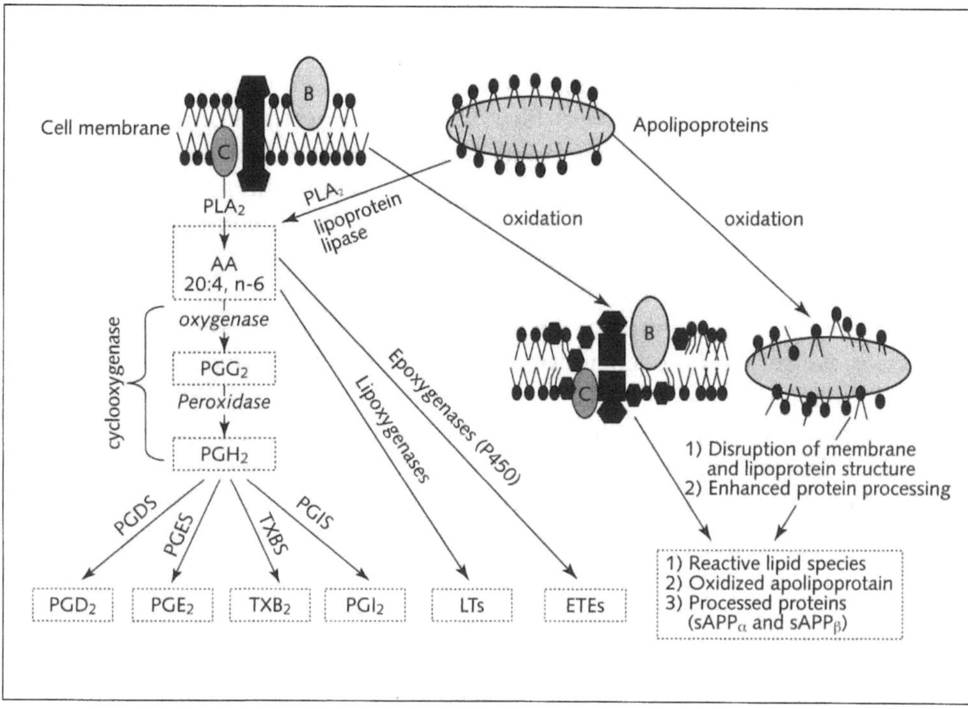

Figure 2
Formation of mediators of inflammation and membrane perturbation in AD
AA in phospholipid membranes or in lipoproteins is mobilized by PLA_2 or lipoprotein lipase activity, respectively. Release of fatty acid from cellular membranes or from lipoproteins disturbs their structure and function. In addition to lipases, autoxidation of PUFAs generates reactive phospholipid species that also destroy membrane integrity. Free AA is converted to lipid mediators of inflammation by various oxygenases and the perturbed cellular membrane exposes amyloid precursor proteins (A) to secretases (B and C), making them more susceptible to the amyloidogenic processing. Generation of lipid mediators of inflammation coupled with the destruction of cellular membranes may result in apoptosis of brain cells and the propagation of AD pathology that is typified by increased amyloid peptide cleavage and deposition of plaques in the brain.

Similar to LO products, little is known about the production of CYP products in the brain. However, oxidized EETs have been shown conjugated to phospholipids and these could potentially alter lipoprotein and cell membrane structure and function (Fig. 2), but this is not established for plasma or brain. HDL lipoproteins in patients with dementia showed reduced AA (20:4, n-6) compared to normal controls, likely due to enhanced degradation by enzymes or increased auto peroxidation [101].

Auto-oxidation of PUFAs

Since PUFAs are important in cell membrane structure and function, they can play a role in AD in one of three ways. First, PUFA are essential in brain physiology. Therefore any change in their composition or their distribution in the brain or in the blood cells is likely to have pathophysiological ramifications. A second role of PUFAs in AD reflects their inflammatory properties. In the context of an inflammatory process, PUFAs may be mobilized from cells by lipases [102]. Free PUFAs can be reduced and oxidized by enzymes to form inflammatory products. This inflammatory process is enhanced in AD compared to normal subjects. Thirdly, a decrease in the antioxidant defense system will result in the autoxidation of PUFAs. This auto oxidative process is catalyzed by free radicals and results in the formation of cytotoxic aldehyde products, isoprostanes and oxidized lipoproteins. Montine and colleagues have done extensive studies, which suggest that there is increased oxidative damage, and enhanced lipid peroxidation of PUFA rich lipids in AD brain tissues [54, 104–106]. The following pieces of evidence are cited as evidence:

1) Lipoproteins from AD extracellular fluid are more vulnerable to oxidation than controls.
2) ApoE is a major apolipoprotein in the CNS that is implicated in lipid transport and receptor-mediated regulation of lipid metabolism. Inheritance of the apoE4 allele is the strongest known genetic risk factor for sporadic AD.
3) ApoE isoforms may influence cellular distribution of peroxidation products in the brain by influencing lipoprotein trafficking and lipid oxidation and the convergence of these processes contributes to neurodegeneration in AD brain.
4) Mass spectrometric data show that there is enhanced formation of reactive products of auto-oxidized PUFAs known as isoprostanes [54, 104–106].

Similar to the enzyme-catalyzed breakdown of PUFA-containing lipids, auto-oxidized lipids perturb the cellular structure of cell membranes and lipoproteins leading to dysfunction of these entities and thence AD pathology. Oxidation of PUFAs that are components of lipids that make up lipoproteins may alter lipoprotein function and increase its reactivity towards brain cells (Fig. 2). Autopsy studies and analyses of spinal fluids show that AA and DHA are oxidized to F_2 and F_3 series isoprostanes, respectively in the brain of AD subjects [107–111]. Theoretically, lipid changes would enormously influence many of the hallmarks of Alzheimer's disease (AD). Most prominently, a lipid-depleted plasma membrane would ease production of deadly βA_{1-42}, since a secretase would more easily access a longer stretch of the peptide chain as it protrudes from a shrunken lipid bilayer [101, 103]. Amyloidogenic processing of proteins will generate cytotoxic peptides and result in the formation of protein plaques. Likewise, the exposure of nerve cells to oxidative breakdown products of PUFA or peptide fragment may induce neurodegeneration *via* apoptosis.

Other oxidative processes and AD

The loss of an electron by a compound is known as oxidation. Chemical species known as free radicals initiate oxidation of macromolecules in biological systems. Free radicals contain an unpaired electron and are thus very reactive because they can gain electrons from other molecules. Molecules that donate their electrons are known as oxidizing agents. In cells, cellular metabolism and exposure to environmental factors result in free radical generation. The most common free radicals are superoxide (O_2-), hydroxyl ($OH_2^{\cdot}$) and nitric oxide (NO) [112–114]. Molecules such as hydrogen peroxide (H_2O_2) and peroxynitrate ($ONOO^-$) readily generate free radicals through various chemical reactions. Reactive oxygen species (ROS) is a general term for all free radicals and related molecules. Regular metabolic process such as the formation of excitatory amino acids and neurotransmitters produce ROS. For example, activation of glial cells results in enhanced production of NO *via* five-electron oxidation of L-arginine [115, 116]. NO synthase (NOS) mediates NO production. Because these compounds can cause oxidative damage to lipids, proteins and DNA, cells tightly control production of ROS. An imbalance in ROS production and inability to defend against them is known as oxidative stress.

Because of the ability of ROS to cause cell injury, oxidative stress is now considered to play a major role in the pathogenesis of AD [117]. Cells control ROS using an antioxidant defense system. The major components of this system are enzymes (superoxide dismutase (SOD), catalase, glutathione peroxidase) that destroy ROS, low molecular weight antioxidants (glutathione, tocopherol, ascorbic acid, retinoic acid, melatonin, uric acid, and lipoic acid), cofactors such as coenzyme Q_{10}, precursors of antioxidants (acetylcysteine, caretenoids) and natural plant compounds such as flavonoids [118–120]. Recent studies have shown that amyloid peptides may also have anti-oxidative properties leading to speculations that their occurrence in AD may be the result of the body mounting a defense against ROS produced by activated microglial cells [121]. However, most links between oxidative stress and AD have not been biochemically proven in human subjects. Since their double bonds make them more susceptible to oxidation, a decrease in oxidant defense will destroy PUFAs. Such destruction of PUFA in the brain will result in altered membrane structure and nerve function. One implication of such a membrane change is to contribute to the abnormal cleavage of βA: a lipid-damaged cell membrane might enable greater access by secretases, to a more proximal region of βAPP and this mechanism may participate in the increased production of amyloid fragments. Oxidized PUFA are also cytotoxic and can induce nerve cell death [122]. Various studies showing an increase in lipid peroxidation in AD suggest that the destruction of PUFA may be one mechanism by which oxidative stress is linked to the pathogenesis of AD.

Role of metals and trace elements in the oxidation of PUFAs

Metals and trace elements play important roles in PUFA metabolism. Metals and elements enter the body through food and diet or from respiration. Metals are important cofactors or components involved in antioxidant defense mechanisms. For example, activation of SOD requires copper and zinc [123–125]. Zinc plays an essential biochemical role in participating in the synthesis/degradation of macromolecules including lipids. Copper is widely distributed in metalloproteins/enzymes and is a major component of respiration. It is an essential component of erythrocytes and plasma where it is mostly bound to caeruloplasmin and closely linked to iron metabolism. Selenium is a constituent of glutathione peroxidase that is a component of the antioxidant defense system of the body [126, 127]. Molybdenum, manganese and vanadium are also activators of several enzymes while vanadium plays a role in lipid metabolism [128]. Other metals are known to be neurotoxic (aluminum, lead, cadmium, mercury) and are implicated in AD etiopathogenesis through mechanisms that are not well defined [31, 129–133]. While no biological function has been attributed to aluminum, it is known to interact and reduce the absorption of calcium, iron, manganese and phosphorus [134]. Therefore, any neurological or neurobehavioral effects of aluminum may be due to the alteration of the metabolism of other metals. In addition to being components of enzymes, metals can also serve as scavengers that prevent free radical formation. While trace amounts of elements can be protective, higher levels or alterations in homeostatic control of metals may induce the formation of free radicals that would destroy PUFA.

Clinical trials

There is presently no cure for AD and no treatment to prevent the gradual progression of the disease. However, remedial medication may help symptoms and make patients more comfortable by controlling sleeplessness, agitation, wandering, anxiety, and depression. For example, in the early or middle stages of AD, tacrine may alleviate some cognitive symptoms and acetylcholinesterase inhibitors (donepezil) and rivastigmine) are prescribed for the treatment of mild to moderate dementia related to AD. Various clinical trials have examined the effects of NSAIDs, antioxidants, and metal chelators on AD development [135–138]. Most of these studies are based on circumstantial evidence or on mouse studies that cannot be duplicated in human subjects. Epidemiological and longitudinal studies have identified a reduced risk for AD in patients (< 70 yrs) previously treated with NSAIDs for non-CNS afflictions such as arthritis [139]. Prospective clinical studies targeting COX-2 are now using more specific inhibitors to avoid gastrointestinal side effects that limited initial trials [140]. However, this new generation of COX-2 inhibitors may also have detrimental side effects because they inhibit a whole spectrum of prostanoids needed for normal physiologic actions of organs (heart, kidney) [141–143]. In addition to COX inhibition, mouse studies have shown that NSAIDs also inhibit secretases, thus preventing βA formation [144]. *In vitro* studies also suggest that NSAIDs may directly scavenge superoxide [145]. The overall effects are to prevent inflammation and oxidative damage. Our studies also show that preventing the conversion of PUFAs to lipid mediators by NSAIDs results in the increase of free PUFA levels in cells and a shift in their distribution in glycerolipids subclasses [146]. An understanding of specific pathways and mechanisms will enable us to design interventions that may lessen AD progression and reduce the side effects of current therapy.

Similar to the NSAID studies, vitamins B1, B12, C, E, folate, choline, magnesium, DHA, phosphatidylserine, DHEA, NADH, acetyl-L-carnitine, melatonin and *Ginkgo Biloba* extract supplements have been administered to AD patients without clear biochemical rationales [147–149]. Thus, while minor beneficial outcomes have been reported in some of these studies, it is difficult to scientifically evaluate these approaches without mechanism-based data. Lipids and their products are associated with nearly all of these agents and baseline lipid biochemistry will allow predictions of lipid altering substances. Furthermore, baseline studies will then enable appropriate monitoring of any treatments.

Conclusions and implications

A biochemical baseline for lipid composition and metabolism in the brain, CSF and plasma, including post-translational modifications, will likely predict candidate lipid markers for early diagnosis of AD, determine which patients might respond to specific treatments and lead to a better understanding of the underlying pathophysiology of AD. There may be a need to replenish PUFA in adult brain using dietary manipulations or identify enzymes that destroy PUFAs in AD in order to direct effort to selectively inhibit specific pathways. Determining whether a decrease in PUFA biosynthesis and transport or whether enhanced oxidation accounts for the decrease in PUFA in the adult brain will better define the use of antioxidants as remedies for AD progression and provide assays to monitor their efficacy. It might turn out that agents that cross the blood brain barrier are needed to control PUFA levels in brain. Alternatively, agents that control PUFA levels in plasma may be sufficient in maintaining physiologic levels of PUFA or preventing their destruction. A combination of dietary manipulation with enzyme inhibitors or antioxidants may be required for effective control of PUFA levels in the adult brain. Future studies are needed to validate all enzyme inhibitors and nutraceuticals approaches that are presently used for AD management.

References

1 Brookmeyer R, Gray S, Kawas C (1998) Projections of Alzheimer's disease in the United States and the public health impact of delaying disease onset. *Am J Public Health* 88: 1337–1342

2 Alzheimer A (1907) Ueber eine eigenartige Erkrankung der Hirnrinde. *Allgemeine Zeitschrift fuer Psychiatrie und Psychisch-Gerichtliche Medizin* 64: 146–148

3 Cole GM, Masliah E, Shelton ER, Chan HW, Terry RD, Saitoh T (1991) Accumulation of amyloid precursor fragment in Alzheimer plaques. *Neurobiol Aging* 12: 85–91

4 Sutton ET, Thomas T, Bryant MW, Landon CS, Newton CA, Rhodin JA (1999) Amyloid-beta peptide induced inflammatory reaction is mediated by the cytokines tumor necrosis factor and interleukin-1. *J Submicrosc Cytol Pathol* 31: 313–323

5 Small GW, Ercoli LM, Silverman DH, Huang SC, Komo S, Bookheimer SY, Lavretsky H, Miller K, Siddarth P, Rasgon NL et al (2000) Cerebral metabolic and cognitive decline in persons at genetic risk for Alzheimer's disease. *Proc Natl Acad Sci USA* 97: 6037–6042

6 Guo Z, Cupples LA, Kurz A, Auerbach SH, Volicer L, Chui H, Green RC, Sadovnick AD, Duara R, DeCarli C et al (2000) Head injury and the risk of AD in the MIRAGE study. *Neurology* 54: 1316–1323

7 Graves AB, Rosner D, Echeverria D, Mortimer JA, Larson EB (1998) Occupational exposures to solvents and aluminium and estimated risk of Alzheimer's disease. *Occup Environ Med* 55: 627–633

8 Moceri VM, Kukull WA, Emanuel I, van Belle G, Larson EB (2000) Early-life risk factors and the development of Alzheimer's disease. *Neurology* 54: 415–420

9 Chapman PF, Falinska AM, Knevett SG, Ramsay MF (2001) Genes, models and Alzheimer's disease. *Trends Genet* 17: 254–261

10 Dal Forno G, Carson KA, Brookmeyer R, Troncoso J, Kawas CH, Brandt J (2002) APOE genotype and survival in men and women with Alzheimer's disease. *Neurology* 58: 1045–1050

11 Alvarez A, Toro R, Caceres A, Maccioni RB (1999) Inhibition of tau phosphorylating protein kinase cdk5 prevents beta-amyloid-induced neuronal death. *FEBS Lett* 459: 421–426

12 Koo EH, Lansbury PT, Jr., Kelly JW (1999) Amyloid diseases: abnormal protein aggregation in neurodegeneration. *Proc Natl Acad Sci USA* 96: 9989–9990

13 Nixon RA, Cataldo AM (1994) Free radicals, proteolysis, and the degeneration of neurons in Alzheimer disease: how essential is the beta-amyloid link? *Neurobiol Aging* 15: 463–469

14 Barger SW, Harmon AD (1997) Microglial activation by Alzheimer amyloid precursor protein and modulation by apolipoprotein E. *Nature* 388: 878–881
15 Jantzen PT, Connor KE, DiCarlo G, Wenk GL, Wallace JL, Rojiani AM, Coppola D, Morgan D, Gordon MN (2002) Microglial activation and beta-amyloid deposit reduction caused by a nitric oxide-releasing nonsteroidal anti-inflammatory drug in amyloid precursor protein plus presenilin-1 transgenic mice. *J Neurosci* 22: 2246–2254
16 Klegeris A, Walker DG, McGeer PL (1994) Activation of macrophages by Alzheimer beta amyloid peptide. *Biochem Biophys Res Commun* 199: 984–991
17 Xia W, Zhang J, Ostaszewski BL, Kimberly WT, Seubert P, Koo EH, Shen J, Selkoe DJ (1998) Presenilin 1 regulates the processing of beta-amyloid precursor protein C-terminal fragments and the generation of amyloid beta-protein in endoplasmic reticulum and Golgi. *Biochemistry* 37: 16465–16471
18 Cataldo AM, Peterhoff CM, Troncoso JC, Gomez-Isla T, Hyman BT, Nixon RA (2000) Endocytic pathway abnormalities precede amyloid beta deposition in sporadic Alzheimer's disease and Down syndrome: differential effects of APOE genotype and presenilin mutations. *Am J Pathol* 157: 277–286
19 Beffert U, Danik M, Krzywkowski P, Ramassamy C, Berrada F, Poirier J (1998) The neurobiology of apolipoproteins and their receptors in the CNS and Alzheimer's disease. *Brain Res Brain Res Rev* 27: 119–142
20 Montine KS, Bassett CN, Ou JJ, Markesbery WR, Swift LL, Montine TJ (1998) Apolipoprotein E allelic influence on human cerebrospinal fluid apolipoproteins. *J Lipid Res* 39: 2443–2451
21 Laffont I, Shuvaev VV, Briand O, Lestavel S, Barbier A, Taniguchi N, Fruchart JC, Clavey V, Siest G (2002) Early-glycation of apolipoprotein E: Effect on its binding to LDL receptor, scavenger receptor A and heparan sulfates. *Biochim Biophys Acta* 1583: 99–107
22 Shuvaev VV, Fujii J, Kawasaki Y, Itoh H, Hamaoka R, Barbier A, Ziegler O, Siest G, Taniguchi N (1999) Glycation of apolipoprotein E impairs its binding to heparin: Identification of the major glycation site. *Biochim Biophys Acta* 1454: 296–308
23 Du YS, Zhu H, Fu J, Yan SF, Roher A, Tourtellotte WW, Rajavashisth T, Chen X, Godman GC, Stern D et al (1997) Amyloid-beta peptide-receptor for advanced glycation endproduct interaction elicits neuronal expression of macrophage-colony stimulating factor: A pro-inflammatory pathway in Alzheimer disease. *Proc Natl Acad Sci USA* 94: 5296–5301
24 Sasaki N, Fukatsu R, Tsuzuki K, Hayashi Y, Yoshida T, Fujii N, Koike T, Wakayama I, Yanagihara R, Garruto R et al (1998) Advanced glycation end products in Alzheimer's disease and other neurodegenerative diseases. *Am J Pathol* 153: 1149–1155
25 Shuvaev VV, Laffont I, Serot JM, Fujii J, Taniguchi N, Siest G (2001) Increased protein glycation in cerebrospinal fluid of Alzheimer's disease. *Neurobiol Aging* 22: 397–402
26 Tabaton M, Perry G, Smith M, Vitek M, Angelini G, Dapino D, Garibaldi S, Zaccheo D, Odetti P (1997) Is amyloid beta-protein glycated in Alzheimer's disease? *Neuroreport* 8: 907–909

27 Cummings JL, Mendez MF (1997) Alzheimer's disease: Cognitive and behavioral pharmacotherapy. *Conn Med* 61: 543–552
28 Floyd RA (1999) Antioxidants, oxidative stress, and degenerative neurological disorders. *Proc Soc Exp Biol Med* 222: 236–245
29 Heininger K (2000) A unifying hypothesis of Alzheimer's disease. IV. Causation and sequence of events. *Rev Neurosci* 11: 213–328
30 Kril JJ, Halliday GM (2001) Alzheimer's disease: its diagnosis and pathogenesis. *Int Rev Neurobiol* 48: 167–217
31 Mantyh PW, GhilardiJR, Rogers S, DeMaster E, Allen CJ, Stimson ER, Maggio JE (1993) Aluminum, iron, and zinc ions promote aggregation of physiological concentrations of beta-amyloid peptide. *J Neurochem* 61: 1171–1174
32 Martyn CN (1992) The epidemiology of Alzheimer's disease in relation to aluminium. *Ciba Found Symp* 169: 69–79
33 McLachlan DR, Fraser PE, Dalton AJ (1992) Aluminium and the pathogenesis of Alzheimer's disease: A summary of evidence. *Ciba Found Symp* 169: 87–98
34 Montine KS, Reich E, Neely MD, Sidell KR, Olson SJ, Markesbery WR, Montine TJ (1998) Distribution of reducible 4-hydroxynonenal adduct immunoreactivity in Alzheimer disease is associated with APOE genotype. *J Neuropathol Exp Neurol* 57: 415–425
35 Polidori MC, Marvardi M, Cherubini A, Senin U, Mecocci P (2001) Heart disease and vascular risk factors in the cognitively impaired elderly: Implications for Alzheimer's dementia. *Aging (Milano)* 13: 231–239
36 Storey E, Masters CL (1995) Amyloid, aluminium and the aetiology of Alzheimer's disease. *Med J Aust* 163: 256–259
37 Wallin A, Sjogren M, Edman A, Blennow K, Regland B (2000) Symptoms, vascular risk factors and blood-brain barrier function in relation to CT white-matter changes in dementia. *Eur Neurol* 44: 229–235
38 Buckwalter JG, Sobel E, Dunn ME, Diz MM, Henderson VW (1993) Gender differences on a brief measure of cognitive functioning in Alzheimer's disease. *Arch Neurol* 50: 757–760
39 Borri P, Macchi G, Taramelli, M (1969) Fatty acid composition of brain cholesterol esters in some neurological disorders with myelin involvement. (Preliminary communication). *Neuropatol Pol* 7: 287–296
40 Dietschy JM, Turley SD (2001) Cholesterol metabolism in the brain. *Curr Opin Lipidol* 12: 105–112
41 Dobiasova M, Frohlich JJ (1999) Advances in understanding of the role of lecithin cholesterol acyltransferase (LCAT) in cholesterol transport. *Clin Chim Acta* 286: 257–271
42 Hill SA, McQueen MJ (1997) Reverse cholesterol transport – A review of the process and its clinical implications. *Clin Biochem* 30: 517–525
43 Koldamova RP, Lefterov IM, Ikonomovic MD, Skoko J, Lefterov PI, Isanski BA, DeKosky ST, Lazo JS (2003) 22R-Hydroxycholesterol and 9-cis-retinoic acid induce

ATP-binding cassette transporter A1 expression and cholesterol efflux in brain cells and decrease amyloid beta secretion. *J Biol Chem* 278: 13244–13256
44 Bretillon L, Siden A, Wahlund LO, Lutjohann D, Minthon L, Crisby M, Hillert J, Groth CG, Diczfalusy U, Bjorkhem I (2000) Plasma levels of 24S-hydroxycholesterol in patients with neurological diseases. *Neurosci Lett* 293: 87–90
45 Schonknecht P, Lutjohann D, Pantel J, Bardenheuer H, Hartmann T, Von Bergmann K, Beyreuther K, Schroder J (2002) Cerebrospinal fluid 24S-hydroxycholesterol is increased in patients with Alzheimer's disease compared to healthy controls. *Neurosci Lett* 324: 83–85
46 Ansari KA, Shoeman D (1988) Rapid separation of free fatty acids from neutral lipids in brain extracts. *J Chromatogr* 439: 453–458
47 Bazan NG Jr (1971) Changes in free fatty acids of brain by drug-induced convulsions, electroshock and anaesthesia. *J Neurochem* 18: 1379–1385
48 Bazan NG Jr, Rakowski H (1970) Increased levels of brain free fatty acids after electroconvulsive shock. *Life Sci* 9: 501–507
49 Kishimoto Y, Agranoff BW, Radin NS, Burton RM (1969) Comparison of the fatty acids of lipids of subcellular brain fractions. *J Neurochem* 16: 397–404
50 Mead JF, Dhopeshwarkar GA (1971) Types of fatty acids in brain lipids, their derivation and function. Lipids, malnutrition & the developing brain. *Ciba Found Symp* 59–72
51 Rodriguez de Turco EB, Bazan NG (1983) Changes in free fatty acids and diglycerides in mouse brain at birth and during anoxia. *J Neurochem* 41: 794–800
52 Harrington MG, Fonteh AN, Biringer RG, Amato H, Huhmer AF (2003) Defining cerebrospinal fluid proteins by electrophoresis, chromatography and mass spectrometry. *ICES 2003: Proteomics: Current Perspectives and Future Challenges* P37
53 Hagen AA, Gerber JN, Sweeley CC, White RP, Robertson JT (1977) Levels and disappearance of prostaglandin F2alpha in cerebral spinal fluid: a clinical and experimental study. *Stroke* 8: 672–675
54 Montine TJ, Sidell KR, Crews BC, Markesbery WR, Marnett LJ, Roberts LJ, Morrow JD (1999) Elevated CSF prostaglandin E2 levels in patients with probable AD. *Neurology* 53: 1495–1498
55 Westcott JY, Murphy RC, Stenmark K (1987) Eicosanoids in human ventricular cerebrospinal fluid following severe brain injury. *Prostaglandins* 34: 877–887
56 White RP, Leffler CW, Bada HS (1990) Eicosanoid levels in CSF of premature infants with post-hemorrhagic hydrocephalus. *Am J Med Sci* 299: 230–235
57 McIlwain H, Bachelard HS (eds) (1985) *Cerebral lipids in biochemistry and the central nervous system.* 5th ed, Churchill Livingstone, New York, 282–335
58 Jump DB (2002) The biochemistry of n-3 polyunsaturated fatty acids. *J Biol Chem* 277: 8755–8758
59 Horrocks LA, Yeo YK (1999) Health benefits of docosahexaenoic acid (DHA). *Pharmacol Res* 40: 211–225

60 Moore SA (1993) Cerebral endothelium and astrocytes cooperate in supplying docosahexaenoic acid to neurons. *Adv Exp Med Biol* 331: 229–233
61 Katz R, Hamilton JA, Spector AA, Moore SA, Moser HW, Noetzel MJ, Watkins, PA (2001) Brain uptake and utilization of fatty acids: Recommendations for future research. *J Mol Neurosci* 16: 333–335
62 Cao Y, Murphy KJ, McIntyre TM, Zimmerman GA, Prescott SM (2000) Expression of fatty acid-CoA ligase 4 during development and in brain. *FEBS Lett* 467: 263–267
63 Chilton FH, Fonteh AN, Surette ME, Triggiani M, Winkler JD (1996) Control of arachidonate levels within inflammatory cells. *Biochim Biophys Acta* 1299: 1–15
64 Fonteh AN, LaPorteT, Swan D, McAlexander MA (2001) A decrease in remodeling accounts for the accumulation of arachidonic acid in murine mast cells undergoing apoptosis. *J Biol Chem* 276: 1439–1449
65 Chilton FH, Murphy RC (1986) Remodeling of arachidonate-containing phosphoglycerides within the human neutrophil. *J Biol Chem* 261: 7771–7777
66 MacDonald JI, Sprecher H (1991) Phospholipid fatty acid remodeling in mammalian cells. *Biochim Biophys Acta* 1084: 105–121
67 Fonteh AN, Chilton FH (1993) Mobilization of different arachidonate pools and their roles in the generation of leukotrienes and free arachidonic acid during immunologic activation of mast cells. *J Immunol* 150: 563–570
68 Surette ME, Winkler JD, Fonteh AN, Chilton FH (1996) Relationship between arachidonate-phospholipid remodeling and apoptosis. *Biochemistry* 35: 9187–9196
69 Yamashita A, Sugiura T, Waku K (1997) Acyltransferases and transacylases involved in fatty acid remodeling of phospholipids and metabolism of bioactive lipids in mammalian cells. *J Biochem (Tokyo)* 122: 1–16
70 Edmond J (2001) Essential polyunsaturated fatty acids and the barrier to the brain: the components of a model for transport. *J Mol Neurosci* 16: 181–193
71 Dobiasova M, Frohlich J (1999) Advances in understanding of the role of lecithin cholesterol acyltransferase (LCAT) in cholesterol transport. *Clinica Chimica Acta* 286: 257–271
72 Pitas RE, Boyles JK, Lee SH, Hui D, Weisgraber KH (1987) Lipoproteins and their receptors in the central nervous system. Characterization of the lipoproteins in cerebrospinal fluid and identification of apolipoprotein B,E(LDL) receptors in the brain. *J Biol Chem* 262: 14352–14360
73 Albers JJ, Marcovina SM, Christenson RH (1992) Lecithin cholesterol acyltransferase in human cerebrospinal fluid: reduced level in patients with multiple sclerosis and evidence of direct synthesis in the brain. *Int J Clin Lab Res* 22: 169–172
74 Deeb SS, Failor RA, Brown BG, Brunzell JD, Albers JJ, Motulsky AG, Wijsman E (1992) Association of apolipoprotein B gene variants with plasma apoB and low density lipoprotein (LDL) cholesterol levels. *Hum Genet* 88: 463–470
75 Bjorkhem I, Andersson U, Ellis E, Alvelius G, Ellegard L, Diczfalusy U, Sjovall J, Einarsson C (2001) From brain to bile. Evidence that conjugation and omega-hydroxylation

are important for elimination of 24S-hydroxycholesterol (cerebrosterol) in humans. *J Biol Chem* 276: 37004–37010

76 Isch K, Lutjohann D, Von Bergmann K, Heun R (2003) The Role of 24S-Hydroxycholesterol in Alzheimer's disease. *J Nutr Health Aging* 7: 458–462

77 Lutjohann D, Papassotiropoulos A, Bjorkhem I, Locatelli S, Bagli M, Oehring RD, Schlegel U, Jessen F, Rao ML, Von Bergmann K et al (2000) Plasma 24S-hydroxycholesterol (cerebrosterol) is increased in Alzheimer and vascular demented patients. *J Lipid Res* 41: 195–198

78 Glomset JA (1968) The plasma lecithins: Cholesterol acyltransferase reaction. *J Lipid Res* 9: 155–167

79 Drayna D, Jarnagin AS, McLean J, Henzel W, Kohr W, Fielding C, Lawn R (1987) Cloning and sequencing of human cholesteryl ester transfer protein cDNA. *Nature* 327: 632–634

80 McLean J, Wion K, Drayna D, Fielding C, Lawn R (1986) Human lecithin-cholesterol acyltransferase gene: Complete gene sequence and sites of expression. *Nucleic Acids Res* 14: 9397–9406

81 Yang CY, Manoogian D, Pao Q, Lee FS, Knapp RD, Gotto AM, Jr., Pownall HJ (1987) Lecithin: Cholesterol acyltransferase. Functional regions and a structural model of the enzyme. *J Biol Chem* 262: 3086–3091

82 Yang CY, Kim TW, Pao Q, Chan L, Knapp RD, Gotto AM Jr, Pownall HJ (1989) Structure and conformational analysis of lipid-associating peptides of apolipoprotein B-100 produced by trypsinolysis. *J Protein Chem* 8: 689–699

83 Funke H, von Eckardstein A, Pritchard PH, Albers JJ, Kastelein JJ, Droste C, Assmann G (1991) A molecular defect causing fish eye disease: An amino acid exchange in lecithin-cholesterol acyltransferase (LCAT) leads to the selective loss of alpha-LCAT activity. *Proc Natl Acad Sci USA* 88: 4855–4859

84 Klein HG, Santamarina-Fojo S, Duverger N, Clerc M, Dumon MF, Albers JJ, Marcovina S, Brewer HB Jr (1993) Fish eye syndrome: A molecular defect in the lecithin-cholesterol acyltransferase (LCAT) gene associated with normal alpha-LCAT-specific activity. Implications for classification and prognosis. *J Clin Invest* 92: 479–485

85 Vaisman BL, Klein HG, Rouis M, Berard AM, Kindt MR, Talley GD, Meyn SM, Hoyt RF Jr, Marcovina SM, Albers JJ et al (1995) Over-expression of human lecithin cholesterol acyltransferase leads to hyper-alpha-lipoproteinemia in transgenic mice. *J Biol Chem* 270: 12269–12275

86 Albers JJ, Tu AY, Paigen B, Chen H, Cheung MC, Marcovina SM (1996) Transgenic mice expressing human phospholipid transfer protein have increased HDL/non-HDL cholesterol ratio. *Int J Clin Lab Res* 26: 262–267

87 Lukiw WJ, Bazan NG (2000) Neuroinflammatory signaling up-regulation in Alzheimer's disease. *Neurochem Res* 25: 1173–1184

88 Stephenson D, Rash K, Smalstig B, Roberts E, Johnstone E, Sharp J, Panetta J, Little S, Kramer R, Clemens J (1999) Cytosolic phospholipase A_2 is induced in reactive glia following different forms of neurodegeneration. *Glia* 27: 110–128

89 Smith WL, DeWitt DL, Garavito RM (2000) Cyclooxygenases: structural, cellular, and molecular biology. *Annu Rev Biochem* 69: 145–182

90 Kitamura Y, Shimohama S, Koike H, Kakimura J, Matsuoka Y, Nomura Y, Gebicke-Haerter PJ, Taniguchi T (1999) Increased expression of cyclo-oxygenases and peroxisome proliferator-activated receptor-gamma in Alzheimer's disease brains. *Biochem Biophys Res Commun* 254: 582–586

91 Urade Y, Watanabe K, Hayaishi O (1995) Prostaglandin D, E, and F synthases. *J Lipid Mediat Cell Signal* 12: 257–273

92 Smith WL, Meade EA, DeWitt DL (1997) Interaction of PGH synthase isozymes-1 and -2 with nonsteroidal anti-inflammatory drugs. *Adv Exp Med Biol* 400A: 189–196

93 Fonteh AN, Harrington MG (2002) Importance of proteins that metabolize polyunsaturated fatty acids in the pathogenesis of Alzheimer's disease. *From Genome to Proteome: Functional Proteomics, 5th Sienna meeting*, 91

94 Ho L, Purohit D, Haroutunian V, Luterman JD, Willis F, Naslund J, Buxbaum JD, Mohs RC, Aisen PS, Pasinetti GM (2001) Neuronal cyclo-oxygenase 2 expression in the hippocampal formation as a function of the clinical progression of Alzheimer disease. *Arch Neurol* 58: 487–492

95 Funk CD, Matsumoto T, Samuelsson B (1990) Molecular biology and cloning of archidonate 5-lipoxygenase. *Methods Enzymol* 187: 491–501

96 Samuelsson B (2000) The discovery of the leukotrienes. *Am J Respir Crit Care Med* 161: S2–S6

97 Murphy S, Rich G, Orgren KI, Moore SA, Faraci FM (1994) Astrocyte-derived lipoxygenase product evokes endothelium-dependent relaxation of the basilar artery. *J Neurosci* Res 38: 314–318

98 Nishiyama M, Watanabe T, Ueda N, Tsukamoto H, Watanabe K (1993) Arachidonate 12-lipoxygenase is localized in neurons, glial cells, and endothelial cells of the canine brain. *J Histochem Cytochem* 41: 111–117

99 Wu S, Moomaw CR, Tomer KB, Falck JR, Zeldin DC (1996) Molecular cloning and expression of CYP2J2, a human cytochrome P450 arachidonic acid epoxygenase highly expressed in heart. *J Biol Chem* 271: 3460–3468

100 Zeldin DC (2001) Epoxygenase pathways of arachidonic acid metabolism. *J Biol Chem* 276: 36059–36062

101 Buchet R, Pikula S (2000) Alzheimer's disease: its origin at the membrane, evidence and questions. *Acta Biochim Pol* 47: 725–733

102 Cummings BS, McHowat J, Schnellmann RG (2000) Phospholipase A(2)s in cell injury and death. *J Pharmacol Exp Ther* 294: 793–799

103 Bazan NG, Rodriguez de Turco EB (1980) Membrane lipids in the pathogenesis of brain edema: phospholipids and arachidonic acid, the earliest membrane components changed at the onset of ischemia. *Adv Neurol* 28: 197–205

104 MontineTJ, Markesbery WR, Morrow JD, Roberts LJ (1998) Cerebrospinal fluid F_2-isoprostane levels are increased in Alzheimer's disease. *Ann Neurol* 44: 410–413

105 Montine TJ, Kaye JA, Montine KS, McFarland L, Morrow JD, Quinn JF (2001) Cere-

brospinal fluid abeta42, tau, and f_2-isoprostane concentrations in patients with Alzheimer disease, other dementias, and in age-matched controls. *Arch Pathol Lab Med* 125: 510–512
106 MontineTJ, Beal MF, Cudkowicz ME, O'Donnell H, Margolin RA, McFarland L, Bachrach AF, Zackert WE, Roberts LJ, Morrow JD (1999) Increased CSF F_2-isoprostane concentration in probable AD. *Neurology* 52: 562–565
107 Liu T, Stern A, Roberts LJ, Morrow JD (1999) The isoprostanes: Novel prostaglandin-like products of the free radical-catalyzed peroxidation of arachidonic acid. *J Biomed Sci* 6: 226–235
108 Morrow JD, Roberts LJ (1997) The isoprostanes: Unique bioactive products of lipid peroxidation. *Prog Lipid Res* 36: 1–21
109 Nourooz-Zadeh J, Liu EH, Anggard E, Halliwell B (1998) F_4-isoprostanes: A novel class of prostanoids formed during peroxidation of docosahexaenoic acid (DHA). *Biochem Biophys Res Commun* 242: 338–344
110 Reich EE, Markesbery WR, Roberts LJ, Swift LL, Morrow JD, Montine TJ (2001) Brain regional quantification of F-ring and D-/E-ring isoprostanes and neuroprostanes in Alzheimer's disease. *Am J Pathol* 158: 293–297
111 Reich EE, Markesbery WR, Roberts LJ, Swift LL, Morrow JD, Montine TJ (2001) Quantification of F-ring and D-/E-ring isoprostanes and neuroprostanes in Alzheimer's disease. *Adv Exp Med Biol* 500: 253–256
112 Facchinetti F, Dawson VL, Dawson TM (1998) Free radicals as mediators of neuronal injury. *Cell Mol Neurobiol* 18: 667–682
113 Gutteridge JM, Halliwell B (2000) Free radicals and antioxidants in the year 2000. A historical look to the future. *Ann NY Acad Sci* 899: 136–147
114 Halliwell B (2001) Role of free radicals in the neurodegenerative diseases: therapeutic implications for antioxidant treatment. *Drugs Aging* 18: 685–716
115 Abe T, Tohgi H, Murata T, Isobe C, Sato C (2001) Reduction in asymmetrical dimethylarginine, an endogenous nitric oxide synthase inhibitor, in the cerebrospinal fluid during aging and in patients with Alzheimer's disease. *Neurosci Lett* 312: 177–179
116 Xia Y, Dawson VL, Dawson TM, Snyder SH, Zweier JL (1996) Nitric oxide synthase generates superoxide and nitric oxide in arginine-depleted cells leading to peroxynitrite-mediated cellular injury. *Proc Natl Acad Sci USA* 93: 6770–6774
117 Colton C, Wilt S, Gilbert D, Chernyshev O, Snell J, Dubois-Dalcq M (1996) Species differences in the generation of reactive oxygen species by microglia. *Mol Chem Neuropathol* 28: 15–20
118 Di Mascio P, Murphy ME, Sies H (1991) Antioxidant defense systems: The role of carotenoids, tocopherols, and thiols. *Am J Clin Nutr* 53: 194S–200S
119 Vendemiale G, Grattagliano I, Altomare E (1999) An update on the role of free radicals and antioxidant defense in human disease. *Int J Clin Lab Res* 29: 49–55
120 Wilson JX (1997) Antioxidant defense of the brain: a role for astrocytes. *Can J Physiol Pharmacol* 75: 1149–1163
121 Kontush A, Donarski N, Beisiegel U (2001) Resistance of human cerebrospinal fluid to

in vitro oxidation is directly related to its amyloid-beta content. *Free Radic Res* 35: 507–517
122 MontineTJ, Markesbery WR, Zackert W, Sanchez SC, Roberts LJ, Morrow JD (1999) The magnitude of brain lipid peroxidation correlates with the extent of degeneration but not with density of neuritic plaques or neurofibrillary tangles or with APOE genotype in Alzheimer's disease patients. *Am J Pathol* 155: 863–868
123 Gutteridge JM, Bannister JV (1986) Copper + zinc and manganese superoxide dismutases inhibit deoxyribose degradation by the superoxide-driven Fenton reaction at two different stages. Implications for the redox states of copper and manganese. *Biochem J* 234: 225–228
124 Kasarskis EJ, Ehmann WD, Markesbery WR (1993) Trace metals in human neurodegenerative diseases. *Prog Clin Biol Res* 380: 299–310
125 Takeda A, Akiyama T, Sawashita J, Okada S (1994) Brain uptake of trace metals, zinc and manganese, in rats. *Brain Res* 640: 341–344
126 Hock A, Demmel U, Schicha H, Kasperek K, Feinendegen LE (1975) Trace element concentration in human brain. Activation analysis of cobalt, iron, rubidium, selenium, zinc, chromium, silver, cesium, antimony and scandium. *Brain* 98: 49–64
127 Ince PG, Shaw PJ, Candy JM, Mantle D, Tandon L, Ehmann WD, Markesbery WR (1994) Iron, selenium and glutathione peroxidase activity are elevated in sporadic motor neuron disease. *Neurosci Lett* 182: 87–90
128 Sasi MM, Haider SS, el Fakhri M, Ghwarsha KM (1993) Micro-chromatographic analysis of lipids, protein, and occurrence of lipid peroxidation in various brain areas of vanadium exposed rats: A possible mechanism of vanadium neurotoxicity. *Neurotoxicology* 14: 57–63
129 Carney JM (2000) Oxidative stress leading to loss of critical proteases in Alzheimer's disease. An alternative view of the etiology of AD. Ann NY Acad Sci 924: 160–163
130 Exley C (1996) Amyloid, aluminum and the aetiology of Alzheimer's disease. *Med J Aust* 164: 252–253
131 Savory J, Garruto RM (1998) Aluminum, tau protein, and Alzheimer's disease: an important link? *Nutrition* 14: 313–314
132 Yokel RA (2000) The toxicology of aluminum in the brain: A review. *Neurotoxicology* 21: 813–828
133 Zapatero MD, Garcia DJ, Pascual F, Calvo ML, Escanero J, Marro A (1995) Serum aluminum levels in Alzheimer's disease and other senile dementias. *Biol Trace Elem Res* 47: 235–240
134 Hussein AS, Cantor AH, Johnson TH, Yokel RA (1990) Effect of dietary aluminum sulfate on calcium and phosphorus metabolism of broiler chicks. *Poult Sci* 69: 985–991
135 Aisen PS (2002) Evaluation of Selective COX-2 Inhibitors for the Treatment of Alzheimer's disease. *J Pain Symptom Manage* 23: S35–S40
136 O'Banion MK (1999) COX-2 and Alzheimer's disease: Potential roles in inflammation and neurodegeneration. *Expert Opin Investig Drugs* 8: 1521–1536
137 Xiang Z, Ho L, Valdellon J, Borchelt D, Kelley K, Spielman L, Aisen PS, Pasinetti GM

(2002) Cyclooxygenase (COX)-2 and cell cycle activity in a transgenic mouse model of Alzheimer's disease neuropathology. *Neurobiol Aging* 23: 327–334
138 Grundman M (2000) Vitamin E and Alzheimer disease: the basis for additional clinical trials. *Am J Clin Nutr* 71: 630S–636S
139 Halliday G, Robinson SR, Shepherd C, Kril J (2000) Alzheimer's disease and inflammation: A review of cellular and therapeutic mechanisms. *Clin Exp Pharmacol Physiol* 27: 1–8
140 Pasinetti GM (2001) Cyclooxygenase and Alzheimer's disease: Implications for preventive initiatives to slow the progression of clinical dementia. *Arch Gerontol Geriatr* 33: 13–28
141 Bing RJ, Lomnicka M (2002) Why do cyclo-oxygenase-2 inhibitors cause cardiovascular events? *J Am Coll Cardiol* 39: 521–522
142 Miyataka M, Rich KA, Ingram M, Yamamoto T, Bing RJ (2002) Nitric oxide, anti-inflammatory drugs on renal prostaglandins and cyclooxygenase-2. *Hypertension* 39: 785–789
143 Whelton A (1999) Nephrotoxicity of non-steroidal anti-inflammatory drugs: physiologic foundations and clinical implications. *Am J Med* 106: 13S–24S
144 Weggen S, Eriksen JL, Das P, Sagi SA, Wang R, Pietrzik CU, Findlay KA, Smith TE, Murphy MP, Bulter T et al (2001) A subset of NSAID's lower amyloidogenic Abeta42 independently of cyclo-oxygenase activity. *Nature* 414: 212–216
145 Ikeda Y, Matsumoto K, Dohi K, Jimbo H, Sasaki K, Satoh K (2001) Direct superoxide scavenging activity of non-steroidal anti-inflammatory drugs: Determination by electron spin resonance using the spin trap method. *Headache* 41: 138–141
146 Trimboli AJ, Waite BM, Atsumi G, Fonteh AN, Namen AM, Clay CE, Kute TE, High KP, Willingham MC, Chilton FH (1999) Influence of coenzyme A-independent transacylase and cyclo-oxygenase inhibitors on the proliferation of breast cancer cells. *Cancer Res* 59: 6171–6177
147 Bowman BA (1992) Acetyl-carnitine and Alzheimer's disease. *Nutr Rev* 50: 142–144
148 Maurer K, Ihl R, Dierks T, Frolich L (1997) Clinical efficacy of *Ginkgo biloba* special extract EGb 761 in dementia of the Alzheimer type. *J Psychiatr Res* 31: 645–655
149 Gilgun-Sherki Y, Melamed E, Offen D (2001) Oxidative stress induced-neurodegenerative diseases: The need for antioxidants that penetrate the blood brain barrier. *Neuropharmacology* 40: 959–975

Lipoxins and resolvins: Local mediators in endogenous anti-inflammation and resolution

Charles N. Serhan and Nan Chiang

The Center for Experimental Therapeutics and Reperfusion Injury, Department of Anesthesiology, Perioperative and Pain Medicine, Brigham and Women's Hospital, Harvard Medical School, Boston, MA 02115, USA

Introduction

Since the early concepts on the importance of the phagocyte in the host defense and inflammation as introduced by Metnivos more than 100 years ago (for which he was awarded the Nobel prize) the focus of research in inflammation was maintained with the credo – elucidate the chemical mediators that could cause/mediate the cardinal signs of inflammation – heat, redness, swelling, pain and loss of function. Inflammation initially involves the recognition of self and non-self by leukocytes. It is now clear that a diverse range of endogenous chemical mediators control these events and orchestrate the host response [1]. These small chemical signals regulate leukocyte traffic as well as the cardinal signs of inflammation. It is well established that the classic eicosanoids such as prostaglandins (PG) and leukotrienes (LT) play important roles and exert a wide range of actions in responses of interest in inflammation [2]. In recent years, the scope and range of chemical mediators identified has expanded considerably [1] to include novel lipid mediators, many new cytokines and chemokines, gases (i.e., nitric oxide and carbon monoxide), and reactive oxygen species as well as new roles for nucleotides as mediators such as adenosine [3–5] and the most recently uncovered of this class, namely inosine monophosphate (IMP), that also regulates neutrophil (PMN) trafficking [6].

A body of evidence demonstrated that endogenous mediators are generated to dampen the host response and orchestrate resolution [1, 7, 8]. In this regard, the lipoxins (LX) were the first to be identified and recognized as endogenous anti-inflammatory lipid mediators relevant in resolution in that they can function as "braking signals" or chalones in inflammation (Fig. 1) [9]. It is of particular interest that aspirin (ASA), a widely used non-steroidal anti-inflammatory drug with many beneficial properties [10] in addition to its well-appreciated ability to inhibit PG [11], also triggers the endogenous generation of 15-epimeric LX, termed aspirin-triggered LX (ATL). This occurs *via* acetylation of cyclooxygenase (COX)-2 at sites of inflammation *in vivo* [12] (*vide infra*) that carry anti-inflammatory and anti-pro-

Arachidonate Remodeling and Inflammation, edited by Alfred N. Fonteh and Robert L. Wykle

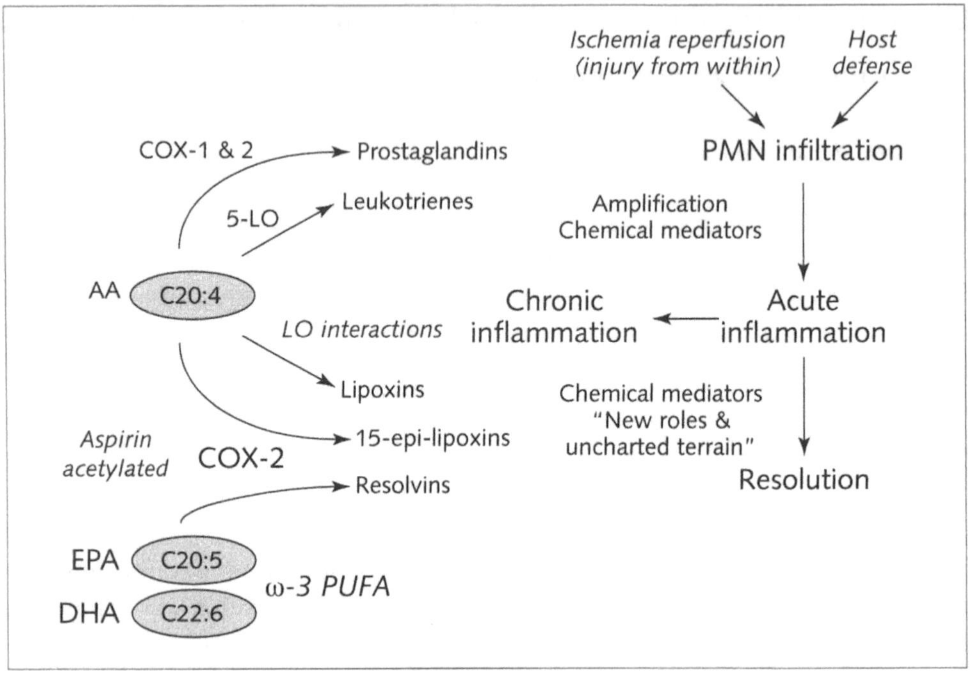

Figure 1
Resolution circuits in inflammation: Novel lipid mediators
During cell-cell interactions, lipid-derived mediators can be amplified (e.g., PG and LT) to enhance the actions of signal molecules, or braking signals (e.g., LX, ATL and resolvins) that can be generated via *cell-cell interactions to limit further PMN recruitment and promote resolution.*

liferative actions [13, 14]. This is a previously unappreciated and novel mechanism of drug action that has intriguing implications for targeted drug design. But more importantly, they help to further illustrate the importance of endogenous generation of lipid mediators with anti-inflammatory properties.

As a class, LX, ATL and their analogs possess physiologic, pathophysiologic and pharmacological actions in several target tissues. Each action of lipoxins is stereoselective in that changes in potencies accompany double bond isomerization and change in alcohol chirality (R or S) at key positions, as well as selective dehydrogenation of alcohols and reduction of double bonds. The self-limited impact of LX in the local micro-environment suggests that they contribute to resolution of injury sites and/or resolve inflammatory loci by regulating further recruitment of PMN and stimulating monocyte migration to promote healing and remodeling. LXA_4 stimulates rapid lipid remodeling within seconds and releases arachidonic acid within PMN but without oxygenation, which is sensitive to pertussis toxin (PTX) treatment [15, 16]; findings that pointed to the involvement of a G protein-coupled receptor (GPCR) in the actions of LX on human leukocytes. This specific GPCR was identified and cloned in human and mouse, and denoted LXA_4 receptor (ALX). Together, they were identified as the first cloned lipoxygenase (LO)-derived eicosanoid receptors.

The traditional approach to develop anti-inflammatory drugs, as in other human conditions amenable to pharmacologic interventions, is the use of biosynthesis inhibitors and receptor antagonists of pro-inflammatory mediators, which indeed have enjoyed both considerable clinical and commercial successes [1, 17], but are not without significant unwanted side effects [18–20]. Hence, the emergence of endogenous pathways and cellular mechanisms involved in counter-regulation of responses that can lead to tissue injury and acute inflammation not only charts relatively unappreciated sides of human biology [21, 22], but also provides an opportunity to explore new therapeutic approaches based on these novel endogenous mechanisms that may reduce the possibilities for unwanted toxic side effects and help control inflammation with a high degree of precision.

Biosynthesis of lipoxins and aspirin-triggered lipoxins

Transcellular biosynthesis of LXs: The role of cell-cell interactions

Formation of LXs are promoted during platelet-leukocyte interactions and/or platelet-leukocyte micro-aggregates [23] by transcellular conversion of the leukocyte 5-lipoxygenase (LO) epoxide product LTA_4 (Fig. 2, right side). Once thought to be solely an intracellular intermediate in LT production, it is now clear that LTA_4 released by activated leukocytes is available for enzymatic conversion by neighboring cell types [5, 24]. When platelets are adherent, their 12-LO converts LTA_4 to lipoxin A_4 and B_4. For a review and mechanistic details with recombinant 12-LO, see [22]. Hence it is important to note that human platelets, which do not produce LX on their own, become a major source of LX, given their abundance *in vivo* and their highly active 12-LO.

The second classical pathway for LX production is initiated by 15-LO (Fig. 2, left side) in airway epithelial cells, monocytes or eosinophils, which up-regulate their 15-LO when exposed to cytokines such as IL-4 or IL-13 [25, 26]. The 15-LO, by definition, inserts molecular oxygen at the carbon 15 position of, for example, arachidonic acid, in the "S" configuration. When these cell types are activated, they generate and release 15S-HETE, which is rapidly taken up and converted by PMN to LX *via* the action of their 5-LO. This event not only leads to LX biosynthesis, but also "turns off" LT formation.

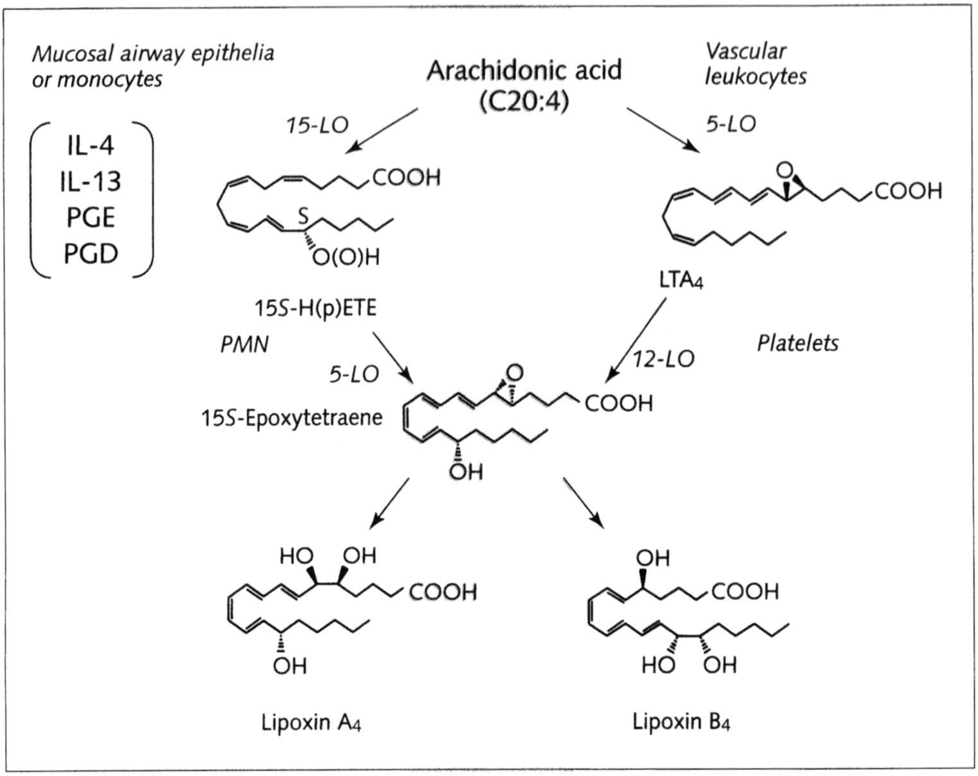

Figure 2
Transcellular lipoxin biosynthesis
During cell-cell interactions, lipoxins can be amplified by transcellular biosynthesis via *the interactions of two or more cell types. Two main pathways appear to be used in human cells and tissues to generate LX. This event also blocks LT formation and therefore regulates leukocytes.*

Biosynthesis of aspirin-triggered lipoxins *via* aspirin-acetylated COX-2

We sought evidence for alternate explanations for ASA's therapeutic actions because many beneficial new actions have been documented in recent clinical studies. These new potential therapeutic indicators for ASA include decreasing incidence of lung, colon, and breast cancer (reviewed by Levy [27]), and prevention of cardiovascular diseases [28]. Inhibition of cyclo-oxygenase and biosynthesis of prostaglandins can account for many of ASA's therapeutic properties [29]; however, its ability to regulate neutrophil-mediated inflammation or cell proliferation remains of interest. Along these lines, we uncovered a new action of aspirin that involves COX-2-bearing cells such as vascular endothelial cells or epithelial cells and their co-activation with PMN (Fig. 3). Hence, inflammatory stimuli (i.e., TNF-α, LPS, etc.) induce COX-2 to generate 15R-HETE when ASA is administered [13]. This intermediate carries a carbon-15 alcohol in the R configuration that is rapidly converted by 5-LO in activated PMN to 15 epimeric-LX, or LX that carry their 15 position alcohol in the R configuration [22] rather than 15S native LX, which in humans can result from LO:LO interaction. 5-LO conversion of 15R-HETE also results in inhibition of LT biosynthesis [14]. 15R-HETE is a major product of arachidonic acid in several cell types when COX-2 is up-regulated after acetylation by ASA. Thus, it is possible that aspirin can regulate the *in vivo* production of LT by 15R-HETE conversion to 15-epi-LX, and 15-epi-LX can in turn also regulate the cellular actions of LT.

LXB_4 is a positional isomer of LXA_4, carrying alcohol groups at carbon 5S, 14R, and 15S positions, instead of the C-5S, 6R, and 15S positions present in LXA_4. Aspirin-triggered LXB_4 carries a 15R alcohol, hence 15-epi-LXB_4. Although LXA_4 and LXB_4 show similar activities in some biologic systems [30], in many others each shows distinct actions ([31] and reviewed in [22]). 15-epi-LXB_4, for example, is a more potent inhibitor of cell proliferation than LXA_4 or 15-epi-LXA_4 [22].

Generation of LXA_4 and 15-epi-LXA_4 in animal models and in human diseases

LXA_4 is produced *in vivo* during the course of inflammation such as in an experimental immune complex glomerulonephritis model [32, 33] and in pleural exudate upon allergen challenge in rats [34]. Also, endogenous LXA_4 was also produced in ischemic lungs and elevated by reperfusion in a hind limb ischemia reperfusion model [35]. A recent report demonstrated that LXA_4 is generated during microbial infection reported in a *T. gondii*-exposed murine model [36, 37] as well as in a murine model of asthma [38]. In addition, LXA_4 is formed in rat brain and elevated in focal cerebral ischemia [39]. In human subjects, a reduction and alteration in LX generation was found in patients with chronic liver disease [40] and chronic myelogenous leukemia [41–45]. These diseases contrast with recent findings that LXA_4 production is up-regulated in localized juvenile periodontitis [46] and mild asthma [47], as well as following atherosclerotic plaque rupture [48], and with nasal polyps [49].

Recently, using a newly developed specific enzyme-linked immunosorbent assay (ELISA) method and liquid chromatography tandem mass spectrometry (LC-MS-MS) system [12], 15-epi-LXA_4 could be detected *in vivo*. For examples, 15-epi-LXA_4 was generated in murine peritonitis [12] and murine dorsal air-pouches [50], and also was detected in rat kidney [33] and liver [51] in an aspirin-dependent manner. Along these lines, it was reported that ASA rapidly up-regulates COX-2 expression in the stomach and causes a significant increase in gastric 15-epi-LXA_4 pro-

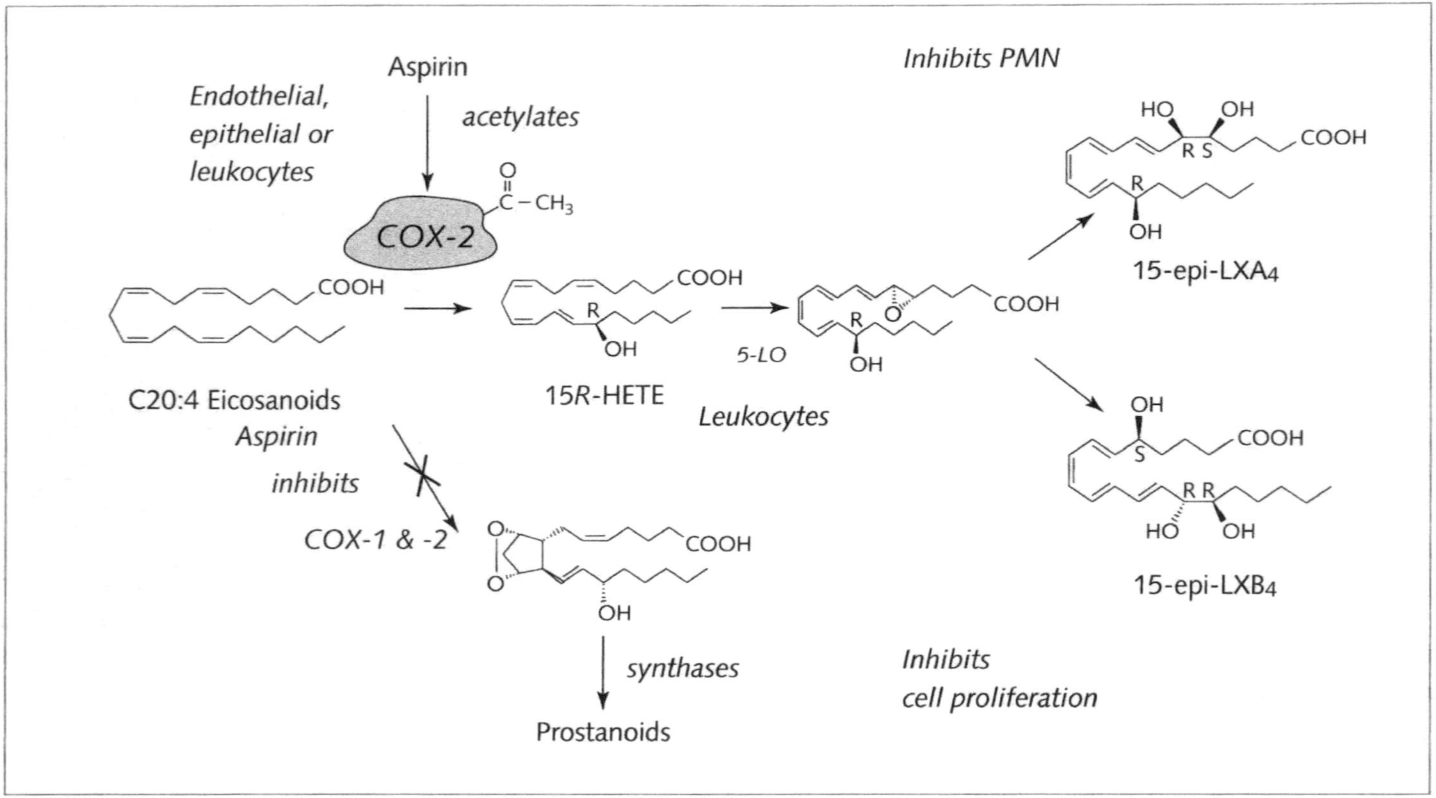

Figure 3
Generation of aspirin-triggered lipid mediators
15-epi-LXs are generated when COX-2 is up-regulated and acetylated by ASA. Irreversible acetylation of COX-2 by aspirin changes the enzyme's product from prostaglandin intermediate to precursors of ATL. The acetylated COX-2 remains catalytically active to generate 15R-HETE.

Table 1 - Lipoxin A_4 and human diseases

Organ/System	**Impact** *in vivo*	**Refs.**
Hematologic	Defect in LX production with cells from chronic myeloid leukemia patients in blast crisis	[42]
	LX stimulate nuclear form of PKC in erythroleukemia cells	[121]
	Formation of LX by granulocytes from eosinophilic donors	[122]
Cardiovascular	Angioplasty-induced plaque rupture triggers LX formation	[48]
Dermatologic	LXA_4 regulates delayed hypersensitive reactions in skin	[123]
Pulmonary	LXA_4 detected in bronchoalveolar lavage fluids from patients with pulmonary disease and asthma	[124]
	Production of LX by nasal polyps and bronchial tissue	[49]
	LXA_4 inhalation shifts and reduces LTC_4-induced contraction is asthmatic patients	[125]
	Aspirin-intolerant asthmatics display a lower biosynthetic capacity than aspirin-tolerant patients	[53]
	LXA_4 inhibits IL-8 release by monocytes from asthma patients	[47]
Hepatic	LX generation decreased in cirrhotic patients	[40]
Rheumatoid arthritis	LX levels increase with recovery	[126]
Oral	LXA_4 production is up-regulated in localized juvenile periodontitis	[46]

duction in rats [52]. These methods (e.g., LC/MS/MS and ELISA) were used to evaluate ATL and LXA_4 formation in ASA-tolerant and ASA-intolerant asthmatics and their relation to leukotriene C_4. Of interest, the ASA-tolerant subjects generated both LX and ATL, but the ASA-intolerant patients proved to have a diminished capacity to generate ATL and LX upon ASA challenge [53]. The lower levels of these potentially protective mediators could contribute to the pathobiology of this chronic disorder in that the disease state is not only characterized by the overproduction of pro-inflammatory mediators but the loss or reduction in LX and ATL that may keep inflammation in check. Together, this result indicates that alterations in LX and ATL levels may be linked to the pathophysiology of several human diseases and may display local organ-specific functions that stand apart from their roles in inflammation and within local inflammatory lesions (Tab. 1).

Stable analogs of LXA_4 and 15-epi-LXA_4 resist rapid metabolic inactivation

Enzymatic inactivation of LXs: Structure requirements for LXA_4 anti-inflammatory actions

As other autacoids, lipoxins are rapidly generated in response to stimuli, act locally and then are rapidly enzymatically inactivated. The major route of LX inactivation is through dehydrogenation by monocytes that convert LXA_4 to 15-oxo-LXA_4, followed by specific reduction of the double bond adjacent to the ketone [21]. 15-hydroxy/oxo-eicosanoid oxidoreductase (15-PGDH) catalyzes the oxidation of LXA_4 to 15-oxo-LXA_4 (Fig. 4). This compound is biologically inactive and is further converted to 13,14-dihydro-15-oxo-LXA_4 by the action of LXA_4/PGE 13,14-reductase/LTB_4 12-hydroxydehydrogenase (PGR/LTB_4DH). Moreover, reduction of the 15-oxo-group by 15-PGDH yields 13,14-dihydro-LXA_4, revealing an additional catalytic activity for this enzyme [54]. LXB_4 can also be dehydrogenated by 15-PGDH at carbon-5 to produce 5-oxo-LXB_4, therefore sharing a common route of inactivation [55]. It has recently been shown that 15-oxo-LXA_4 is also produced from LXA_4 in mouse whole blood [56] suggesting that the mouse shares with the human a common pathway for LXA_4 inactivation.

Each action of lipoxins is stereo-selective in that changes in potencies accompany double bond isomerization and change in alcohol chirality (R or S) at key positions as well as selective dehydrogenation of alcohols and reduction of double bonds. For example, the 15-hydroxyl group is important for anti-inflammatory properties since aspirin-triggered LXA_4 (15R-LXA_4) with the 15-hydroxyl group in the R-configuration as well as 15(R/S)-methyl-LXA_4 have been established in several experimental settings to be more potent than native LXA_4 (15S-LXA_4) *in vitro* and *in vivo* [57, 58]. Also, both 15-oxo-LXA_4 [54] and 15-deoxy-LXA_4 [21] are biologically inactive in inhibiting superoxide anion generation and transmigration in PMN, respectively. The 13,14-double bond is important since 13,14-dihydro-LXA_4 proved to be inactive in inhibiting superoxide anion generation [54]. These pharmacophores for LX's anti-inflammatory action are also required for their interaction with ALX since these biologically inactive isomers (e.g., 15-oxo-LXA_4, 15-deoxy-LXA_4 and 13,14-dihydro-LXA_4) did not bind to ALX, whereas the active ones (e.g., 15R-LXA_4 and 15(R/S)-methyl-LXA_4) give specific binding to ALX, as demonstrated by specific [^{3}H]-LXA_4 binding (see Fig. 5).

Design of stable analogs of LXA_4 and 15-epi-LXA_4

Since LXs are rapidly transformed and inactivated by monocytes, and, potentially, other cells *in vivo*, it was highly desirable to design LX analogs that could resist this form of metabolism, maintain their structural integrity, and potentially enhance

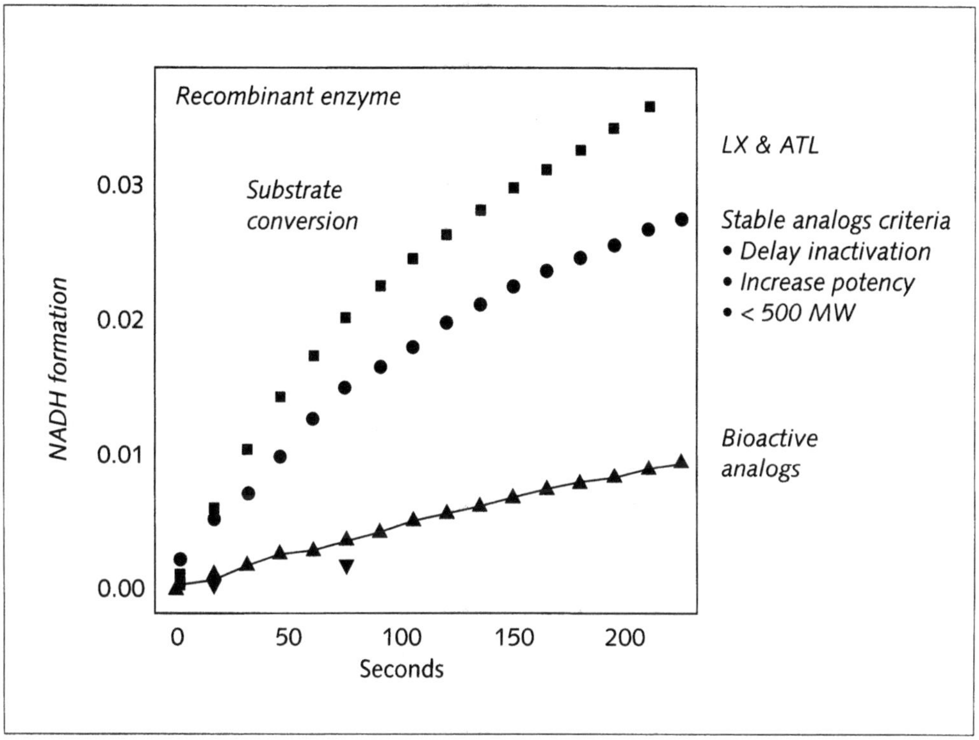

Figure 4
LX and ATL stable analogs resist rapid inactivation
The initial step in LXA_4 inactivation is dehydrogenation of the 15-hydroxyl group, catalyzed by 15-hydroxy prostaglandin dehydrogenase (15-PGDH). Metabolic stable analogs of both LXA_4 and 15-epi-LXA_4 were designed to resist rapid inactivation at carbon 15 as well as the ω-end of the molecule.

beneficial bioactions. LX analogs were constructed with specific modifications of the native structures of LXA_4 and LXB_4, such as the addition of methyl groups on carbon-15 and carbon-5 of LXA_4 and LXB_4 structures, respectively, to block dehydrogenation by 15-PGDH. For example, 15(R/S)-methyl-LXA_4 is a racemic stable analog of both LXA_4 and 15-epi-LXA_4. Additional analogs of LXA_4 were synthesized with a phenoxy group bonded to carbon-16 and replacing the ω-end of the molecule. This design permits 16-phenoxy-LXA_4 to resist potential ω-oxidation and to be protected from dehydrogenation *in vivo*. Fluoride was added to the para-position of the phenoxy ring to make 16-(para-fluoro)-phenoxy-LXA_4 to hinder degradation of the phenoxy ring. The aspirin-triggered 15-epi counterpart of 16-(para-fluoro)-phenoxy-LXA_4, namely 15-epi-16-(para-fluoro)-phenoxy-LXA_4, was

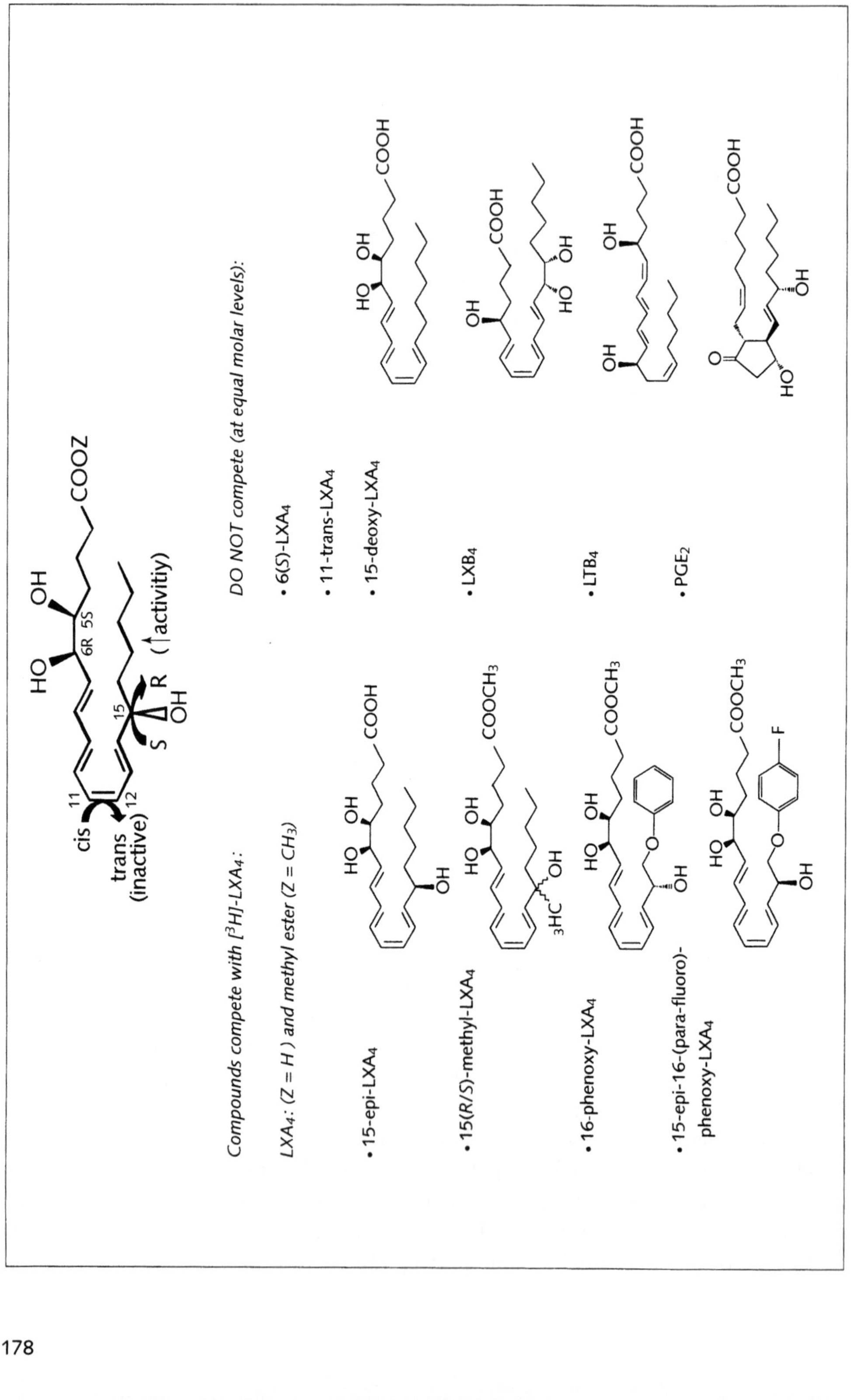
HO
OH
6R
5S
COOZ
11
12
cis
trans
(inactive)
15
S
R
OH
(↑activitiy)
Compounds compete with [3H]-LXA4:
LXA4: (Z = H) and methyl ester (Z = CH3)
• 15-epi-LXA4
• 15(R/S)-methyl-LXA4
• 16-phenoxy-LXA4
• 15-epi-16-(para-fluoro)-phenoxy-LXA4
DO NOT compete (at equal molar levels):
• 6(S)-LXA4
• 11-trans-LXA4
• 15-deoxy-LXA4
• LXB4
• LTB4
• PGE2
COOH
COOCH3
3HC
F
O

also synthesized. These modifications not only prolong the half-life of the compounds in blood but also enhance their bioavailabilities as well as bioactivities (Fig. 4) [56].

When compared to the native LXs, the ATLs are less effectively converted *in vitro* to their 15-oxo-metabolite [21]. This indicates that the dehydrogenation step is highly stereo-specific and suggests that, when ATLs are generated *in vivo*, their biologic half-life is increased by about two-fold greater than that of native LXA_4, thereby enhancing their ability to evoke bioactions. Hence, biologically stable analogs of LX and ATL can be engineered to enhance their bioactions, which suggests that they are useful tools, and offers leads for developing novel therapeutic modalities. These analogs proved to be active and also interact directly with ALX (Fig. 5).

Biological properties of LX and ATL

Actions of LX and ATL *in vitro*

LXA_4 and ATL display counter-regulatory roles in various cell types *in vitro* (Fig. 6, Tabs. 2 and 3). With human peripheral blood leukocytes (Tab. 2), LXA_4 inhibits both isolated PMN and eosinophil chemotaxis *in vitro* in the nanomolar range [59, 60] and blocks human natural killer (NK) cell cytotoxicity in a stereo-selective fashion [22] as well as stimulate myeloid bone marrow-derived progenitors [45]. In cell-cell interaction systems, LXA_4 inhibits PMN transmigration across both endothelial and epithelial monolayers [30, 61] *via* actions on both cell types (i.e., PMN and endothelial cells, PMN and epithelial cells). LXA_4 stimulates chemotaxis and adherence in monocytes but no apparent "pro-inflammatory" responses of these cells *in vitro* or *in vivo*, findings that may relate to the recruitment of monocytes to sites of

Figure 5
Ligand binding specificity of human ALX
LXA_4 interaction with ALX is highly stereo-specific, that is the 5S, 6R-orientation of the two hydroxyl groups as well as 11-cis double bond conformation are essential for bioactions. 15-epi-LXA_4 (an aspirin-triggered lipoxin, ATL) carries a C-15 alcohol at the R configuration, opposite to the S configuration in native LXA_4, and was shown to have higher potency than native LXA_4 in certain bioassays. In 15(R/S)-methyl-LXA_4, hydrogen at C-15 was replaced by a methyl group as a racemate at C-15. 16-phenoxy-LXA_4 has a phenoxyl group at C-16. These compounds, which are more resistant to rapid dehydrogenation by 15-PGDH than native LXA_4, compete with [^{3}H]-LXA_4 specific binding on PMN as well as recombinant ALX and are potent inhibitors for PMN functions in vitro *and* in vivo.

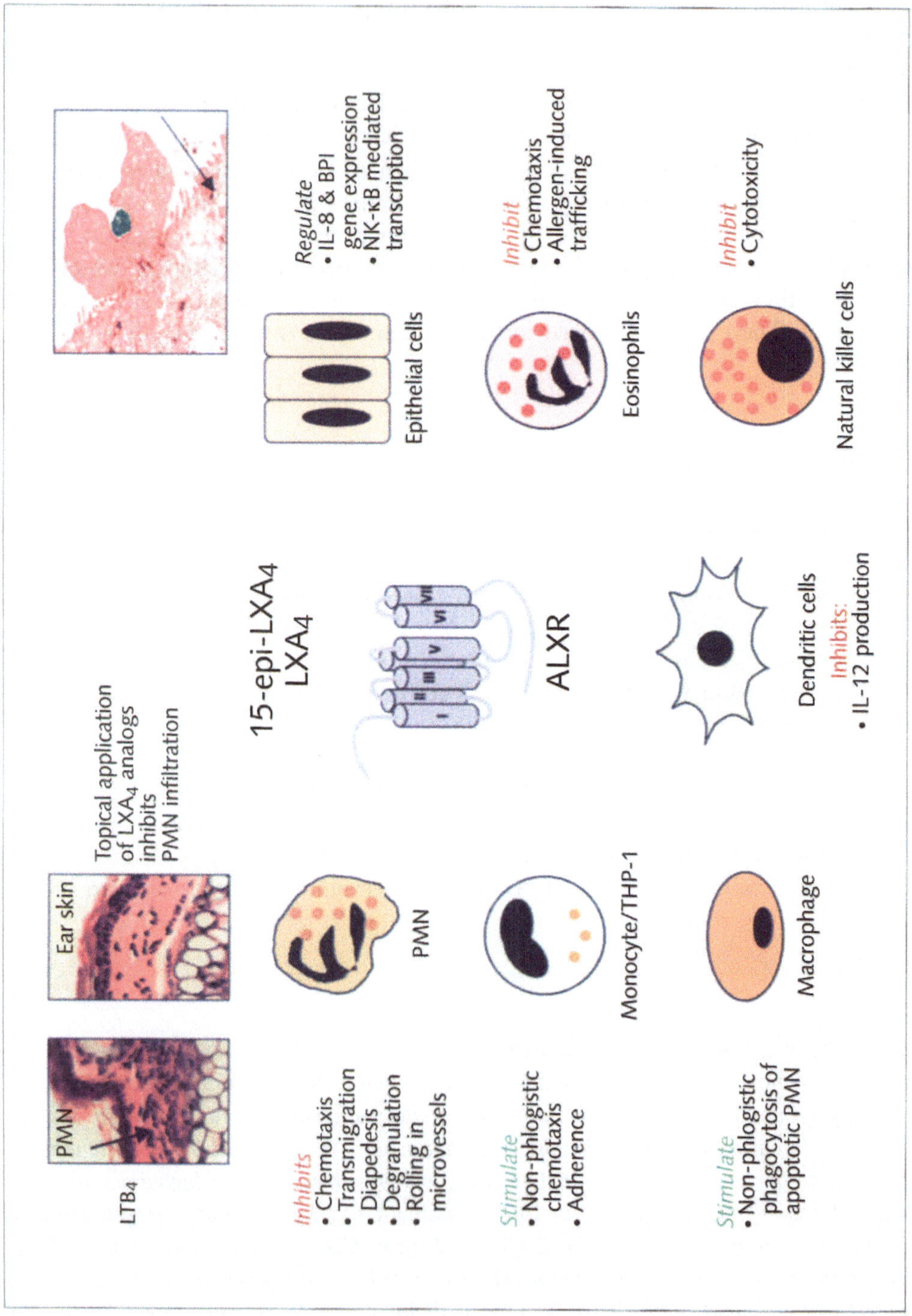
LTB4
PMN
Ear skin
Topical application of LXA4 analogs inhibits PMN infiltration
15-epi-LXA4
LXA4
ALXR
Inhibits
• Chemotaxis
• Transmigration
• Diapedesis
• Degranulation
• Rolling in microvessels
PMN
Stimulate
• Non-phlogistic chemotaxis
• Adherence
Monocyte/THP-1
Stimulate
• Non-phlogistic phagocytosis of apoptotic PMN
Macrophage
Dendritic cells
Inhibits:
• IL-12 production
Epithelial cells
Regulate
• IL-8 & BPI gene expression
• NK-κB mediated transcription
Eosinophils
Inhibit
• Chemotaxis
• Allergen-induced trafficking
Natural killer cells
Inhibit
• Cytotoxicity

wound healing and clearance. Indeed, LX and ATL stimulate the uptake of apoptotic PMN by macrophages in a non-phlogistic fashion [62].

With human enterocytes and fibroblast, LXA_4 regulates pro-inflammatory cytokine release as well as gene expression (see below). These counter-regulatory actions are initiated *via* unique cell surface receptors on leukocytes and enterocytes as well as fibroblasts. In hepatocytes, ATL significantly reduces PPARα and cytokine-induced neutrophil chemo-attractant (CINC)-1 [63]. With other cell types such as endothelium and mesangial cells, LXA_4 evokes bioactions and interacts with a subclass of peptido-LT receptors (CysLT1) [reviewed recently in Ref. 22]. The leukocyte receptors are physiologically and pharmacologically distinct and evoke selective actions on each type of leukocyte tested to date.

Actions of LX and ATL *in vivo*

The metabolically stable analogs of LX and ATL have been examined in various experimental animal models and summarized in Table 4.

Acute inflammation

In dermal inflammation, these LX stable analogues when applied topically to mouse ears inhibit both PMN infiltration and vascular permeability changes in a concentration-dependent fashion [57, 58]. Also, the fluorinated analog of ATL, denoted ATLa, at levels as low as ~24 nmol/mouse, potently inhibited TNF-α-induced leukocyte recruitment into the dorsal air-pouch [56]. Inhibition was evident by either local intra-air-pouch delivery (~77% inhibition) or *via* systemic delivery by intravenous injection (~85% inhibition) and proved more potent than local delivery of ASA. Recently, using a thioglycollate-induced peritonitis, ATL analogs were shown to rapidly promote macrophage phagocytosis of apoptotic PMN, supporting a role for LXs as pro-resolution signals in inflammation [64].

Figure 6
LXA_4 actions via *ALX in leukocytes and enterocytes*
Actions of LXA_4 in leukocytes (reviewed in [22]) and human epithelial cells [74, 102]. (Upper left panel) Ear biopsies: Inhibition of LTB_4-induced PMN infiltration into mouse ear by topical application of LXA_4 analogs in acute skin inflammation [58]. PMN is indicated by an arrow. (Upper right panel) Photomicrograph: Internalization of Salmonella typhimurium *(shown in green) by intestinal epithelium (indicated by an arrow). In response to this gastrointestinal pathogen, intestinal epithelium secretes chemokines, which promote neutrophil infiltration. This chemokine (IL-8) secretion can be down-regulated by LXA_4 analogs.*

Table 2 - Lipoxin A_4 and ATL on myeloid cells

Cell type/tissue	Action	Refs.
Whole blood	Down-regulate CD11/CD18, prevent shedding of L-selectin and reduce peroxynitrite generation on PMN, monocytes and lymphocytes	[98, 127]
Neutrophils	Inhibit chemotaxis, adherence and transmigration	[21, 59]
	Inhibit PMN-epithelial and endothelial cell interactions	[30, 61]
	Block superoxide anion generation	[97]
	Inhibit CD11b/CD18 expression and IP_3 formation	[15, 93];
	Inhibit peroxynitrite generation	[98]
	Attenuate AP-1 and NF-κB accumulation and inhibit IL-8 gene expression	[98]
Monocytes	Stimulate chemotaxis and adhesion to laminin without increase in cytotoxicity	[82]
	Inhibit peroxynitrite generation	[98]
	Inhibit IL-8 release by cells obtained from asthma patients	[47]
Macrophages	Stimulate non-phlogistic phagocytosis of apoptotic PMN	[62, 64]
Dendritic cells	Inhibit IL-12 production	[36, 37]
Eosinophils	Inhibit migration/chemotaxis	[65]
NK cells	Block cytotoxicity	[128]
Myeloid progenitors	Stimulate myeloid bone marrow-derived progenitors	[45]

Table 3 - Actions of lipoxin A_4 and ATL on resident cell types

Cell type/tissue	Action	Refs.
Enterocytes	Inhibit TNF-α-induced IL-8 expression and release	[74]
	Inhibit *Salmonella typhimurium*-induced IL-8	[102]
Fibroblasts	Inhibit IL-1β-induced IL-6, IL-8 and MMP-3 production	[83]
Endothelia (HUVEC)	Stimulate protein kinase C-dependent prostacyclin formation	[129]
	Block P-selectin expression	[68]
Mesangial cells	Inhibit LTD4-induced proliferation	[108]
Pulmonary artery	Induce relaxation and reverses pre-contraction by PGF_2 or endothelin-1	[130]
Hepatocytes	Reduce PPARα and CINC-1 levels	[63]
Bronchi	Relaxation after pre-contraction by blocking peptido-leukotrienes in human airway	[125]

Table 4 - Actions of lipoxin A_4 and ATL in vivo *in disease models*

System	Model	Action	Refs.
Acute inflammation (mouse)	Peritonitis	Promote macrophage phagocytosis of PMN	[64]
	Dorsal air pouch	Inhibit TNF-α-induced PMN infiltration	[56]
	Dermal inflammation	Inhibit LTB_4 induced PMN infiltration into ear skin	[58]
Lung (mouse)	Asthma	Inhibit airway hyper-responsiveness and pulmonary inflammation	[38]
	Ischemia and reperfusion	Inhibit PMN infiltration into lungs	[35]
Kidney (rat)	Glomerulonephritis	Antagonize the effects of LTC_4 and LTD_4	[139]
	Ischemia and reperfusion	Reduce PMN infiltration and is protective against acute renal failure	[69]
Microbial infection	*T. gondii* infection (mouse)	Stun and block dendritic cell migration and IL-12 production	[36, 37]
	A. costaricenssis infection (rat)	Shorten the duration of pleural exudation	[34]
Endothelium (mouse)	Granuloma	Reduce angiogenic phenotype	[71]
Eye (rabbit)		Lower intraocular pressure	[72, 73]
Gastrointestinal tract	Colitis (mouse)	Attenuate induction of pro-inflammatory gene expression and reduce severity of DSS-induced colitis	[70]
	Aspirin-induced gastric damage (rat)	Reduce the severity of gastric damage and suppress aspirin-induced leukocyte adherence	[52]

Lung (endogenous *versus* exogenous stimuli)

Since LXA_4 and ATL selectively regulate leukocyte responses, they were tested in BLT1 transgenic mice that give dramatically increased PMN trafficking to lungs after hind limb ischemia-reperfusion. Despite excessive PMN recruitment in BLT1 transgenic mice, intravenous injection of ATL sharply diminished reperfusion-initiated PMN trafficking to lungs, revealing a novel protective role for LX and ATL in stress responses that have applications in perioperative medicine [35]. With exogenous allergen challenge in a murine model of asthma, LXA_4 biosynthesis and ALX expression was increased. Transgenic expression of human ALX leads to significant inhibition of pulmonary inflammation and eicosanoid-initiated eosinophil infiltration. Administration of a stable analog of ATL blocks both airway hyper-responsiveness and pulmonary inflammation, highlighting a unique counter-regulatory role for LX in airway responses [38]. Along these lines, LXA_4/ATL analogues dramatically blocked allergic pleural eosinophil influx on allergen-induced eosinophilic pleurisy in sensitized rats [65].

Kidney

LXA_4 and ATL display physiological and pathophysiological roles in kidney including regulation of renal functions by increasing glomerular filtration rate and renal plasma flow rate [66, 67]. LXA_4 also antagonizes the effects of LTC_4 and LTD_4 on the glomerular microcirculation [139]. In ischemic acute renal failure, ATL stable analog shows functional and morphologic protection and reduces PMN infiltration. In addition, ATL-treated mice also display increased renal mRNA levels for suppressors of cytokine signaling (SOCS)-1 and SOCS-2 [69].

Gastrointestinal tract

During aspirin-induced gastric damage in rats, LXA_4 exhibits potent protective action on gastric mucosa by reducing the severity of gastric damage and suppressing aspirin-induced leukocyte adherence [52]. Also, ATL is protective in intestinal inflammation in a mouse model of dextran sodium sulfate-induced colitis. Oral administration of ATL analog (10 μg/day) significantly reduced the weight loss, hematochezia and mortality that characterize DSS-induced colitis [70].

Microbial infection

In *Angiostrongylus costaricensis* infected rats, two stable LXA_4 analogues did not alter the magnitude of pleural exudation response, but clearly shortened its duration. These results indicate that the early resolution of allergic pleural edema observed during *A. costaricensis* infection coincided with a selective local eosinophilia and seemed to be mediated by COX-2-derived PGE_2 and LXA_4 [34]. Along these lines, ATL stable analog shows reduced splenic dendritic cell mobilization and IL-12 response in *T. gondii*-infected mice, demonstrating a novel role for LXs in regulating pro-inflammatory responses during microbial infection [36, 37].

Endothelium

ATL analog inhibits endothelial cell proliferation and VEGF-induced endothelial cell chemotaxis. In a granuloma *in vivo* model of inflammatory angiogenesis, ATL treatment (10 μg/mouse) reduces the angiogenic phenotype, as assessed by both vascular casting and fluorescence. Together, these results identify a novel and potent previously unappreciated action of ATL in angiogenesis [71].

Eye

Both LXA_4 and LXB_4 and their stable analogs lower intraocular pressure in rabbits and that may underlie their role in the physiology of ocular pressure regulation within the eye [72, 73]. In human eye tissues, the receptor ALX is indeed present and appears to be associated with corneal epithelial cells [74].

Cell surface receptors for LXA_4 and ATL

Human and mouse ALX: Molecular cloning, receptor expression and up-regulation

The specific LXA_4 binding sites were first characterized on human PMN that are likely to mediate many of its selective actions on these cells [75]. Intact PMN demonstrate specific and reversible [11, 12-^{3}H]-LXA_4 [described in 76] binding (Kd ~0.5 nM and Bmax ~1,830 sites/PMN), which is modulated by guanosine stable analogs. These LXA_4 binding sites are inducible in promyelocytic lineage (HL-60) cells exposed to differentiating agents (e.g., retinoic acid, DMSO and PMA) and confer LXA_4-stimulated phospholipase activation [77]. Together, these findings provided further evidence that LXA_4 interacts with specific membrane-associated receptors on human leukocytes that belong to the classical GPCR. Based on our finding that functional LXA_4 receptors are inducible in HL-60 cells, several putative receptor cDNAs that are also induced within this temporal frame, cloned earlier from myeloid lineages and designated orphans [78, 79], were systematically examined for their ability to specifically bind and signal with LXA_4.

One of the orphans (denoted previously as pINF114, also known as FPRL1 and FPR2) when transfected into Chinese hamster ovary (CHO) cells displays specific [^{3}H]-LXA_4 binding with high affinity (Kd =1.7 nM) and demonstrated selectivity when compared to LXB_4, LTB_4, LTD_4 and PGE_2 (Fig. 5) [80]. These transfected CHO cells transmit signal with LXA_4, activating both GTPase and the release of arachidonic acid (C20:4) from membrane phospholipid, indicating that this cDNA encodes a functional receptor for LXA_4 in myeloid cells (Tab. 5). The mouse LXA_4

Table 5 - Lipoxin A_4 and ATL signal transduction with human ALX

Cell type	LXA_4 and ATL-evoked signal transduction	Kinase associated	Gene expression
HL-60 (differentiated)	PLD activation (lipid remodeling)	protein kinase C (staurosporine sensitive)	
PMN	PLD activation GTPase activity C20:4 release PIPP signal (↑PSDP accumulation) (with second signal) No increase of cAMP and proton efflux Weak $[Ca^{2+}]_i$	tyrosine kinase (genistein sensitive)	Inhibit IL-8 mRNA expression Up-regulate NAB1
Monocyte	Increase of $[Ca^{2+}]_i$ (PTX sensitive) No increase of cAMP and proton efflux		
Enterocyte	No proton efflux		Reduce IL-8 mRNA level Reduce NF-κB mediated transcription activation Regulate bactericidal/ permeability-increasing protein (BPI)
Synovial fibroblast	PLD activation Inhibit NF-κB binding		Stimulate tissue inhibitor of metalloproteinase (TIMP)-1 and TIMP-2 transcription

receptor cDNA was cloned from a spleen cDNA library and displays specific [^{3}H]-LXA_4 binding and LXA_4-initiated GTPase activity when transfected into CHO cells [58]. The human and mouse LXA_4 receptors represented the first cloned LO-derived eicosanoid receptors.

Both human [80] and mouse [58] ALX cDNA contain an open reading frame of 1051 nucleotides, which encode a protein of 351 amino acids. Northern blot analysis demonstrated that ALX mRNA is ~1.4 Kb in both human and mouse [58]. Chromosome mapping revealed that the gene encoding ALX [80] is located on chromosome 19q [81], denoted as FPRH1 in this early report of the orphan receptor. Northern blot analysis of multiple murine tissues demonstrated that ALX mRNA is most abundant in PMN, spleen and lung with lesser amounts in heart and liver [58]. In the absence of disease, the pattern is similar in human tissues. In humans, ALX mRNA is also abundant in PMN, followed by spleen, lung, placenta and liver [58, 80].

To date, ALX is identified by function and direct actions, and cloned in both human and mouse PMN [58, 80], human monocytes [82] and human enterocytes [74], as well as synovial fibroblasts [83]. In human PMN, results of subcellular fractionation experiments revealed that [^{3}H]-LXA$_4$ binding sites are associated with plasma membrane and endoplasmic reticulum (42.1%) and granule (34.5%) as well as nuclear-enriched fractions (23.3%), a distribution distinct from [^{3}H]-LTB$_4$ binding [75]. The finding that LXA$_4$ blocks both PAF and fMLP-stimulated eosinophil chemotaxis [60] suggests that functional ALX is also present on eosinophils. In human enterocytes, ALX is present in crypt and brush border colonic epithelial cells [74].

Retinoic acid, PMA and DMSO, which lead to granulocytic phenotypes in HL-60 cells, induce a ~3–5-fold increase in the expression of ALX as monitored by specific [^{3}H]-LXA$_4$ binding [77] (see Tab. 5). Also, ALX transcription is dramatically up-regulated by cytokines in human enterocytes, with lymphocyte-derived IL-13 and interferon (IFN)-γ being most potent, followed by IL-4 and IL-6. IL-1β and LPS also showed moderate induction of ALX mRNA [74]. In view of the cytokine regulation of ALX, it is likely that the expression of these receptors will change dramatically in disease states, which in turn, might attenuate mucosal inflammatory and allergic responses.

Structure-function relationships of ALX

ALX belongs to the GPCR superfamily characterized by seven putative transmembrane segments (TMS) with N-terminus on the extracellular side of the membrane and C-terminus on the intracellular side [84]. The overall homology between human and mouse ALXs is 76% in nucleotide sequence and 73% in deduced amino acid [58]. An especially high homology is evident for their second intracellular loop (100%) and between their sixth TMS (97%) followed by the second, third and seventh TMS as well as the first extracellular loop (87–89%), suggesting essential roles for these regions in ligand recognition and G protein coupling. Molecular evolution analysis (Fig. 7) suggests that ALX is only distantly related to prostanoid receptors and belongs to the cluster of chemoattractic peptide receptors exemplified by fMLP, C5a and IL-8 receptors [85] and now known to also include BLTs as well as the recently cloned cysteinyl-leukotriene receptors (CysLTs). BLT1 was obtained from human HL-60 cells [86] and mouse eosinophils [87] and found to share an overall

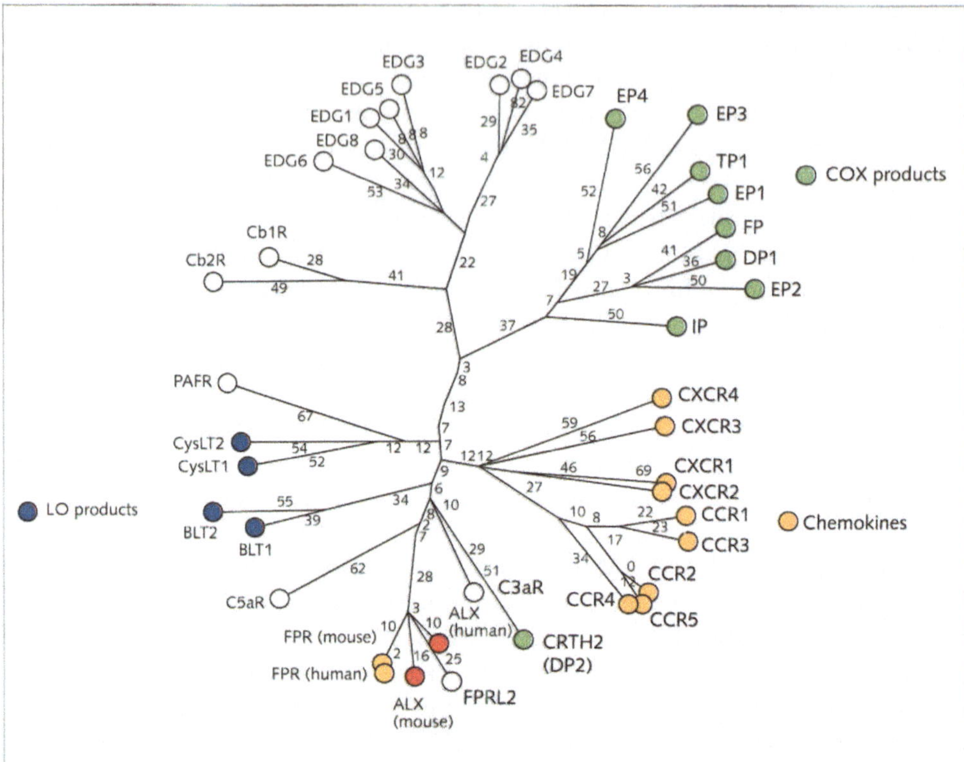

Figure 7
Phylogenetic tree of GPCRs for chemokines and lipid mediators
This tree is constructed with deduced amino acid sequences of human eicosanoid and chemokine receptors using the "All–All Program" at the Computational Biochemistry Server at ETHZ (http://cbrg.inf.ethz.ch/Server/AllAll.html).
Abbreviations: TP, thromboxane A_2 receptor; DP, prostaglandin D_2 receptor; EP1, EP2, EP3, EP4, subtypes of prostaglandin E_2 receptor; FP, prostaglandin F_2 receptor; IP, prostacyclin receptor; Cb1R, Cb2R, cannabinoid receptors; ALXR, lipoxin A_4 receptor and BLT, leukotriene B_4 receptor. BLT, CXCR-4 and CCR-5 were identified as co-receptors for HIV-1 entry.

~30% homology with ALX in deduced amino acid sequences. High homologous region (~46%) is present within the second transmembrane segments in both ALX and BLT1 with the amino acid sequence LNLALAD. Prostanoids interact with their receptors *via* carboxyl group interacting with an arginine residue within the seventh transmembrane segment [88]. Neither ALX nor BLT1 share this Arg (in 7th transmembrane segment) requirement [80, 86], yet both ligands contain COOH, which at physiological pH could present as a counter-anion. Together, these findings further provide evidence that the origin of receptors for LT and LX is distinct from that of receptors for prostanoids.

Several conserved motifs and amino acid residues important for post-translational modification are found in ALX. For example, conserved N-glycosylation sites are present on Asn-4 and Asn-179 of human ALX and N-glycosylation was shown to be important for ligand specificity of this receptor [89]. Site-directed mutagenesis study demonstrates that Ser-236, Ser-237 and Tyr-302 are essential for human ALX phosphorylation and signaling [90]. In addition, the NPXXY motif in the TMS-7, which plays a role in receptor desensitization and/or resensitization [91], and DRY motif in the TMS-3 that is essential for β-arrestin binding and G protein activation [92] are also present in ALX, suggesting a conserved mechanism in regulation of ALX signaling.

Ligand binding specificity: peptide and lipid ligands

ALX is stereo-selective for its eicosanoid-based ligands. Intact human PMN and retinoic acid-differentiated HL-60 cells demonstrate specific and reversible [^{3}H]-LXA_4 binding with Kds ~0.5 and ~0.6 nM, respectively [75, 77]. Several isomers of LXA_4 tested, namely 11-trans-LXA_4, 6S-LXA_4 and LXB_4, did not compete for these recognition sites, consistent with their functional responses in these systems. Results from Scatchard analyses indicate that [^{3}H]-LXA_4 binds PMN granule membrane-enriched fractions with comparable Kd (0.8 nM) but with a larger B_{max} (4.1×10^{-11} M) than plasma membrane fractions (Kd = 0.7 nM, $B_{max} = 2.1 \times 10^{-11}$ M) [80]. Hence, it appears that additional receptors can be mobilized by granule fusion to the plasma membrane of PMN. [^{3}H]-LXA_4 specific binding is stereo-selective, since LTB_4, LXB_4, 6S-LXA_4, 11-trans-LXA_4 or SKF104353 (a CysLT1 antagonist) does not compete for [^{3}H]-LXA_4 in human PMN. Human and mouse ALX cDNA, each transfected into CHO cells, display specific binding with [^{3}H]-LXA_4, with Kd values of 1.7 nM [80] and 1.5 nM [58], respectively. Human ALX-transfected CHO cells were also tested for binding with other eicosanoids, including LXB_4, LTD_4, LTB_4 and PGE_2. Only LTD_4 shows competition with [^{3}H]-LXA_4 binding, giving a K_i of 80 nM [80].

It is of interest to note that, although ALX shares ~70% homology with FPR, ALX binds [^{3}H]-fMLP with only low affinity (Kd ~5 μM) and proves to be selective for LXA_4 by three log orders of magnitude [93]. More recently, it was reported that certain peptides/proteins can also interact with ALX in *in vitro* and *in vivo* settings (Tab. 6). For examples, MHC binding peptide (a potent necrotactic peptide derived from NADH dehydrogenase subunit 1 from mitochondria) directly binds to human ALX and evokes PMN chemotaxis that is inhibited by ATL analog [89]; it also stimulates macrophage phagocytosis of PMN [64]. In addition, naturally produced cleaved form (i.e., D2D3(88–274)) of urokinase-type plasminogen activator

Table 6 - Comparison of LXs versus *potential peptide ligands for human ALX*

Ligand	Binding and signaling	Bioaction	EC_{50}
LXA_4 and ATLa	See Table 5 for details	Inhibit PMN See Table 2, 3 and 4 for details	~ 1 nM
MHC peptide [64, 89]	Displace $[^3H]$-LXA_4 binding	Induce PMN chemotaxis Induce phagocytosis of apoptotic PMN	~ 1 nM
SAA [131]	Partially displace $[^3H]$-LXA_4 binding Ca^{2+} mobilization Activate ERK-1/2 and P38	Induce PMN and monocyte chemotaxis Induce IL-8 and TNF-α secretion Induce NF-κB and IL-8 gene expression	~ 100 nM
uPAR (aa 88–274) fragment [94]	Direct $[^{125}I]$-peptide binding Activate Hck tyrosine kinase	Induce PMN and monocyte chemotaxis	~ 5 pM
Annexin 1 and derived peptides [50]	Direct $[^{125}I]$-peptide binding	Inhibit PMN infiltration into murine dorsal air pouches	~ 1 μM
HIV-1 gp120 peptides (F and V3 peptides) [132]	Ca^{2+} mobilization	Induce chemotaxis Down-regulate CCR5 and CXCR4	~ 1–10 μM
HIV-1 gp41 peptides (T21 and N36 peptides) [133]	Ca^{2+} mobilization	Induces chemotaxis	~ 0.1–10 μM
Amyloid β_{42} [134]	Ca^{2+} mobilization	Induces chemotaxis	~ 1 μM
LL-37 [135]	Ca^{2+} mobilization	Induces chemotaxis	~ 1 μM
Hp (2–20) *H. pylori* peptide [136]		Induce PMN chemotaxis Up-regulate integrins (Mac-1) Activate NADPH oxidase	~ 0.3 μM
Prion (aa 106–126) [137]		Induces chemotaxis in monocytes	~ 25 μM
MMK-1 [64, 89]	Evoke proton efflux Displace $[^3H]$-LXA_4 binding	Induce PMN chemotaxis Evoke PMN infiltration into murine dorsal air pouches Induces phagocytosis of apoptotic PMN	~ 1 nM
WKYMVm [138]	Ca^{2+} mobilization phosphoinositide hydrolysis	Induces chemotaxis Down-regulates CCR5 and CXCR4 Stimulates superoxide generation Enhances monocyte survival by inhibiting caspase-3 activity	~1 pM

(uPAR) directly binds to ALX and is a unique endogenous chemotactic agonist for ALX providing the first direct link between the fibrinolytic machinery and the inflammatory response [94]. These endogenously generated peptides evoke their bioaction *via* ALX in the subnanomolar range.

We recently reported that glucocorticoid-induced annexin 1 (ANXA1)-derived peptides (e.g., Ac2-26) are generated *in vivo* and act at the ALX to halt PMN diapedesis. These peptides specifically interact directly with recombinant human ALX, demonstrated by radioligand binding and function as well as immunoprecipitation of PMN receptors. In addition, the combination of both ATL and ANXA1-derived peptides limited PMN infiltration and reduced production of inflammatory mediators in murine dorsal air pouches [50]. These results demonstrate spatial and temporal separation in endogenous lipid and peptide anti-inflammatory circuits where both ATL and specific ANXA1-derived peptides act at ALX to down-regulate PMN recruitment to inflammatory loci.

Along these lines, additional peptides in the micromolar range can also interact with ALX in some *in vitro* model systems [95] including HIV envelope peptides (e.g., T21, N36, V3 and F peptides) and bacterial-derived peptides (e.g., Hp2–20 from *H. pylori*), as well as host-derived peptides (e.g., SAA, PrP106-126 and Aβ). These findings are summarized in Table 6. The functional role(s) of these peptides in human biology, pertaining to their ability to activate FPRL-1/ALX, albeit at μM levels, remains of interest.

These new findings suggest that small peptides as well as bioactive lipids can function as ligands for the same receptor, however with different affinity and/or distinct interaction sites within the receptor and separate intracellular signaling depending on the cell type and model system. Hence, it appears likely that the intracellular protein interactions following ligand-receptor binding are different for peptide *versus* lipid ligands of this receptor because different conformations of the ligand–receptors are likely to be formed. Taken together, the finding that specific LXA_4-related structures and certain peptides interact with this receptor may reflect the need for multi-recognition and receptor redundancies in the immune system.

Signal transduction of ALX

PMN *versus* monocytes

The intracellular signaling cascade of ALX appears to be highly cell type-specific. For example, in human PMN, LXA_4 stimulates rapid lipid remodeling (within seconds) with release of arachidonic acid that is evoked *via* PTX-sensitive G proteins [16] without formation of either LT or PG. Only a modest Ca^{2+} mobilization was observed. Also, LXA_4 was reported to block intracellular generation of IP3 [15] as well as Ca^{2+} mobilization in response to other stimuli [59]. In human peripheral blood monocytes and cultured THP-1 cells, LXA_4 triggers intracellular Ca^{2+} release and adherence to laminin [82, 96]. Thus, different intracellular signaling pathways are present in PMN *versus* monocytes despite identical receptor sequences (see Tab. 5). It is of interest that Ca^{2+} is not the second messenger for LX actions in monocytes, since LXA_4-stimulated monocyte adherence to laminin is not dependent on a LX-stimulated increase in $[Ca^{2+}]_i$. The EC_{50} value for LXA_4-stimulated increase in $[Ca^{2+}]_i$ is > 100 nM in monocytes, which is more than two log orders of magnitude higher than that required for LXA_4-stimulated adherence (EC_{50} < 1 nM). In view of G-protein coupling events in monocytes, both Ca^{2+} mobilization and adherence are PTX-sensitive. This indicates that receptor coupling in monocytes and PMN is similar to this point, although there could be different PTX-sensitive G-protein subtypes that couple to the intracellular domains of the receptors and diverge downstream in the signal transduction pathways leading to chemotaxis of monocytes and inhibition of PMN.

Novel anti-inflammatory signaling (not just desensitization)

Recently, additional results from this laboratory indicated that, with PMN, ALX interaction with LX and ATL analogs regulates a newly described polyisoprenyl phosphate (PIPP) signaling pathway [97]. ALX activation reverses LTB_4-initiated polyisoprenyl phosphate remodeling, leading to accumulation of presqualene diphosphate (PSDP), a potent negative intracellular signal in PMN which inhibits recombinant PLD and superoxide anion generation. Along these lines, LXA_4 reduces peroxynitrite formation and thus can oppose peroxynitrite signaling in leukocytes [98]. Recently findings by Maderna et al. [99] demonstrated that LXs induces cytoskeleton reorganization and an increase in membrane-associated RhoA GTPase as well as RhoA activity in monocytes and macrophages *via* inhibition of cAMP, revealing a potential mechanism of LXs in promoting resolution.

In retinoic acid-differentiated HL-60 cells, LXA_4 stimulated PLD activation that is staurosporine sensitive, suggesting the involvement of PKC in signal transduction in these cells [77]. It was also demonstrated that LXA_4 blocks LTB_4 or fMLP-stimulated PMN transmigration or adhesion by regulation of β_2 integrin-dependent PMN adhesion [93]. This modulatory action is partially reversed by prior exposure to genistein, a tyrosine kinase inhibitor [30].

ALX regulate pro-inflammatory gene expression

Leukocytes

Using differential display RT-PCR, a subset of genes was identified that was selectively up-regulated upon short exposure of PMN to ATLa. Among them, a transcriptional co-repressor NAB1 identified previously as a glucocorticoid-responsive gene in hamster smooth muscle cells was further investigated and also found to be up-regulated by ATLa in murine lung vascular smooth muscle *in vivo* [100]. These findings provide evidence for rapid transcriptional induction of a cassette of genes *via* an ATLa-stimulated G protein-coupled receptor pathway.

In addition, ATLa attenuates nuclear accumulation of activator protein-α and NF-κB in both PMN and monocytes and inhibits IL-8 mRNA expression [98].

Epithelial cells

Microarray analysis reveals that epithelial cells of wide origin (oral, pulmonary, and gastrointestinal mucosa) express bactericidal/permeability-increasing protein (BPI), an antibacterial and endotoxin-neutralizing molecule and that is transcriptionally regulated by ATL. A BPI-neutralizing anti-serum revealed that surface BPI blocks endotoxin-mediated signaling in epithelia and kills *Salmonella typhimurium* [101]. These studies identify a previously unappreciated "molecular shield" for protection of mucosal surfaces against Gram-negative bacteria and their endotoxin. Along these lines, in human enterocytes, ALX activation by LXA_4 and LX analogs diminishes *Salmonella typhimurium*-induced IL-8 transcription [102]. The reduction of IL-8 mRNA level parallels decrements in IL-8 secretion, indicating that in these cells ALX's mechanism of action for blocking this chemokine is at the gene transcriptional level. In an effort to elucidate the mechanism by which these lipid mediators modulate cellular pro-inflammatory programs, global epithelial gene expression was surveyed using microarray analysis. ATL analog pretreatment attenuates induction of approximately 50% of *Salmonella typhimurium*-induced gene expression [70]. A major subset of genes whose induction was reduced by ATL analog pretreatment is regulated by NF-κB, suggesting that ATL analog was influencing the activity of this transcription factor. Nanomolar concentrations of ATL analog reduced NF-κB-mediated transcriptional activation in an ALX dependent manner and inhibited induced degradation of IκBα.

Resident cell types

In human synovial fibroblasts, LXA_4 inhibits IL-1β responses with reduction of IL-6 and IL-8 synthesis and prevented IL-1β-induced MMP-3 synthesis at nanomolar concentrations [83]. Also, LXA_4 induces a two-fold increase of tissue inhibitor of metalloproteinase (TIMP)-1 and an approximately three-fold increase of TIMP-2 protein levels that is abrogated by pretreatment with LXA_4 receptor antiserum. LXA_4-induced changes of IL-6 and TIMP were accompanied by parallel changes in mRNA levels. These findings suggest that LXA_4 may be involved in a negative feedback loop opposing inflammatory cytokine-induced activation of synovial fibroblasts. In addition, LXA_4 induces tissue factor activity by increasing its mRNA level in EC304 cells (non-endothelial parenchymal cells) *via* a PTX-sensitive and PKC-dependent mechanism [103]. The ability of LXA_4 to induce tissue factor is an intriguing result. Its physiological role remains to be established in relation to LX generation and proximity to tissue factor releasing cells *in vivo*.

Additional sites of action for LX and ATL – CysLTs

In several tissues and cell types other than leukocytes, results from pharmacological experiments indicate that LXA_4 acts *via* interacting with a subclass of peptido-LT receptors ($CysLT_1$) as a partial agonist to mediate its actions [75, 104]. Along these lines, both LTC_4 and LXA_4, albeit at high concentrations (> 1 μM), induce contractions of guinea pig lung parenchyma and release of thromboxane A_2 that is sensitive to $CysLT_1$-receptor antagonists [105], which is not likely to be a physiologic action of LXA_4. In certain cell types, LXA_4 (in the nanomolar range) blocks LTD_4 actions, and in this regard blocks specific [^{3}H]-LTD_4 binding to mesanglial cells [104] and human umbilical vein endothelial cells (HUVEC) [58, 77]. HUVEC specifically bind [^{3}H]-LXA_4 at a Kd of 11 nM, which can be inhibited by LTD_4 and SKF104353 [77]. Therefore, it appears that LXA_4 interacts with at least two classes of cell surface receptors, one specific for LXA_4, which is present on leukocytes and enterocytes (ALX), the other shared by LTD_4, which is present on HUVEC and mesanglial cells ($CysLT_1$). Along these lines, an inducible $CysLT_1$ was recently identified and cloned from HUVEC [106].

Recombinant $CysLT_1$ receptor gave stereo-specific binding with both [^{3}H]-LTD_4 and a novel labeled mimetic of ATL ([^{3}H]-ATLa) that was displaced with LTD_4 and ATLa (~ IC_{50} 0.2–0.9 nM), and not with a biologically inactive ATL/LX isomer. In sharp contrast, LTD_4 was an ineffective competitive ligand for recombinant ALX with [^{3}H]-ATLa. Endogenous murine $CysLT_1$ receptors also gave specific [^{3}H]-ATLa binding that was displaced with essentially equal affinity by LTD_4 or ATLa. Systemic ATLa proved to be a potent inhibitor (> 50%) of $CysLT_1$-mediated vascular leakage in murine skin (200 μg/kg) in addition to its ability to block PMN recruitment to dorsal air-pouch (4 μg/kg). These results indicate that ATL and LTD_4 bind and compete with equal affinity at $CysLT_1$, providing a molecular basis for ATL serving as a local damper of both vascular $CysLT_1$ signals as well as ALX-regulated PMN traffic [106].

In human renal mesangial cells, LXA_4 inhibits PDGF and LTD_4-stimulated proliferation *via* regulation of PDGFRβ [107]. Also, LXA_4 stimulates MAP kinase superfamily *via* two distinct receptors: one *via* a PTX-sensitive G protein, leading to p38 activation, and the other *via* a PTX-insensitive G protein, leading to ERK activation [108]. Also, LXA_4 modulates MAP kinase activities on mesangial cells in a

PTX-insensitive manner [109], suggesting the presence of additional novel LXA_4 receptor subtypes and/or signaling pathways in these cells.

LXs and ATL exhibit unique molecular mechanism in anti-inflammation

LXA_4, ATL and their stable analogs activate ALX, which then modulates PMN responses *in vitro*, such as chemotaxis, transmigration, adhesion, degranulation, cytokine release and functions, as well as inhibit PMN recruitment in several murine models. For example, ATL analog inhibits TNF-α-initiated PMN infiltration in murine dorsal air-pouch [56] and LTB_4-induced PMN influx during dermal inflammation [58] as well as PMN-mediated second organ injury [35]. When compared to other eicosanoids of COX, LO and p450 products reported in the literature to display potential anti-inflammatory properties, LXs and ATL stand apart both in mechanism and amount range for action. For example, PGE_2 reduced the antigen response [110] and inhibited macrophage phagocytosis [111], presumably *via* increasing intracellular cAMP levels, which in turn inhibits MAPK activation by stimulating PKA-dependent phosphorylation of Raf-1 [112]. In contrast, LXA_4 does not give significant increase of cAMP levels in PMN [96]. In addition, cyclopentenone prostaglandins such as 15-deoxy-$\Delta^{12,14}$-PGJ_2, in relatively high amounts, give anti-inflammatory action in adjuvant-induced arthritis in rats [113]. However, it appears that 15-deoxy-$\Delta^{12,14}$–PGJ_2 acts *via* mechanisms that are independent of cell surface GPCRs [114, 115]. To date, LXA_4 and ATL are the only lipid mediators that possess anti-inflammatory and pro-resolution properties acting in the nanomolar range, since they regulate leukocyte trafficking and contribute to the early resolution.

Resolvins: The novel omega-3 PUFA derived endogenous local autacoids in anti-inflammation and pro-resolution

A recent report demonstrates that inflammatory exudates from mice treated with omega-3 polyunsaturated fatty acid (PUFA) and aspirin (ASA) generate a novel array of bioactive lipid signals (Fig. 8) [116]. Human endothelial cells with up-regulated COX-2 treated with ASA converted C20:5 omega-3 PUFA, namely eicosapentaenoic acid (EPA), to 18R-hydroxyeicosapentaenoic acid (HEPE) and 15R-HEPE. Each was used by PMN to generate separate classes of novel trihydroxy-containing mediators, including 5-series 15R-LX(5) and 5,12,18R-triHEPE. These new compounds proved to be potent inhibitors of human PMN transendothelial migration and infiltration *in vivo* (ATL analogue > 5,12,18R-triHEPE > 18R-HEPE) (Fig. 9). Acetaminophen and indomethacin also permitted 18R-HEPE and 15R-HEPE generation with recombinant COX-2 as well as omega-5 and omega-

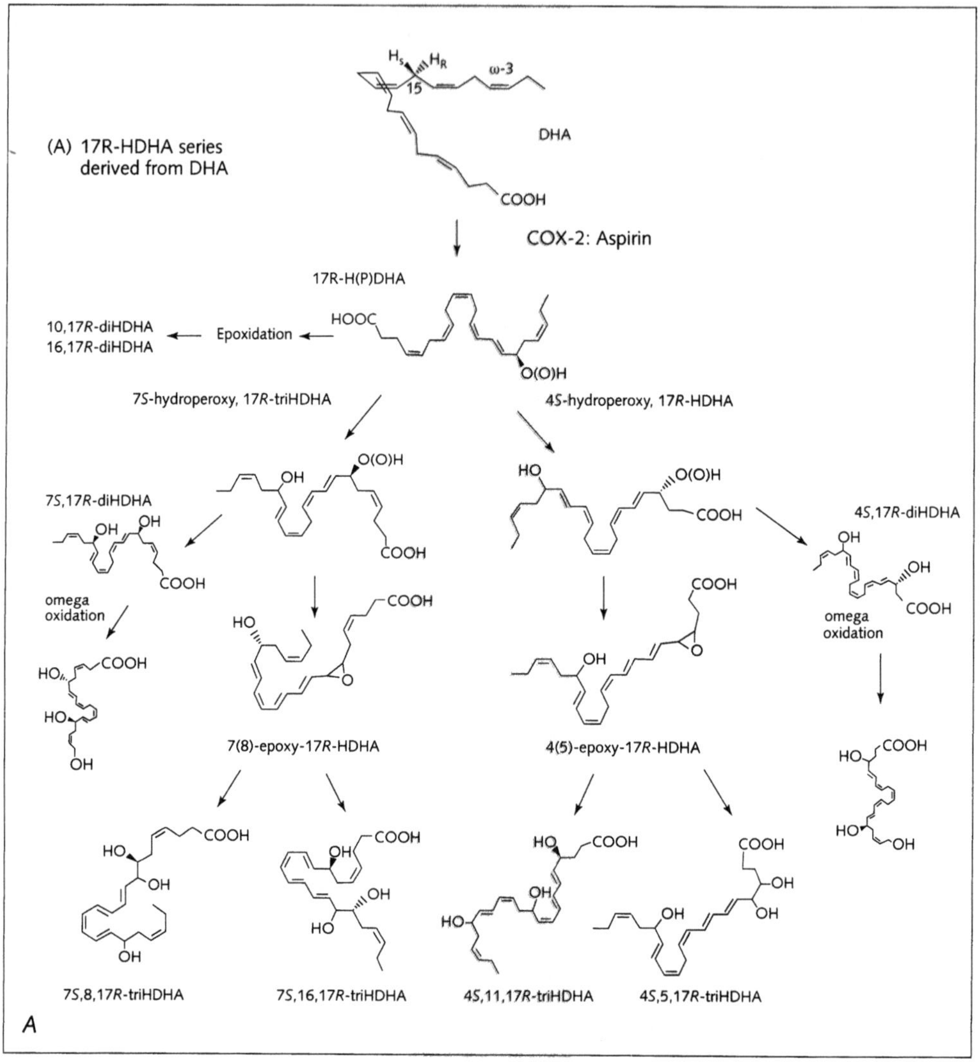

Figure 8
Proposed biosynthetic scheme for resolvins
Interactions of omega-3 polyunsaturated fatty acids (ω-3 PUFA) with aspirin (ASA)-acetylated COX-2 generate novel arrays of bioactive compounds, which inhibits PMN transmigration in vitro *and inflammation* in vivo*. A prototypic oxygenation with DHA (panel A) and EPA (panel B) is depicted as an omega-3-containing fatty acid.*

9 oxygenations of other fatty acids that act on hematologic cells. These findings establish new transcellular routes for producing arrays of bioactive lipid mediators

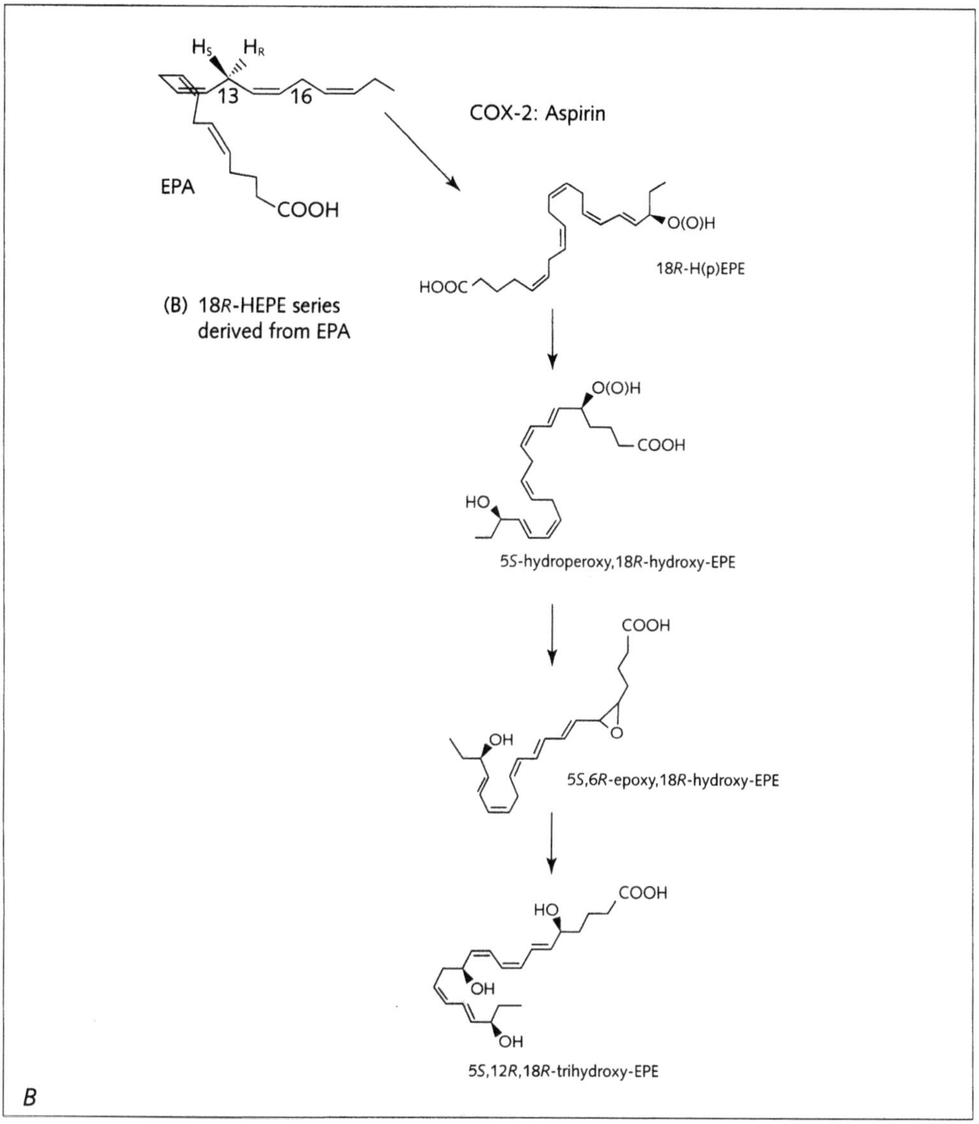

via COX-2-non-steroidal anti-inflammatory drug-dependent oxygenations and cell-cell interactions that impact micro-inflammation. The generation of these and related compounds provides a novel mechanism(s) for the therapeutic benefits of omega-3 dietary supplementation, which may be important in inflammation, neoplasia, and vascular diseases.

Along these lines, lipidomic analysis of exudates obtained in the resolution phase from mice treated with ASA and docosahexaenoic acid (DHA) (C22:6) produced a novel family of bioactive 17R-hydroxy-containing di- and tri-hydroxy-docosanoids

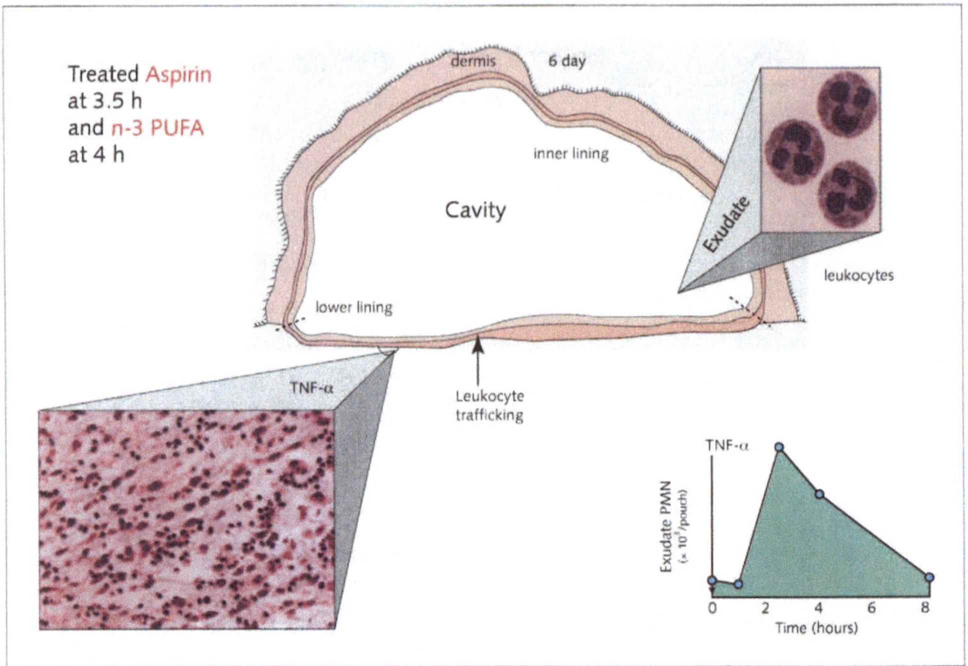

Figure 9
Murine dorsal air pouch
The 6-day murine dorsal air-pouch is characterized by the presence of a nascent lining that encloses the air cavity. TNF-α induces leukocyte infiltration, predominantly PMN, which is inhibited by iv injection of aspirin and ω-3 PUFA (25–60%).

[117]. Murine brain treated with aspirin produced endogenous 17R-hydroxydocosahexaenoic acid, as did human microglial cells. Human COX-2 converted DHA to 13-hydroxy-DHA that switched with ASA to 17R-HDHA, which also proved a major route in hypoxic endothelial cells. Human neutrophils transformed COX-2-ASA-derived 17R-hydroxy-DHA into two sets of novel di- and tri-hydroxy products; one initiated *via* oxygenation at carbon 7 and the other at carbon 4. These compounds inhibited (IC_{50} approximately 50 pM) microglial cell cytokine expression and *in vivo* dermal inflammation and peritonitis at nanogram doses, reducing leukocytic exudates by 40–80%. These results indicate that exudates, vascular cells, leukocytes and neural cells treated with aspirin convert DHA to a novel 17R-hydroxy series of docosanoids that are potent regulators. Together, these biosynthetic pathways utilize omega-3 DHA and EPA during multicellular events in resolution to produce a family of protective compounds, i.e., resolvins, which enhance pro-resolution status.

Summary

Lipoxins are the trihydroxy-tetraene-containing eicosanoids that are primarily generated by cell-cell interactions *via* transcellular biosynthesis that serve as local endogenous anti-inflammatory mediators. These "stop signals" in inflammation and other related processes may be involved in switching the cellular response from additional PMN recruitment toward monocytes (in a non-phlogistic fashion) that could lead to resolution of the inflammatory response and/or promotion of repair and healing. Aspirin impinges on this homeostatic system and evokes the endogenous biosynthesis of the carbon 15 epimers of lipoxins, namely aspirin-triggered-lipoxins, which mimic the bioactions of native LX in several biological systems and can thus modulate in part the beneficial actions of ASA in humans. The activation of a LX biosynthetic circuit *in vivo* requires up-regulation of key enzymes by cytokines such as IL-4 and IL-13 that also control the expression of the receptor ALX [74]. Moreover, both the temporal and spatial components in LX formation and actions are important determinants in their bio-impact during an acute inflammatory reaction [118]. In this regard, generation of lipid (i.e., ATL) *versus* protein (i.e., ANXA1) mediators during the host inflammatory response displays different time courses. This temporal difference suggests that ALX/FPRL1 could regulate PMN by interacting with each class of ligands within specific phases of the inflammatory response.

LXA_4 elicits biological actions *via* at least two main classes of receptor systems known to date: (1) ALX on leukocytes and enterocytes, and (2) a shared CysLT1 subtype on endothelial and mesangial cells.

ALX belongs to the classical G-protein-coupled receptors (GPCR) and was identified in mammalian tissues and characterized using direct evidence obtained with specific $[^3H]$-LXA_4 binding and activation of functional responses with LXA_4. ALX is the first cloned lipoxygenase-derived eicosanoid receptor. ALX and BLT are more akin to chemokine receptors in their deduced amino acid sequences than the currently known prostanoid receptors. The signaling pathways and bioactions of ALX are cell type-specific. In agreement with *in vitro* results, ALX agonists, namely LXA_4 and 15-epi-LXA_4 as well as their stable analogs, are topically active in inhibiting PMN infiltration and vascular permeability during murine dermal inflammation. In addition, it appears that LX also display organ-specific actions in addition to host defense and immune roles such as in the eye, kidney, lung, oral and GI tract and within bone marrow progenitors, possibly involving stem cells. The development of these relatively few synthetic stable analogs has already provided valuable tools to evaluate the biological roles, significance and pharmacological actions of ALX as well as provided a novel means to selective therapies for inflammatory diseases.

The relationship between LX generation and current NSAID therapies is more intertwined than currently appreciated [119] in that aspirin inhibits COX-1 and converts COX-2 into an ASA-triggered lipid mediator-generating system that produces an array of novel endogenous local autacoids from dietary omega-3 PUFA;

some of which display potent anti-inflammatory or anti-neutrophil recruitment activity [116, 117] as well as impinge on the role of these compounds in resolution and thus termed resolvins. Hence, it is not surprising that others have recently found a protective action for COX-2 in cardiovascular disease [120]. Together with the LXs and 15-epi-LXs, the identification of these novel endogenous anti-inflammatory lipid mediators [116] gives us new avenues of approach in considering therapeutics for inflammation, cardiovascular diseases and cancer.

Acknowledgments
The author's laboratory is supported in part by National Institutes of Health grant nos. GM38765, DK50305, and P01-DE13499. The authors thank Mary Halm Small for expert assistance in manuscript preparation.

References

1 Gallin JI, Snyderman R, Fearon DT, Haynes BF, Nathan C (eds) (1999) *Inflammation: Basic Principles and Clinical Correlates.* 3rd ed. Lippincott Williams & Wilkins, Philadelphia

2 Samuelsson B, Dahlén SE, Lindgren JÅ, Rouzer CA, Serhan CN (1987) Leukotrienes and lipoxins: Structures, biosynthesis, and biological effects. *Science* 237: 1171–1176

3 Cronstein BN, Montesinos MC, Weissmann G (1999) Sites of action for future therapy: An adenosine-dependent mechanism by which aspirin retains its anti-inflammatory activity in cyclo-oxygenase-2 and NFκB knockout mice. *Osteoarthritis and Cartilage* 7: 361–363

4 Cronstein BN, Montesinos MC, Weissmann G (1999) Salicylates and sulfasalazine, but not glucocorticoids, inhibit leukocyte accumulation by an adenosine-dependent mechanism that is independent of inhibition of prostaglandin synthesis and p105 of NFκB. *Proc Natl Acad Sci USA* 96: 6377–6381

5 Krump E, Picard S, Mancini J, Borgeat P (1997) Suppression of leukotriene B_4 biosynthesis by endogenous adenosine in ligand-activated human neutrophils. *J Exp Med* 186: 1401–1406

6 Qiu F-H, Wada K, Stahl GL, Serhan CN (2000) IMP and AMP deaminase in reperfusion injury down-regulates neutrophil recruitment. *Proc Natl Acad Sci USA* 97: 4267–4272

7 de Waal Malefyt R (1999) Role of interleukin-10, interleukin-4, and interleukin-13 in resolving inflammatory responses. Third ed. In: Gallin JI, Snyderman R, Fearon DT, Haynes BF, Nathan C, (eds): *Inflammation: Basic Principles and Clinical Correlates.* Lippincott Williams & Wilkins, Philadephia, 837–849

8 Diamond P, McGinty A, Sugrue D, Brady HR, Godson C (1999) Regulation of leukocyte trafficking by lipoxins. *Clin Chem Lab Med* 37: 293–297

9 Serhan CN (1994) Lipoxin biosynthesis and its impact in inflammatory and vascular events. *Biochim Biophys Acta* 1212: 1–25

10 Marcus AJ (1995) Aspirin as prophylaxis against colorectal cancer. *N Engl J Med* 333: 656–658

11 Vane JR (1982) *Adventures and excursions in bioassay: The stepping stones to prostacyclin: Les Prix Nobel: Nobel Prizes, Presentations, Biographies and Lectures*. Almqvist & Wiksell, Stockholm, 181–206

12 Chiang N, Takano T, Clish CB, Petasis NA, Tai H-H, Serhan CN (1998) Aspirin-triggered 15-epi-lipoxin A_4 (ATL) generation by human leukocytes and murine peritonitis exudates: Development of a specific 15-epi-LXA_4 ELISA. *J Pharmacol Exp Ther* 287: 779–790

13 Clària J, Serhan CN (1995) Aspirin triggers previously undescribed bioactive eicosanoids by human endothelial cell-leukocyte interactions. *Proc Natl Acad Sci USA* 92: 9475–9479

14 Clària J, Lee MH, Serhan CN (1996) Aspirin-triggered lipoxins (15-epi-LX) are generated by the human lung adenocarcinoma cell line (A549)-neutrophil interactions and are potent inhibitors of cell proliferation. *Mol Med* 2: 583–596

15 Grandordy BM, Lacroix H, Mavoungou E, Krilis S, Crea AE, Spur BW, Lee TH (1990) Lipoxin A_4 inhibits phosphoinositide hydrolysis in human neutrophils. *Biochem Biophys Res Commun* 167: 1022–1029

16 Nigam S, Fiore S, Luscinskas FW, Serhan CN (1990) Lipoxin A_4 and lipoxin B_4 stimulate the release but not the oxygenation of arachidonic acid in human neutrophils: dissociation between lipid remodeling and adhesion. *J Cell Physiol* 143: 512–523

17 Showell HJ, Cooper K (1999) Inhibitors and antagonists of cyclo-oxygenase, 5-lipoxygenase, and platelet activating factor. In: Gallin JI, Snyderman R, (eds): *Inflammation: Basic Principles and Clinical Correlates*. Lippincott Williams & Wilkins, Philadelphia, 1177–1193

18 Malaviya R, Abraham SN (2000) Role of mast cell leukotrienes in neutrophil recruitment and bacterial clearance in infectious peritonitis. *J Leukoc Biol* 67: 841–846

19 Swan SK, Rudy DW, Lasseter KC, Ryan CF, Buechel KL, Lambrecht LJ, Pinto MB, Dilzer SC, Obrda O, Sundblad KJ et al (2000) Effect of cyclo-oxygenase-2 inhibition on renal function in elderly persons receiving a low-salt diet: A randomized, controlled trial. *Ann Intern Med* 133: 1–9

20 Cannon GW, Caldwell JR, Holt P, McLean B, Seidenberg B, Bolognese J, Ehrich E, Mukhopadhyay S, Daniels B (2000) Rofecoxib, a specific inhibitor of cyclo-oxygenase 2, with clinical efficacy comparable with that of diclofenac sodium: Results of a one-year, randomized, clinical trial in patients with osteoarthritis of the knee and hip. Rofecoxib Phase III Protocol 035 Study Group. *Arthritis Rheum* 43: 978–987

21 Serhan CN, Maddox JF, Petasis NA, Akritopoulou-Zanze I, Papayianni A, Brady HR, Colgan SP, Madara JL (1995) Design of lipoxin A_4 stable analogs that block transmigration and adhesion of human neutrophils. *Biochemistry* 34: 14609–14615

22 Serhan CN (1997) Lipoxins and novel aspirin-triggered 15-epi-lipoxins (ATL): a jungle of cell-cell interactions or a therapeutic opportunity? *Prostaglandins* 53: 107–137

23 Lehr H-A, Frei B, Arfors K-E (1994) Vitamin C prevents cigarette smoke-induced leukocyte aggregation and adhesion to endothelium *in vivo*. *Proc Natl Acad Sci USA* 91: 7688–7692

24 Fiore S, Serhan CN (1989) Phospholipid bilayers enhance the stability of leukotriene A_4 and epoxytetraenes: stabilization of eicosanoids by liposomes. *Biochem Biophys Res Commun* 159: 477–481

25 Nassar GM, Morrow JD, Roberts LJ, II, Lakkis FG, Badr KF (1994) Induction of 15-lipoxygenase by interleukin-13 in human blood monocytes. *J Biol Chem* 269: 27631–27634

26 Levy BD, Romano M, Chapman HA, Reilly JJ, Drazen J, Serhan CN (1993) Human alveolar macrophages have 15-lipoxygenase and generate 15(S)-hydroxy-5,8,11-cis-13-trans-eicosatetraenoic acid and lipoxins. *J Clin Invest* 92: 1572–1579

27 Levy GN (1997) Prostaglandin H synthases, non-steroidal anti-inflammatory drugs, and colon cancer. *FASEB J* 11: 234–247

28 Savage MP, Goldberg S, Bove AA, Deutsch E, Vetrovec G, Macdonald RG, Bass T, Margolis JR, Whitworth HB, Taussig A et al (1995) Effect of thromboxane A_2 blockade on clinical outcome and restenosis after successful coronary angioplasty. *Circulation* 92: 3194–3200

29 Samuelsson B (1982) *From studies of biochemical mechanisms to novel biological mediators: prostaglandin endoperoxides, thromboxanes and leukotrienes: Les Prix Nobel: Nobel Prizes, Presentations, Biographies and Lectures*. Almqvist & Wiksell, Stockholm, 153–174

30 Papayianni A, Serhan CN, Brady HR (1996) Lipoxin A_4 and B_4 inhibit leukotriene-stimulated interactions of human neutrophils and endothelial cells. *J Immunol* 156: 2264–2272

31 Tamaoki J, Tagaya E, Yamawaki I, Konno K (1995) Lipoxin A_4 inhibits cholinergic neurotransmission through nitric oxide generation in the rabbit trachea. Eur J Pharmacol 287: 233–238

32 Papayianni A, Serhan CN, Phillips ML, Rennke HG, Brady HR (1995) Transcellular biosynthesis of lipoxin A_4 during adhesion of platelets and neutrophils in experimental immune complex glomerulonephritis. *Kidney Int* 47: 1295–1302

33 Munger KA, Montero A, Fukunaga M, Uda S, Yura T, Imai E, Kaneda Y, Valdivielso JM, Badr KF (1999) Transfection of rat kidney with human 15-lipoxygenase suppresses inflammation and preserves function in experimental glomerulonephritis. *Proc Natl Acad Sci USA* 96: 13375–13380

34 Bandeira-Melo C, Serra MF, Diaz BL, Cordeiro RSB, Silva PMR, Lenzi HL, Bakhle YS, Serhan CN, Martins MA (2000) Cyclo-oxygenase-2-derived prostaglandin E_2 and lipoxin A_4 accelerate resolution of allergic edema in Angiostrongylus costaricensis-infected rats: Relationship with concurrent eosinophilia. *J Immunol* 164: 1029–1036

35 Chiang N, Gronert K, Clish CB, O'Brien JA, Freeman MW, Serhan CN (1999) Leuko-

triene B_4 receptor transgenic mice reveal novel protective roles for lipoxins and aspirin-triggered lipoxins in reperfusion. *J Clin Invest* 104: 309–316

36 Aliberti J, Serhan C, Sher A (2002) Parasite-induced Lipoxin A(4) Is an endogenous regulator of IL-12 production and immunopathology in *Toxoplasma gondii* infection. *J Exp Med* 196: 1253–1262

37 Aliberti J, Hieny S, Reis e Sousa C, Serhan CN, Sher A (2002) Lipoxin-mediated inhibition of IL-12 production by DCs: a mechanism for regulation of microbial immunity. *Nature Immunol* 3: 76–82

38 Levy BD, De Sanctis GT, Devchand PR, Kim E, Ackerman K, Schmidt BA, Szczeklik W, Drazen JM, Serhan CN (2002) Multi-pronged inhibition of airway hyper-responsiveness and inflammation by lipoxin A_4. *Nature Med* 8: 1018–1023

39 Kim SJ, Tominaga T (1989) Formation of lipoxins by the brain. *Ann NY Acad* Sci 559: 461–464

40 Clària J, Titos E, Jiménez W, Ros J, Ginès P, Arroyo V, Rivera F, Rodés J (1998) Altered biosynthesis of leukotrienes and lipoxins and host defense disorders in patients with cirrhosis and ascites. *Gastroenterology* 115: 147–156

41 Stenke L, Näsman-Glaser B, Edenius C, Samuelsson J, Palmblad J, Lindgren JÅ (1990) Lipoxygenase products in myeloproliferative disorders: Increased leukotriene C_4 and decreased lipoxin formation in chronic myeloid leukemia. *Adv Prostaglandin Thromboxane Leukot Res* 21B: 883–886

42 Stenke L, Edenius C, Samuelsson J, Lindgren JA (1991) Deficient lipoxin synthesis: a novel platelet dysfunction in myeloproliferative disorders with special reference to blastic crisis of chronic myelogenous leukemia. *Blood* 78: 2989–2995

43 Stenke L, Nasman-Glaser B, Edenius C, Samuelsson J, Palmblad J, Lindgren JA (1991) Lipoxygenase products in myeloproliferative disorders: Increased leukotriene C_4 and decreased lipoxin formation in chronic myeloid leukemia. *Adv Prostaglandin Thromboxane Leukot Res* 21B: 883–886

44 Stenke L, Mansour M, Edenius C, Reizenstein P, Lindgren JA (1991) Formation and proliferative effects of lipoxins in human bone marrow. *Biochem Biophys Res Commun* 180: 255–261

45 Stenke L, Reizenstein P, Lindgren JA (1994) Leukotrienes and lipoxins – New potential performers in the regulation of human myelopoiesis. *Leukemia Res* 18: 727–732

46 Pouliot M, Clish CB, Petasis NA, Van Dyke TE, Serhan CN (2000) Lipoxin A_4 analogues inhibit leukocyte recruitment to Porphyromonas gingivalis: A role for cyclo-oxygenase-2 and lipoxins in periodontal disease. *Biochemistry* 39: 4761–4768

47 Bonnans C, Vachier I, Chavis C, Godard P, Bousquet J, Chanez P (2002) Lipoxins are potential endogenous antiinflammatory mediators in asthma. *Am J Respir Crit Care Med* 165: 1531–1535

48 Brezinski DA, Nesto RW, Serhan CN (1992) Angioplasty triggers intra-coronary leukotrienes and lipoxin A_4. Impact of aspirin therapy. *Circulation* 86: 56–63

49 Edenius C, Kumlin M, Björk T, Anggard A, Lindgren JA (1990) Lipoxin formation in human nasal polyps and bronchial tissue. *FEBS Lett* 272: 25–28

50 Perretti M, Chiang N, La M, Fierro IM, Marullo S, Getting SJ, Solito E, Serhan CN (2002) Endogenous lipid- and peptide-derived anti-inflammatory pathways generated with glucocorticoid and aspirin treatment activate the lipoxin A(4) receptor. *Nat Med* 8: 1296–1302

51 Titos E, Chiang N, Serhan CN, Romano M, Gaya J, Pueyo G, Clària J (1999) Hepatocytes are a rich source of novel aspirin-triggered 15-epi-lipoxin A_4 (ATL). *Am J Physiol* 277: C870–877

52 Fiorucci S, De Lima OM, Jr., Mencarelli A, Palazzetti B, Distrutti E, McKnight W, Dicay M, Ma L, Romano M, Morelli A et al (2002) Cyclo-oxygenase-2-derived lipoxin A_4 increases gastric resistance to aspirin-induced damage. *Gastroenterology* 123: 1598–1606

53 Sanak M, Levy BD, Clish CB, Chiang N, Gronert K, Mastalerz L, Serhan CN, Szczeklik A (2000) Aspirin-tolerant asthmatics generate more lipoxins than aspirin-intolerant asthmatics. *Eur Respir J* 16: 44–49

54 Clish CB, Levy BD, Chiang N, Tai H-H, Serhan CN (2000) Oxidoreductases in lipoxin A_4 metabolic inactivation. *J Biol Chem* 275: 25372–25380

55 Maddox JF, Colgan SP, Clish CB, Petasis NA, Fokin VV, Serhan CN (1998) Lipoxin B_4 regulates human monocyte/neutrophil adherence and motility: design of stable lipoxin B_4 analogs with increased biologic activity. *FASEB J* 12: 487–494

56 Clish CB, O'Brien JA, Gronert K, Stahl GL, Petasis NA, Serhan CN (1999) Local and systemic delivery of a stable aspirin-triggered lipoxin prevents neutrophil recruitment *in vivo*. *Proc Natl Acad Sci USA* 96: 8247–8252

57 Takano T, Clish CB, Gronert K, Petasis N, Serhan CN (1998) Neutrophil-mediated changes in vascular permeability are inhibited by topical application of aspirin-triggered 15-epi-lipoxin A_4 and novel lipoxin B_4 stable analogues. *J Clin Invest* 101: 819–826

58 Takano T, Fiore S, Maddox JF, Brady HR, Petasis NA, Serhan CN (1997) Aspirin-triggered 15-epi-lipoxin A_4 and LXA_4 stable analogs are potent inhibitors of acute inflammation: Evidence for anti-inflammatory receptors. *J Exp Med* 185: 1693–1704

59 Lee TH, Horton CE, Kyan-Aung U, Haskard D, Crea AE, Spur BW (1989) Lipoxin A_4 and lipoxin B_4 inhibit chemotactic responses of human neutrophils stimulated by leukotriene B_4 and N-formyl-L-methionyl-L-leucyl-L-phenylalanine. *Clin Sci* 77: 195–203

60 Soyombo O, Spur BW, Lee TH (1994) Effects of lipoxin A_4 on chemotaxis and degranulation of human eosinophils stimulated by platelet-activating factor and N-formyl-L-methionyl-L-leucyl-L-phenylalanine. *Allergy* 49: 230–234

61 Colgan SP, Serhan CN, Parkos CA, Delp-Archer C, Madara JL (1993) Lipoxin A_4 modulates transmigration of human neutrophils across intestinal epithelial monolayers. *J Clin Invest* 92: 75–82

62 Godson C, Mitchell S, Harvey K, Petasis NA, Hogg N, Brady HR (2000) Cutting edge: Lipoxins rapidly stimulate non-phlogistic phagocytosis of apoptotic neutrophils by monocyte-derived macrophages. *J Immunol* 164: 1663–1667

63 Planagumà A, Titos E, López-Parra M, Gaya J, Pueyo G, Arroyo V, Clària J (2002)

Aspirin (ASA) regulates 5-lipoxygenase activity and peroxisome proliferator-activated receptor a-mediated CINC-1 release in rat liver cells: Novel actions of lipoxin A_4 (LXA_4) and ASA-triggered 15-epi-LXA_4. *FASEB J* 16: 1937–1939
64 Mitchell S, Thomas G, Harvey K, Cottell D, Reville K, Berlasconi G, Petasis NA, Erwig L, Rees AJ, Savill J et al (2002) Lipoxins, aspirin-triggered epi-lipoxins, lipoxin stable analogues, and the resolution of inflammation: Stimulation of macrophage phagocytosis of apoptotic neutrophils *in vivo*. *J Am Soc Nephrol* 13: 2497–2507
65 Bandeira-Melo C, Bozza PT, Diaz BL, Cordeiro RSB, Jose PJ, Martins MA, Serhan CN (2000) Cutting edge: Lipoxin (LX) A_4 and aspirin-triggered 15-epi-LXA_4 block allergen-induced eosinophil trafficking. *J Immunol* 164: 2267–2271
66 Badr KF, Serhan CN, Nicolaou KC, Samuelsson B (1987) The action of lipoxin-A on glomerular microcirculatory dynamics in the rat [published erratum appears in *Biochem Biophys Res Commun* (1987) 147 (3): 1300]. *Biochem Biophys Res Commun* 145: 408–414
67 Katoh T, Takahashi K, DeBoer DK, Serhan CN, Badr KF (1992) Renal hemodynamic actions of lipoxins in rats: A comparative physiological study. *Am J Physiol* 263: F436–442
68 Scalia R, Gefen J, Petasis NA, Serhan CN, Lefer AM (1997) Lipoxin A_4 stable analogs inhibit leukocyte rolling and adherence in the rat mesenteric microvasculature: Role of P-selectin. *Proc Natl Acad Sci USA* 94: 9967–9972
69 Leonard MO, Hannan K, Burne MJ, Lappin DW, Doran P, Coleman P, Stenson C, Taylor CT, Daniels F, Godson C et al (2002) 15-epi-16-(para-fluorophenoxy)-lipoxin A_4-methyl ester, a synthetic analogue of 15-epi-lipoxin A_4, is protective in experimental ischemic acute renal failure. *J Am Soc Nephrol* 13: 1657–1662
70 Gewirtz AT, Collier-Hyams LS, Young AN, Kucharzik T, Guilford WJ, Parkinson JF, Williams IR, Neish AS, Madara JL (2002) Lipoxin A_4 analogs attenuate induction of intestinal epithelial proinflammatory gene expression and reduce the severity of dextran sodium sulfate-induced colitis. *J Immunol* 168: 5260–5267
71 Fierro IM, Kutok JL, Serhan CN (2002) Novel lipid mediator regulators of endothelial cell proliferation and migration: Aspirin-triggered-15R-lipoxin A_4 and lipoxin A_4. *J Pharmacol Exp Ther* 300: 385–392
72 Cotran PR, Hsu C, Fokin VV, Petasis NA, Serhan CN (1995) Lipoxin analogs reduce intraocular pressure in rabbits. *International Symposium on Experimental and Clinical Ocular Pharmacology and Pharmaceutics, Geneva, Switzerland*
73 Cotran PR, Hsu C, Serhan CN (1995) Lipoxin derivatives reduce intraocular pressure in rabbits. *Association for Research in Vision and Ophthalmology*
74 Gronert K, Gewirtz A, Madara JL, Serhan CN (1998) Identification of a human enterocyte lipoxin A_4 receptor that is regulated by IL-13 and IFN-γ and inhibits TNF-α-induced IL-8 release. *J Exp Med* 187: 1285–1294
75 Fiore S, Ryeom SW, Weller PF, Serhan CN (1992) Lipoxin recognition sites. Specific binding of labeled lipoxin A_4 with human neutrophils. *J Biol Chem* 267: 16168–16176
76 Brezinski DA, Serhan CN (1991) Characterization of lipoxins by combined gas chro-

matography and electron-capture negative ion chemical ionization mass spectrometry: formation of lipoxin A_4 by stimulated human whole blood. *Biol Mass Spectrom* 20: 45–52
77 Fiore S, Romano M, Reardon EM, Serhan CN (1993) Induction of functional lipoxin A_4 receptors in HL-60 cells. *Blood* 81: 3395–3403
78 Nomura H, Nielsen BW, Matsushima K (1993) Molecular cloning of cDNAs encoding a LD78 receptor and putative leukocyte chemotactic peptide receptors. *Int Immunol* 5: 1239
79 Perez HD, Holmes R, Kelly E, McClary J, Andrews WH (1992) Cloning of a cDNA encoding a receptor related to the formyl peptide receptor of human neutrophils. *Gene* (Amst) 118: 303–304
80 Fiore S, Maddox JF, Perez HD, Serhan CN (1994) Identification of a human cDNA encoding a functional high affinity lipoxin A_4 receptor. *J Exp Med* 180: 253–260
81 Bao L, Gerard NP, Eddy RL Jr, Shows TB, Gerard C (1992) Mapping of genes for the human C5a receptor (C5AR), human FMLP receptor (FPR), and two FMLP receptor homologue orphan rceptors (FPRH1, FPRH2) to chromosome 19. *Genomics* 13: 437–440
82 Maddox JF, Hachicha M, Takano T, Petasis NA, Fokin VV, Serhan CN (1997) Lipoxin A_4 stable analogs are potent mimetics that stimulate human monocytes and THP-1 cells *via* a G-protein linked lipoxin A_4 receptor. *J Biol Chem* 272: 6972–6978
83 Sodin-Semrl S, Taddeo B, Tseng D, Varga J, Fiore S (2000) Lipoxin A_4 inhibits IL-1 beta-induced IL-6, IL-8, and matrix metalloproteinase-3 production in human synovial fibroblasts and enhances synthesis of tissue inhibitors of metalloproteinases. *J Immunol* 164: 2660–2666
84 Baldwin JM (1993) The probable arrangement of the helices in G protein-coupled receptors. *EMBO J* 12: 1693–1703
85 Toh H, Ichikawa A, Narumiya S (1995) Molecular evolution of receptors for eicosanoids. *FEBS Lett* 361: 17–21
86 Yokomizo T, Izumi T, Chang K, Takuwa T, Shimizu T (1997) A G-protein-coupled receptor for leukotriene B_4 that mediates chemotaxis. *Nature* 387: 620–624
87 Huang W-W, Garcia-Zepeda EA, Sauty A, Oettgen HC, Rothenberg ME, Luster AD (1998) Molecular and biological characterization of the murine leukotriene B_4 receptor expressed on eosinophils. *J Exp Med* 188: 1063–1074
88 Ushikubi F, Hirata M, Narumiya S (1995) Molecular biology of prostanoid receptors; an overview. *Journal of Lipid Mediators and Cell Signalling* 12: 343–359
89 Chiang N, Fierro IM, Gronert K, Serhan CN (2000) Activation of lipoxin A_4 receptors by aspirin-triggered lipoxins and select peptides evokes ligand-specific responses in inflammation. *J Exp Med* 191: 1197–1207
90 Kang Y, Taddeo B, Varai G, Varga J, Fiore S (2000) Mutations of serine 235-237 and tyrosine 302 residues in the human lipoxin A_4 receptor intracellular domains result in sustained signaling. *Biochemistry* 39: 13551–13557
91 Barak LS, Menard L, Ferguson SS, Colapietro AM, Caron MG (1995) The conserved

seven-transmembrane sequence NP(X)2,3Y of the G-protein-coupled receptor superfamily regulates multiple properties of the beta 2-adrenergic receptor. *Biochemistry* 34: 15407–15414

92 Bennett TA, Maestas DC, Prossnitz ER (2000) Arrestin binding to the G protein-coupled N-formyl peptide receptor is regulated by the conserved 'DRY' sequence. *J Biol Chem* 275: 24590–24594

93 Fiore S, Serhan CN (1995) Lipoxin A_4 receptor activation is distinct from that of the formyl peptide receptor in myeloid cells: inhibition of CD11/18 expression by lipoxin A4-lipoxin A4 receptor interaction. *Biochemistry* 34: 16678–16686

94 Resnati M, Pallavicini I, Wang JM, Oppenheim JJ, Serhan CN, Romano M, Blasi F (2002) The fibrinolytic receptor for urokinase activates the G protein-coupled chemotactic receptor FPRL1/LXA_4R. *Proc Natl Acad Sci USA* 99: 1359–1364

95 Le Y, Murphy PM, Wang JM (2002) Formyl-peptide receptors revisited. *Trends Immunol* 23: 541–548

96 Romano M, Maddox JF, Serhan CN (1996) Activation of human monocytes and the acute monocytic leukemia cell line (THP-1) by lipoxins involves unique signaling pathways for lipoxin A_4 *versus* lipoxin B_4. *J Immunol* 157: 2149–2154

97 Levy BD, Fokin VV, Clark JM, Wakelam MJO, Petasis NA, Serhan CN (1999) Polyisoprenyl phosphate (PIPP) signaling regulates phospholipase D activity: A 'stop' signaling switch for aspirin-triggered lipoxin A_4. *FASEB J* 13: 903–911

98 Jozsef L, Zouki C, Petasis NA, Serhan CN, Filep JG (2002) Lipoxin A_4 and aspirin-triggered 15-epi-lipoxin A_4 inhibit peroxynitrite formation, NF-kappa B and AP-1 activation, and IL-8 gene expression in human leukocytes. *Proc Natl Acad Sci USA* 99: 13266–13271

99 Maderna P, Cottell DC, Berlasconi G, Petasis NA, Brady HR, Godson C (2002) Lipoxins induce actin reorganization in monocytes and macrophages but not in neutrophils: differential involvement of rho GTPases. *Am J Pathol* 160: 2275–2283

100 Qiu F-H, Devchand PR, Wada K, Serhan CN (2001) Aspirin-triggered lipoxin A_4 and lipoxin A_4 up-regulate transcriptional co-repressor NAB1 in human neutrophils. *FASEB J* 15: 2736–2738

101 Canny G, Levy O, Furuta GT, Narravula-Alipati S, Sisson RB, Serhan CN, Colgan SP (2002) Lipid mediator-induced expression of bactericidal/permeability-increasing protein (BPI) in human mucosal epithelia. *Proc Natl Acad Sci USA* 99: 3902–3907

102 Gewirtz AT, McCormick B, Neish AS, Petasis NA, Gronert K, Serhan CN, Madara JL (1998) Pathogen-induced chemokine secretion from model intestinal epithelium is inhibited by lipoxin A_4 analogs. *J Clin Invest* 101: 1860–1869

103 Maderna P, Godson C, Hannify G, Murphy M, Brady HR (2000) Influence of lipoxin A(4) and other lipoxygenase-derived eicosanoids on tissue factor expression. *Am J Physiol Cell Physiol* 279: C945–C953

104 Badr KF, DeBoer DK, Schwartzberg M, Serhan CN (1989) Lipoxin A_4 antagonizes cellular and *in vivo* actions of leukotriene D_4 in rat glomerular mesangial cells: evidence for competition at a common receptor. *Proc Natl Acad Sci USA* 86: 3438–3442

105 Wikström Jonsson E (1998) Functional characterization of receptors for cysteinyl leukotrienes in muscle. *Doctoral, Karolinska Institute*
106 Gronert K, Martinsson-Niskanen T, Ravasi S, Chiang N, Serhan CN (2001) Selectivity of recombinant human leukotriene D_4, leukotriene B_4, and lipoxin A_4 receptors with aspirin-triggered 15-epi-LXA_4 and regulation of vascular and inflammatory responses. *Am J Pathol* 158: 3–9
107 McMahon B, Mitchell D, Shattock R, Martin F, Brady HR, Godson C (2002) Lipoxin, leukotriene, and PDGF receptors cross-talk to regulate mesangial cell proliferation. *Faseb J* 16: 1817–1819
108 McMahon B, Stenson C, McPhillips F, Fanning A, Brady HR, Godson C (2000) Lipoxin A_4 antagonizes the mitogenic effects of leukotriene D_4 in human renal mesangial cells: Differential activation of MAP kinases through distinct receptors. *J Biol Chem* 275: 27566–27575
109 McMahon B, McPhilips F, Fanning A, Brady HR, Godson C (1998) Modulation of mesangial cell MAP kinase activities by leukotriene D_4 and lipoxin A_4. J *Am Soc Nephrol* 9: 355A
110 Raud J, Dahlén S-E, Sydbom A, Lindbom L, Hedqvist P (1988) Enhancement of acute allergic inflammation by indomethacin is reversed by prostaglandin E_2: apparent correlation with *in vivo* modulation of mediator release. *Proc Natl Acad Sci USA* 85: 2315–2319
111 Rossi AG, McCutcheon JC, Roy N, Chilvers ER, Haslett C, Dransfield I (1998) Regulation of macrophage phagocytosis of apoptotic cells by cAMP. *J Immunol* 160: 3562–3568
112 Pillinger MH, Capodici C, Han G, Weissmann G (1997) Inflammation and anti-inflammation: gating of cell/cell adhesion at the level of mitogen-activated protein kinases. *Ann NY Acad Sci* 832: 1–12
113 Kawahito Y, Kondo M, Tsubouchi Y, Hashiramoto A, Bishop-Bailey D, Inoue K-i, Kohno M, Yamada R, Hla T, Sano H (2000) 15-deoxy-$\Delta^{12,14}$-PGJ_2 induces synoviocyte apoptosis and suppresses adjuvant-induced arthritis in rats. *J Clin Invest* 106: 189–197
114 Serhan CN, Devchand PR (2001) Novel anti-inflammatory targets for asthma: A role for PPARγ? *Am J Respir Cell Mol Biol* 24: 658–661
115 Rossi A, Kapahi P, Natoli G, Takahashi T, Chen Y, Karin M, Santoro MG (2000) Anti-inflammatory cyclopentenone prostaglandins are direct inhibitors of IkappaB kinase. *Nature* 403: 103–108
116 Serhan CN, Clish CB, Brannon J, Colgan SP, Chiang N, Gronert K (2000) Novel functional sets of lipid-derived mediators with anti-inflammatory actions generated from omega-3 fatty acids *via* cyclo-oxygenase 2-nonsteroidal anti-inflammatory drugs and transcellular processing. *J Exp Med* 192: 1197–1204
117 Serhan CN, Hong S, Gronert K, Colgan SP, Devchand PR, Mirick G, Moussignac R-L (2002) Resolvins: a family of bioactive products of omega-3 fatty acid transformation circuits initiated by aspirin treatment that counter pro-inflammation signals. *J Exp Med* 196: 1025–1037

118 Levy BD, Clish CB, Schmidt B, Gronert K, Serhan CN (2001) Lipid mediator class switching during acute inflammation: signals in resolution. *Nature Immunol* 2: 612–619
119 FitzGerald GA, Patrono C (2001) The coxibs, selective inhibitors of cyclo-oxygenase-2. *N Engl J Med* 345: 433–442
120 Shinmura K, Tang X-L, Wang Y, Xuan Y-T, Liu S-Q, Takano H, Bhatnagar A, Bolli R (2000) Cyclo-oxygenase-2 mediates the cardioprotective effects of the late phase of ischemic preconditioning in conscious rabbits. *Proc Natl Acad Sci USA* 97: 10197–10202
121 Beckman BS, Despinasse BP, Spriggs L (1992) Actions of lipoxins A_4 and B_4 on signal transduction events in Friend erythroleukemia cells. *Proc Soc Exp Biol Med* 201: 169–173
122 Serhan CN, Hirsch U, Palmblad J, Samuelsson B (1987) Formation of lipoxin A by granulocytes from eosinophilic donors. *FEBS Lett* 217: 242–246
123 Feng Z, Godfrey HP, Mandy S, Strudwick S, Lin K-T, Heilman E, Wong PY-K (1996) Leukotriene B_4 modulates *in vivo* expression of delayed-type hypersensitivity by a receptor-mediated mechanism: regulation by lipoxin A_4. *J Pharmacol Exp Ther* 278: 950–956
124 Chavis C, Chanez P, Vachier I, Bousquet J, Michel FB, Godard P (1995) 5,15-diHETE and lipoxins generated by neutrophils from endogenous arachidonic acid as asthma biomarkers. *Biochem Biophys Res Commun* 207: 273–279
125 Christie PE, Spur BW, Lee TH (1992) The effects of lipoxin A_4 on airway responses in asthmatic subjects. *Am Rev Respir Dis* 145: 1281–1284
126 Thomas E, Leroux JL, Blotman F, Chavis C (1995) Conversion of endogenous arachidonic acid to 5,15-diHETE and lipoxins by polymorphonuclear cells from patients with rheumatoid arthritis. *Inflamm Res* 44: 121–124
127 Filep JG, Zouki C, Petasis NA, Hachicha M, Serhan CN (1999) Anti-inflammatory actions of lipoxin A_4 stable analogs are demonstrable in human whole blood: Modulation of leukocyte adhesion molecules and inhibition of neutrophil-endothelial interactions. *Blood* 94: 4132–4142
128 Ramstedt U, Serhan CN, Nicolaou KC, Webber SE, Wigzell H, Samuelsson B (1987) Lipoxin A-induced inhibition of human natural killer cell cytotoxicity: Studies on stereospecificity of inhibition and mode of action. *J Immunol* 138: 266–270
129 Leszczynski D, Ustinov J (1990) Protein kinase C-regulated production of prostacyclin by rat endothelium is increased in the presence of lipoxin A_4. *FEBS Lett* 263: 117–120
130 Dahlén S-E, Serhan CN (1991) Lipoxins: Bioactive lipoxygenase interaction products. In: Wong A, Crooke ST, (eds): *Lipoxygenases and Their Products*. Academic Press, San Diego, 235–276
131 Su SB, Gong W, Gao J-L, Shen W, Murphy PM, Oppenheim JJ, Wang JM (1999) A seven-transmembrane, G protein-coupled receptor, FPRL1, mediates the chemotactic activity of serum amyloid A for human phagocytic cells. *J Exp Med* 189: 395–402
132 Deng X, Ueda H, Su SB, Gong W, Dunlop NM, Gao J-L, Murphy PM, Wang JM (1999) A synthetic peptide derived from human immunodeficiency virus type 1 gp120 down-

regulates the expression and function of chemokine receptors CCR5 and CXCR4 in monocytes by activating the 7-transmembrane G-protein-coupled receptor FPRL1/ LXA_4R. *Blood* 94: 1165–1173
133 Su SB, Gao JI, Gong W, Dunlop NM, Murphy PM, Oppenheim JJ, Wang JM (1999) T21/DP107, a synthetic leucine zipper-like domain of the HIV-1 envelope gp41, attracts and activates human phagocytes by using G-protein-coupled formyl peptide receptors. *J Immunol* 162: 5924–5930
134 Le Y, Gong W, Tiffany HL, Tumanov A, Nedospasov S, Shen W, Dunlop NM, Gao J-L, Murphy PM, Oppenheim JJ et al (2001) Amyloid β42 activates a G-protein-coupled chemoattractant receptor, FPR-like-1. *J Neuroscience* 21: 1–5
135 Yang D, Chen Q, Schmidt AP, Anderson GM, Wang JM, Wooters J, Oppenheim JJ, Chertov O (2000) LL-37, the neutrophil granule- and epithelial cell-derived cathelicidin, utilizes formyl peptide receptor-like 1 (FPRL1) as a receptor to chemoattract human peripheral blood neutrophils, monocytes, and T cells. *J Exp Med* 192: 1069–1074
136 Bylund J, Christophe T, Boulay F, Nystrom T, Karlsson A, Dahlgren C (2001) Proinflammatory activity of a cecropin-like antibacterial peptide from Helicobacter pylori. *Antimicrob Agents Chemother* 45: 1700–1704
137 Le Y, Yazawa H, Gong W, Yu Z, Ferrans VJ, Murphy PM, Wang JM (2001) The neurotoxic prion peptide fragment PrP(106-126) is a chemotactic agonist for the G protein-coupled receptor formyl peptide receptor-like 1. *J Immunol* 166: 1448–1451
138 Dahlgren C, Christophe T, Boulay F, Madianos PN, Rabiet MJ, Karlsson A (2000) The synthetic chemoattractant Trp-Lys-Tyr-Met-Val-DMet activates neutrophils preferentially through the lipoxin A_4 receptor. *Blood* 95: 1810–1818
139 Nassar GM, Badr KF (1995) Role of leukotrienes and lipoxygenases in glomerular injury. *Miner Electrolyte Metab* 21: 262–270

Metabolism and physiological significance of anandamide and 2-arachidonoylglycerol, endogenous cannabinoid receptor ligands

Takayuki Sugiura, Seishi Kishimoto, Saori Oka, Maiko Gokoh and Keizo Waku

Faculty of Pharmaceutical Sciences, Teikyo University, Sagamiko, Tsukui-gun, Kanagawa 199-0195, Japan

Introduction

Arachidonic acid is known to be metabolized into a number of bioactive eicosanoids such as prostaglandins, thromboxanes, leukotrienes, lipoxins and mono- and dihydroxyeicosatetraenoic acids. It is well known that these bioactive eicosanoids are involved in diverse physiological and pathophysiological processes in mammalian tissues. In the last decade of the 20th century, two remarkable derivatives of arachidonic acid, i.e., N-arachidonoylethanolamine (anandamide) and 2-arachidonoylglycerol (2-AG) (Fig. 1) were reported to be new members of the bioactive lipids. Anandamide and 2-AG have unique structural characteristic in that they contain an intact arachidonoyl moiety in their molecules, thus differing from other eicosanoids. Both anandamide and 2-AG have been shown to act as endogenous cannabinoid receptor ligands. In this review, we focused on anandamide and 2-AG and described the metabolism and possible physiological significance of these molecules in mammalian tissues and cells including inflammatory cells and immune competent cells.

Cannabinoid receptors

Marijuana has been used as a traditional medicine and a pleasure-inducing drug for thousands of years around the world, especially in Asia. In the 1960s, Gaoni and Mechoulam [1] demonstrated that Δ^9-tetrahydrocannabinol (Δ^9-THC) (Fig. 1) is the major psychoactive ingredient of marijuana. Δ^9-THC is known to exert a variety of pharmacological effects on experimental animals and humans. For example, the administration of Δ^9-THC induces reduced spontaneous motor activity, immobility, analgesia, heightened sensory awareness, euphoria, hypothermia and impairment of short-term memory [2]. Δ^9-THC is also known to exert profound effects in various biological systems other than in the nervous system such as the suppression of immune responses. The mechanisms of these actions of Δ^9-THC, however, remained

Arachidonate Remodeling and Inflammation, edited by Alfred N. Fonteh and Robert L. Wykle

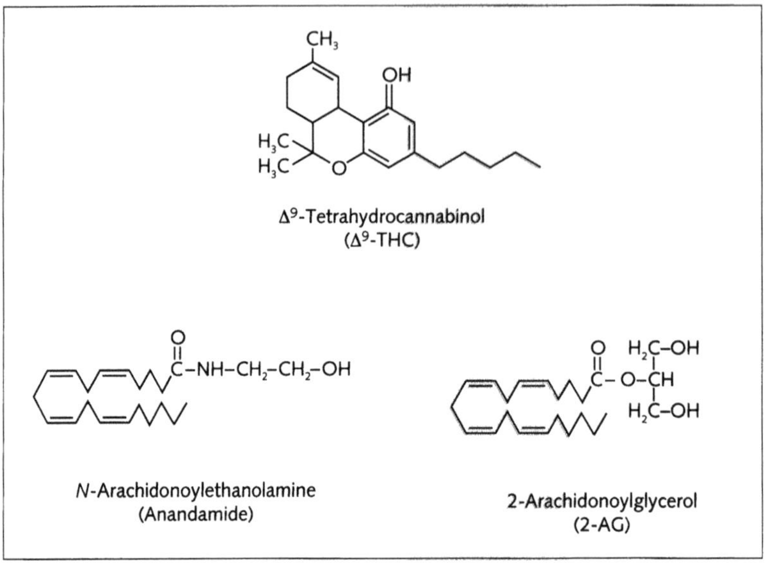

Figure 1
Chemical structures of Δ⁹-tetrahydrocannabinol (Δ⁹-THC), N-arachidonoylethanolamine (anandamide) and 2-arachidonoylglycerol (2-AG)

unclear until recently. The occurrence of specific binding site(s) for cannabinoids has been postulated based on the observation that (–)-Δ^9-THC exhibited potent pharmacological activities whereas its stereoisomer (+)-Δ^9-THC was far less potent. However, binding experiments of radiolabeled cannabinoids were unsuccessful until the late 1980s due to their highly lipophilic properties.

In 1988, Devane et al. [3] succeeded in showing the occurrence of specific binding sites for cannabinoids in rat brain synaptosomes using a radiolabeled synthetic cannabinoid, [^{3}H]CP55940. The K_d was 133 pM and the B_{max} was 1.85 pmol/mg protein. Finally, Matsuda et al. [4] cloned a cDNA encoding a cannabinoid receptor (CB1 receptor) from a rat brain cDNA library in 1990. The CB1 receptor is present in various mammalian tissues, especially in the nervous tissues, and is assumed to be involved in the attenuation of neurotransmission [5]. The CB1 receptor is a seven-transmembrane, G protein-coupled receptor, and contains 472 (human) or 473 (rat) amino acids. It is noteworthy that the whole-brain cannabinoid receptor density is similar to the whole-brain densities of receptors for glutamate and GABA [6]. Indeed, the CB1 receptor is one of the most abundant G protein-coupled receptors expressed in the brain. Among the various brain regions, the CB1 receptor is especially abundant in the substantia nigra, globus pallidus, molecular layer of the cerebellum, hippocampus and cerebral cortex. The CB1 receptor is assumed to be involved in the regulation of cognition, memory and motor activity [5].

In 1993, Munro et al. [7] cloned a cDNA encoding another type of cannabinoid receptor (CB2 receptor) from an HL-60 cell cDNA library. The CB2 receptor is also a seven-transmembrane, G protein-coupled receptor, and consists of 360 amino acid residues. The CB1 receptor and the CB2 receptor share a 44% overall identity (68% identity for the transmembrane domains). The CB2 receptor is abundantly present in various lymphoid tissues such as the marginal zone of the spleen, the cortex of the lymph nodes and the nodular corona of Peyer's patches [6], and is expressed in various types of leukocytes with a rank order of B lymphocytes > natural killer cells > monocytes/macrophages > polymorphonuclear leukocytes > T8 lymphocytes > T4 lymphocytes [8–10]. The CB2 receptor is assumed to participate in the regulation of inflammatory reactions and immune responses [6, 8–10], although the details of the physiological functions of the CB2 receptor still remain to be clarified.

The mechanisms of the diverse actions of the natural and synthetic cannabinoids have not yet been fully elucidated. However, it has been assumed that various cannabinoids including Δ^9-THC exert their biological activities mainly by acting on these cannabinoid receptors (CB1 and CB2) although there remains the possibility that non-CB1 and non-CB2 binding sites for the cannabinoids exist in mammalian tissues.

Identification of endogenous cannabinoid receptor ligands

The discovery of the specific receptors for cannabinoids prompted the search for endogenous ligand(s). In 1992, Devane et al. [11] isolated anandamide from pig brain as the first endogenous cannabinoid receptor ligand. They demonstrated that anandamide binds to the brain cannabinoid receptor (K_i = 52 nM) and inhibits the mouse twitch response. Anandamide also induces the inhibition of the voltage-gated Ca^{2+} channels, the activation of an inwardly rectifying K^+ current, the stimulation of [^{35}S]GTPγS binding to G proteins, the activation of mitogen-activated protein kinase (MAP kinase) and the neural form of focal adhesion kinase, the inhibition of neurotransmitter release, the inhibition of long-term potentiation in hippocampal slices, the impairment of memory, reduced spontaneous motor activities, immobility, hypothermia, analgesia, vasodilation, hypotension, bradycardia, the growth inhibition of human breast and prostate cancer cells and the inhibition of the sperm acrosome reaction [5, 12–16]. It should be noted, however, that anandamide acted as a partial agonist in some cases [16]. It is curious that an endogenous natural ligand acts as a partial agonist at its own receptor; natural ligands usually act as full agonists. Therefore, it seems unlikely that anandamide acts as an endogenous cannabinoid receptor agonist with significant physiological importance, although this still remains a controversial subject.

In addition to acting as an endogenous cannabinoid receptor ligand, anandamide was shown to act as an endogenous ligand for an ion channel-type vanilloid receptor (VR1 receptor). Moreover, anandamide can also bind to several other types of ion channels. Therefore, it seems possible that some of the anandamide effects mentioned above are mediated through receptors or binding sites other than the cannabinoid receptors. It is necessary to keep this in mind when interpreting the experimental results concerning the biological activities of anandamide.

On the other hand, in the mid-1990s, we isolated 2-AG, a unique molecular species of monoacylglycerol, from rat brain, and found that it acts as an endogenous cannabinoid receptor ligand [17, 18]. Independently and concurrently, Mechoulam et al. [19] also isolated 2-AG from the canine gut. It has been shown that 2-AG binds to the cannabinoid receptors (CB1 and CB2), and exhibits several cannabimimetic activities *in vitro* and *in vivo* [5, 12–16, 18–21]. Despite its possible physiological significance, however, much less attention has been paid to 2-AG compared with anandamide until recently. Indeed, the physiological significance of 2-AG has very often been overlooked. Unequivocally, 2-AG is a quite noticeable molecule from a variety of viewpoints. We have focused on 2-AG and intensively investigated the biological activities and physiological roles of 2-AG since 1995.

In 1996, we found by chance that 2-AG induces a rapid transient increase in the intracellular free Ca^{2+} concentrations ($[Ca^{2+}]_i$) in neuroblastoma x glioma hybrid NG108-15 cells expressing the CB1 receptor [22–24]. We confirmed that the response induced by 2-AG was blocked by pretreatment of the cells with SR141716A, a cannabinoid CB1 receptor-specific antagonist, indicating that the response induced by 2-AG was mediated through the CB1 receptor [22, 24]. 2-AG-induced Ca^{2+} transients were also observed in promyelocytic HL-60 cells expressing the CB2 receptor [25]. We confirmed that the response was abolished by pretreatment of the cells with SR144528, a cannabinoid CB2 receptor-specific antagonist, indicating that the response induced by 2-AG was mediated through the CB2 receptor [25]. Free arachidonic acid failed to exhibit any agonistic activity in either case, indicating that arachidonic acid that may be generated from 2-AG during incubation is not involved in the response. We also confirmed that the pretreatment of the cells with indomethacin, a cyclo-oxygenase inhibitor, or nordihydroguaiaretic acid, a lipoxygenase inhibitor, did not affect the response, suggesting that arachidonic acid metabolites are not involved.

We then examined the mechanism underlying the rapid transient increase in $[Ca^{2+}]_i$ induced by 2-AG. We found that pretreatment of the cells with pertussis toxin abolished the response induced by 2-AG, indicating that Gi/Go is involved in the response. We also found that pretreatment of the cells with U73122, a phospholipase C inhibitor abrogated the response induced by 2-AG, whereas pretreatment of the cells with U73343, an inactive analog of U73122, did not affect the response [23], suggesting that phospholipase C is involved in the response induced by 2-AG. We assumed that the 2-AG-induced rapid transient increase in $[Ca^{2+}]_i$ involves:

1) 2-AG binding to the cannabinoid receptors (CB1 and CB2) followed by
2) activation of Gi/Go, and
3) the liberated βγ subunit of Gi/Go then stimulates phospholipase Cβ to enhance the production of inositol 1,4,5-trisphosphate, thereby increasing $[Ca^{2+}]_i$.

We next investigated in detail the structure-activity relationship of 2-AG and other cannabinoid receptor ligands using NG108-15 cells (for the CB1 receptor) and HL-60 cells (for the CB2 receptor). We found that an ether-linked analog of 2-AG (2-AG ether) possesses substantial biological activity, indicating that the structure of 2-AG itself, but not its metabolite, is actually recognized by the receptor molecules (CB1 and CB2) [24, 25]. In contrast to 2-AG, anandamide was found to act as a weak partial agonist toward either the CB1 receptor or the CB2 receptor. The activity of Δ^9-THC, a major psychoactive constituent of marijuana, was also low [23, 25]. We further examined the activities of various 2-AG analogs as well as classical and synthetic cannabinoids. The activities of the monoacylglycerols containing various saturated, monoenoic and dienoic fatty acids were almost inactive. 1(3)-AG, the positional isomers of 2-AG, exhibited appreciable agonistic activities, yet their activities were much lower than those of 2-AG. Among the various naturally occurring analogs, 2-AG acted as the most efficacious agonist toward both types of cannabinoid receptors (CB1 and CB2) [22–25]. This has also been confirmed by several investigators [26–28]. We proposed that 2-AG, but not anandamide, is the intrinsic natural ligand for the cannabinoid receptors (CB1 and CB2) and both receptors are originally and primarily 2-AG receptors [20, 21, 23–25].

Recently, Hanus et al. [29] reported that 2-AG ether is present in pig brain. They described that 2-AG ether is the third endogenous cannabinoid receptor ligand (they called it "noladin ether"). Fezza et al. [30] also reported that a small amount of 2-AG ether is present in rat brain. We examined in detail whether 2-AG ether is actually present in the brain. However, we did not detect any appreciable amounts of the 2-AG ether in the brains of various mammalian species such as mouse, rat, hamster, guinea pig, and pig (< 0.2 pmol/g tissue, if at all present) (S. Oka and T. Sugiura, unpublished results). The ether bond is known to be located exclusively at the 1-position of the glycerol backbone in mammalian tissues [31–33]. It is questionable whether 2-AG ether is a naturally-occurring molecule and acts as an endogenous ligand for the cannabinoid receptors.

Tissue levels of endogenous cannabinoid receptor ligands

Previously, Schmid and co-workers [34] estimated the levels of the N-acylethanolamines including anandamide by GC/MS. They detected anandamide in the brains of several animal species. The levels of anandamide in the sheep brain were negligible, whereas low levels were found in the pig and cow brain (17 and

10 pmol/g tissue, respectively). These values were, however, significantly lower than the levels of anandamide in the pig brain (370 pmol/g tissue) reported by Devane et al. [11]. The exact reason for this difference is uncertain. However, it is possible that a considerable amount of anandamide was generated during the postmortem period in the latter case. Schmid and co-workers [34] demonstrated that the levels of anandamide in isolated pig brain were dramatically augmented when kept at ambient temperature. The generation of anandamide in the brain during the postmortem period was also reported by Kempe et al. [35] and Felder et al. [36].We detected a small amount of anandamide in fresh rat brain (4.3 pmol/g tissue), which accounts for only 0.7% of the total N-acylethanolamine [37]. The predominant species of N-acylethanolamine detected in the rat brain were the N-16:0 and N-18:0 species.

The levels of anandamide in the brains of various mammalian species have also been studied by a number of investigators (reviewed in [21]). The levels of anandamide in the rat brain were 3.37–15 pmol/g tissue and those in mouse brain were 10–15 pmol/g tissue. The level of anandamide was variable depending on the physiological and pathophysiological conditions of the brains. Di Marzo et al. [38] reported that the levels of anandamide in the hypothalamus markedly decreased following the administration of leptin. On the other hand, Giuffrida et al. [39] reported that D2-like dopamine receptor activation evokes the release of anandamide in the dorsal striatum of freely moving rats. Recently, Baker et al. [40] reported that the levels of anandamide in the brain and spinal cord were elevated in spastic mice with chronic relapsing experimental allergic encephalomyelitis.

Anandamide was detected in various peripheral tissues as well (reviewed in [21]), such as the bovine retina (64 pmol/g tissue), the human heart (10 pmol/g tissue), human spleen (15 pmol/g tissue), rat heart (21.2–126 pmol/g tissue), rat liver (19.7–77.1 pmol/g tissue), rat kidney (8–164 pmol/g tissue, 0.32–0.35 pmol/μmol lipid P), rat spleen (6 pmol/g tissue, 0.34 pmol/μmol lipid P), rat thymus (40.6–137 pmol/g tissue), rat testis (2.9–43.5 pmol/g tissue, 0.25–0.31 pmol/μmol P), mouse uterus (2215–20982 pmol/g tissue), rat skin (23 pmol/g tissue, 49 pmol/g tissue), and rat paw skin (0.69 pmol/mg of extracted lipids). Schmid et al. [41] reported that mouse peritoneal macrophages contain a small amount of anandamide (0.25 pmol/μmol lipid P). Anandamide was also detected in rat plasma collected by decapitation (144 pmol/ml) or by cardiac puncture (3.1 pmol/ml) and in the sera from normal donors (4 pmol/ml) and patients with endotoxin shock (18 pmol/ml) (reviewed in [21]). On the other hand, Yang et al. [42] described that the levels of anandamide in the rat brain, spleen, testis, liver, lung and heart were below the levels of quantification achievable (< 0.1 pmol/mg protein).

The available information as to the tissue levels of 2-AG is still limited. Mechoulam et al. [19] detected 2-AG in the canine gut, although they did not quantify it. We detected 3.25 nmol/g tissue of arachidonoylglycerols (2-AG plus 1(3)-AG) [18] and 3.36 nmol/g tissue of 2-AG in rat brain [43]. The level of 2-AG in the brains obtained from rats sacrificed in liquid nitrogen was 0.23 nmol/g tissue [44].

Stella et al. [45] also detected 4.0 nmol/g tissue of 2-AG in the rat brain. Bisogno et al. [46] examined the levels of 2-AG in various brain regions. 2-AG was also detected in the peripheral nervous system. Huang et al. [47] reported that 2-AG is present in the rat sciatic nerve (0.052 nmol/g tissue), lumbar spinal cord (0.432 nmol/g tissue) and lumbar dorsal root ganglion (0.370 nmol/g tissue). 2-AG was also detected in the rat retina (2.97 nmol/g tissue) and bovine retina (1.63 nmol/g tissue). We further estimated the levels of 2-AG in several rat tissues [43]. The levels of 2-AG in the rat liver, spleen, lung, kidney and plasma were 1.15 nmol/g tissue, 1.17 nmol/g tissue, 0.78 nmol/g tissue, 0.98 nmol/g tissue and 0.012 nmol/ml, respectively. The tissue levels of 2-AG in rats have also been reported by Schmid and co-workers [48]: 13.04-6.19 pmol/μmol lipid P (kidney), 3.21-3.05 pmol/μmol lipid P (testis), 10.94 pmol/μmol lipid P (heart), 29.03 pmol/μmol lipid P (spleen) and 5.05 pmol/μmol lipid P (liver). 2-AG was also detected in the sera from normal donors (10 pmol/ml) and patients with endotoxin shock (30 pmol/ml) and in human milk (0.33 μg/ml) and rat paw skin (51.1 pmol/mg of extracted lipids) (reviewed in [21]).

Noticeably, the levels of 2-AG in the brain changed following several treatments of experimental animals. We found that the amount of 2-AG in rat brain was significantly augmented following decapitation [49]. Di Marzo et al. [50] demonstrated that the levels of 2-AG in the globus pallidus were augmented in reserpine-treated rats. Baker et al. [40] demonstrated that the levels of 2-AG in the brain and spinal cord were elevated in spastic mice with chronic relapsing experimental allergic encephalomyelitis. Interestingly, the levels of 2-AG as well as anandamide in the rat hypothalamus were significantly reduced in mice injected with leptin compared with the control [38], suggesting that these endogenous cannabinoid receptor ligands are involved in the regulation of appetite.

Biosynthesis and degradation of endogenous cannabinoid receptor ligands

Evidence is gradually accumulating that anandamide is produced in relatively small amounts in a variety of tissues and cells upon stimulation. Di Marzo et al. [51] demonstrated that rat brain neurons generated anandamide when stimulated with ionomycin or several membrane-depolarizing agents such as kainate, high K^+ and 4-aminopyridine. Di Marzo et al. also demonstrated that anandamide was produced in ionomycin-treated J774 macrophages [52, 53], ionomycin-treated RBL-2H3 cells [53] and phospholipase D-treated N18TG2 neuroblastoma cells [52]. Hansen and co-workers [54] investigated whether N-acylethanolamine is produced in glutamate- or A23187- stimulated mouse cortical neurons in culture. They showed that the generation of N-acylethanolamine takes place in stimulated cells prelabeled with [^{3}H] ethanolamine, although they could not detect the generation of anandamide when the cells were prelabeled with [^{3}H] arachidonic acid. Later, they detected the formation of [^{14}C] anandamide in NaN_3-treated neurons [55]. The generation of anandamide was also observed in ionomycin-stimulated rat macrophages [56], LPS-, platelet-activating factor- and Δ^9-THC-stimulated RAW264.7 mouse macrophages [57], veratridine-, 4-aminopyridine- and A23187-stimulated SK-N-SH neuroblastoma cells [58] and in the periaqueductal gray region of the rat brain following electrical stimulation and the subcutaneous injection of formalin [59]. On the other hand, we found that the level of anandamide was augmented in inflamed rat testis, although the augmentation of the levels of saturated and monoenoic species was much more prominent [60]. Kuwae et al. [61] also demonstrated the generation of anandamide as well as other molecular species of N-acylethanolamine in mouse peritoneal macrophages in culture supplemented with ethanolamine.

Anandamide can be enzymatically formed *via* two independent synthetic pathways (Fig. 2) (reviewed in [21]). One is the direct N-acylation of ethanolamine and the other is transacylase-phosphodiesterase-mediated synthesis (Schmid's pathway). The first pathway is catalyzed by the reverse reaction of an anandamide amidohydrolase/fatty acid amide hydrolase [62–64], suggesting that the formation of anandamide *via* this pathway may not be physiologically relevant, yet there remains the possibility that a significant amount of anandamide can be formed *via* this pathway if high concentrations of arachidonic acid and ethanolamine are co-localized at some sites within the cell. The second pathway (Schmid's pathway) is the formation of anandamide from pre-existing N-arachidonoyl phosphatidylethanolamine (PE) through the action of a phosphodiesterase. This enzyme reaction has been assumed to be the major synthetic route for various N-acylethanolamines such as N-palmitoyl- and N-stearoyl-ethanolamine [65]. Several investigators have demonstrated that anandamide can be formed from N-arachidonoyl PE through the action of a phosphodiesterase [37, 51]. This is probably the major synthetic pathway for anandamide in various mammalian tissues. Nevertheless, this pathway does not appear to be able to generate a large amount of anandamide, because the tissue level of N-arachidonoyl PE is usually very low. The reason for this is that N-arachidonoyl PE is synthesized from PE and arachidonic acid, esterified at the 1-position of the glycerophospholipids through the action of a transacylase [37, 66]. It should be noted that the level of arachidonic acid esterified at the 1-position of glycerophospholipids is usually very low. The absence of efficient synthetic pathways for anandamide coincides with the observation that the tissue levels of anandamide are generally low (in the order of pmol/g tissue) except in a few cases [21].

The degradation of anandamide is catalyzed by two separate types of anandamide amidohydrolase/fatty acid amide hydrolases; the optimal pH of one isoform is 8.5–10 [67] and the optimal pH of the other is 5 [68]. The gene encoding the first enzyme protein has already been cloned by Cravatt et al. [69]. The second enzyme identified and purified by Ueda et al. [68] is less sensitive to several inhibitors than the first anandamide amide hydrolase/fatty acid amide hydrolase having the optimal pH of 8.5–9 and efficiently degraded N-palmitoylethanolamine. The role allotment of these two types of anandamide amidohydrolase/fatty acid amide hydrolase

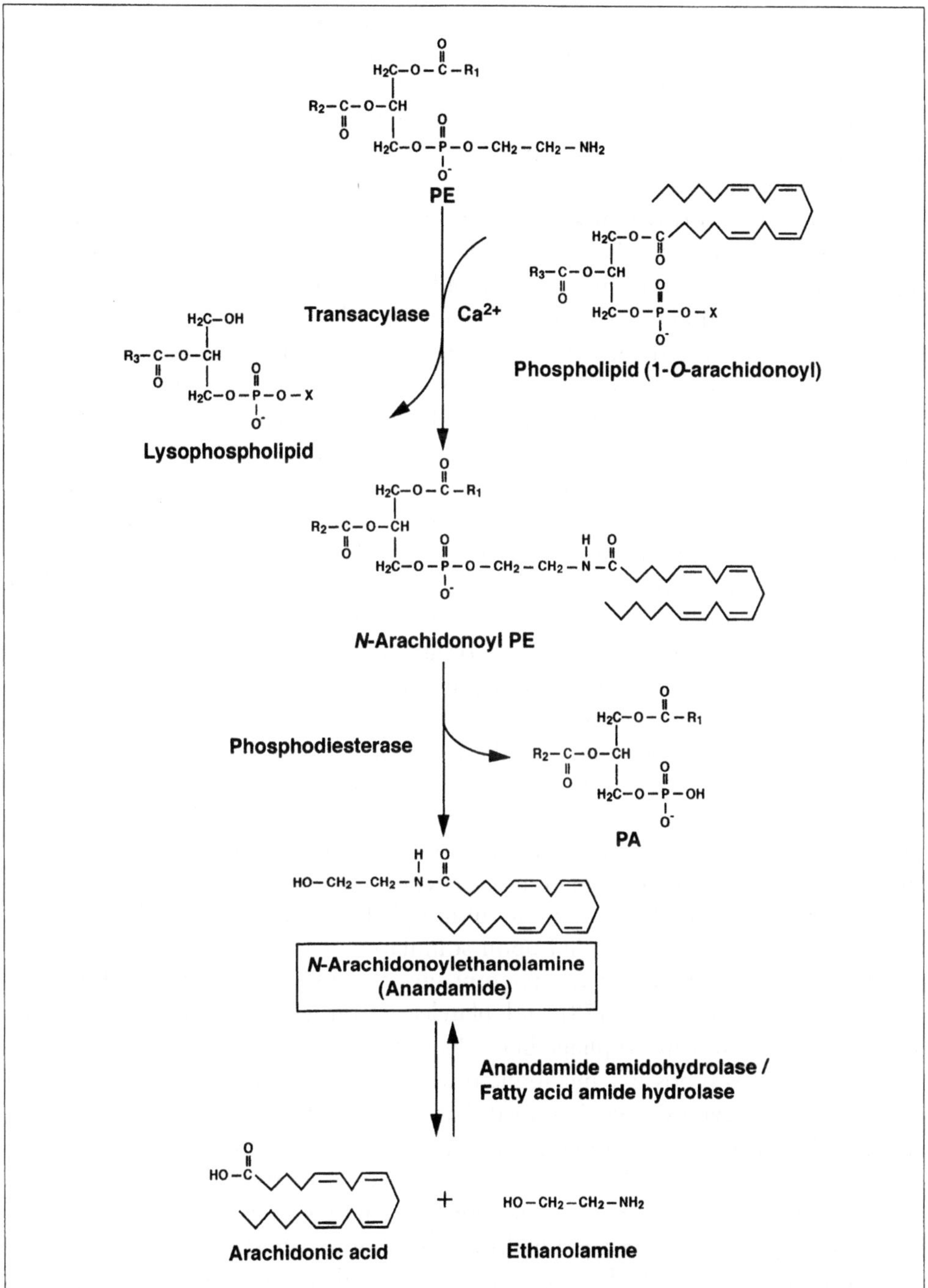

Figure 2
Biosynthesis and degradation of anandamide

remains to be clarified. The anandamide amidohydrolase/fatty acid amide hydrolase is capable of producing anandamide if the concentrations of arachidonic acid and ethanolamine are high enough as already described.

In contrast to anandamide, 2-AG is present in relatively large amounts in various mammalian tissues (the order of nmol/g tissue) [20, 21]. Noticeably, the levels of 2-AG were rapidly elevated in tissues and cells following stimulation [20, 21]. In addition, it is noteworthy that a significant portion of newly formed 2-AG can be released from stimulated cells [20, 21]. These characteristics appear to be favorable in acting as an intercellular mediator derived from stimulated cells.

The first description in the literature concerning the generation of arachidonoylglycerols in stimulated tissues and cells was that in thrombin-stimulated platelets reported by Prescott and Majerus in 1983 [70]. Later, the generation of arachidonoylglycerols in platelet-derived growth factor-stimulated Swiss 3T3 cells [71] and in bradykinin-stimulated rat dorsal ganglion neurons [72] was also demonstrated. However, at that time, the physiological significance of 2-AG as an endogenous cannabinoid receptor ligand was unknown. The generation of 2-AG as an endogenous cannabinoid receptor ligand was first described in ionomycin-stimulated N18TG2 cells [73], and in electrically stimulated rat hippocampal slices and ionomycin-stimulated neurons [45]. We also investigated the generation of 2-AG and found that the rapid generation of 2-AG occurs in the rat brain homogenate during incubation in the presence of Ca^{2+} (T. Sugiura and S. Oka, unpublished results) and in thrombin- or A23187-stimulated human umbilical vein endothelial cells [74]. Furthermore, we recently found that the levels of 2-AG in the rat brain were dramatically elevated (six-fold) following the intraperitoneal injection of picrotoxinin, a central nervous system stimulant [44]. In this case, the generation of monoacylglycerols other than arachidonoylglycerols was negligible or very small. The generation of 2-AG was also observed in the carbachol-treated rat aorta [75] and ethanol-treated cerebellar granule neurons in culture [76]. Several types of blood cells or inflammatory cells produce 2-AG upon stimulation such as in LPS-stimulated rat platelets [77], LPS-stimulated rat macrophages and LPS- or ionomycin-stimulated J774 macrophage-like cells [78] and platelet-activating factor-stimulated human platelets and P388D1 macrophage-like cells [79].

There are several possible metabolic routes for the generation of 2-AG in mammalian tissues. Previously, we described that 2-AG can be formed from arachidonic acid-enriched membrane phospholipids, such as inositol phospholipids, through the combined actions of phospholipase C and diacylglycerol lipase or through the combined actions of phospholipase A_1 and phospholipase C (Fig. 3) [18, 20, 21]. The first pathway, involving the rapid hydrolysis of inositol phospholipids by phospholipase C and subsequent hydrolysis of the resultant diacylglycerol by diacylglycerol lipase, was described two decades ago by Prescott and Majerus as a degradation pathway for arachidonic acid-containing diacylglycerol in platelets [70]. Stella et al. [45] demonstrated that these enzyme activities are involved in the ionomycin-

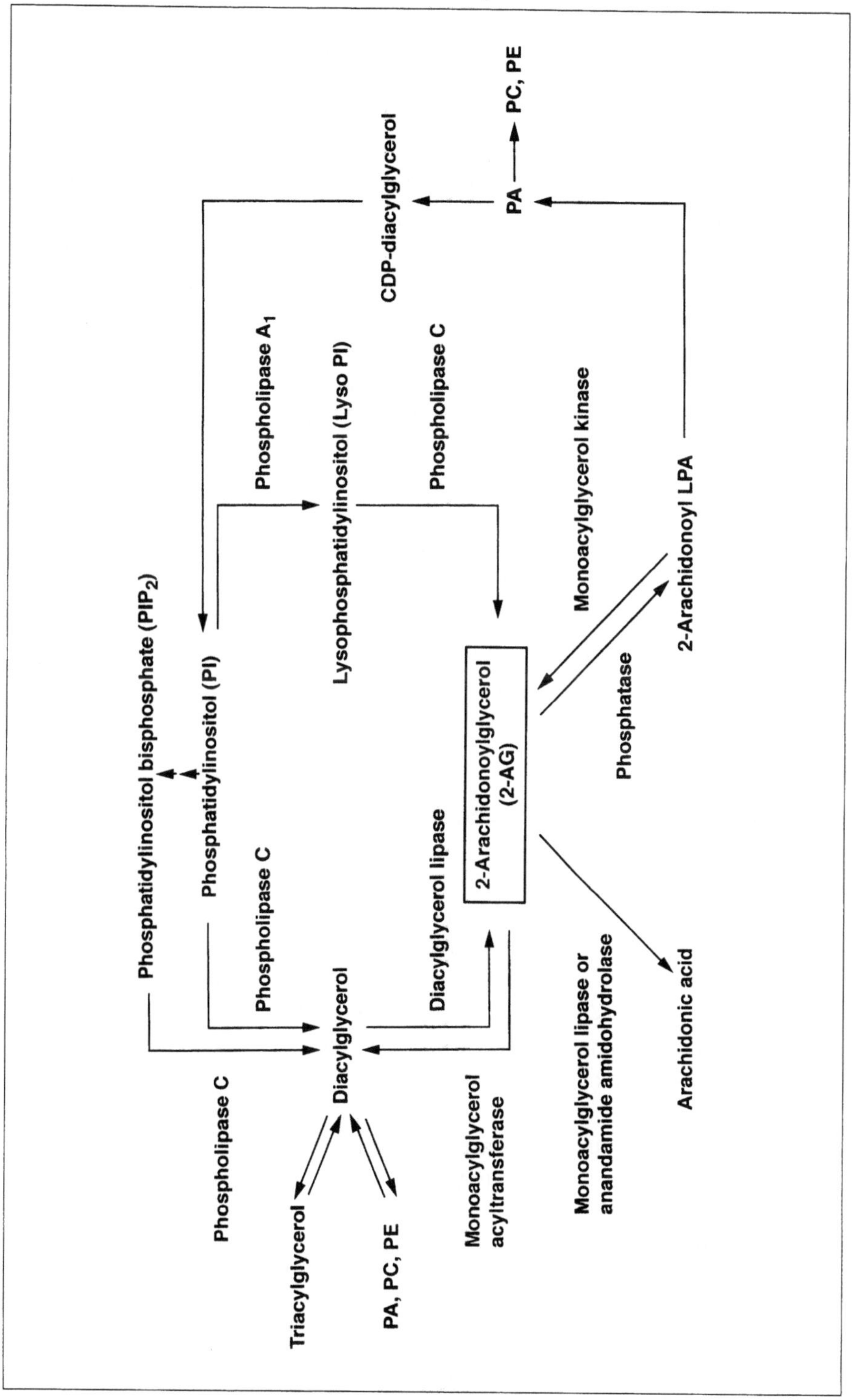

Figure 3
Biosynthesis and degradation of 2-AG

induced generation of 2-AG in cultured neurons using metabolic inhibitors. Recently, we confirmed that this pathway is important for the depolarization-induced generation of 2-AG in rat brain synaptosomes (T. Sugiura and S. Oka, unpublished results).

The second pathway involves the hydrolysis of phosphatidylinositol (PI) by phospholipase A_1 and the hydrolysis of the resultant lysoPI by a specific phospholipase C. Noticeably, lysoPI-specific phospholipase C is distinct from the various types of phospholipase Cs which act on other inositol phospholipids, and is localized in the synaptosomes [80]. Therefore, it is possible that this unique enzyme may also be involved in the generation of 2-AG in synapses. In addition to these two pathways, there are several possible routes for the generation of 2-AG, such as the generation from 2-arachidonoyl LPA [81] or 2-arachidonoyl PA [82]. We obtained evidence that a substantial amount of 2-arachidonoyl LPA is present in rat brain and that the formation of 2-AG from 2-arachidonoyl LPA takes place in the rat brain homogenate [81]. The biosynthetic pathways for 2-AG appear to differ, depending on the types of tissues and cells, and the types of stimuli. A full understanding of the mechanism and regulation of the biosynthesis of 2-AG awaits further investigations.

As for the degradation of 2-AG, 2-AG is metabolized by monoacylglycerol lipase similar to other monoacylglycerols [83]. In addition to this, several investigators have demonstrated that anandamide amidohydrolase/fatty acid amide hydrolase is also able to metabolize 2-AG [84, 85]. A part of 2-AG may be degraded by anandamide amidohydrolase/fatty acid amide hydrolase in some cases.

Physiological roles of endogenous cannabinoid receptor ligands in the nervous system

Anandamide was the first endogenous cannabinoid receptor ligand to be found. To date, it has been established that anandamide possesses strong binding activity toward the cannabinoid receptors and exhibits a variety of cannabimimetic activities *in vitro* and *in vivo* [5, 12–16]. However, evidence has gradually accumulated recently against the physiological significance of anandamide as the natural ligand for the cannabinoid receptors:

1) The levels of anandamide in tissues are usually low except in a few cases.
2) No selective and efficient synthetic pathway for anandamide has hitherto been found.
3) Anandamide, as well as Δ^9-THC, acted as a partial agonist at least in some cases.

Among them, the final issue is particularly important, because it is unusual that an endogenous natural ligand acts as a partial agonist at its own receptor. Hence, it is questionable whether anandamide actually serves as an endogenous cannabinoid receptor ligand of great physiological significance in living animals. Recently, Zygmunt et al. [86] demonstrated that anandamide is a potent agonist of the vanilloid receptor. Anandamide is also known to regulate the background K^+ channel TASK-1 and several other ion channels [21, 87, 88]. It is possible that the physiological significance of anandamide may exist in its being an endogenous ligand of the receptor(s) other than the cannabinoid receptors or as a modulator of ion channels. Alternatively, there may exist some other unknown physiological functions of anandamide. In any case, further studies are necessary to clarify the exact physiological significance and functions of anandamide.

2-AG was the second endogenous cannabinoid receptor ligand to be discovered and was found to act as a full agonist at the cannabinoid CB1 receptor abundantly expressed in the nervous system. Noticeably, 2-AG can be rapidly and selectively produced from a variety of cells upon stimulation. The tissue levels of 2-AG are usually markedly higher than those of anandamide. In contrast to anandamide, 2-AG does not bind to the vanilloid receptor. Based on these observations, we have proposed that 2-AG is the intrinsic natural ligand for the cannabinoid CB1 receptor, and the cannabinoid CB1 receptor is originally and primarily a 2-AG receptor [20, 21, 23–25].

2-AG is present in relatively high amounts in mammalian brains; it seems unlikely that 2-AG induces psychedelic reactions such as heightened sensory awareness, dissociation of ideas, errors in judgment of time and space and hallucinations in living animals. Recently, we obtained evidence that a substantial amount of 2-AG was generated and released from depolarized rat brain synaptosomes (T. Sugiura and S. Oka, unpublished results). We also found that the treatment of the synaptosomes with SR141716A, a cannabinoid CB1 receptor-specific antagonist, enhanced the release of glutamate upon depolarization (T. Sugiura and S. Oka, unpublished results). These results strongly suggest that the endogenous ligand of the cannabinoid CB1 receptor, that is, 2-AG, plays an essential role in the attenuation of neurotransmission. We proposed that the physiological role of 2-AG in the synapse is as follows: 2-AG is generated through increased phospholipid metabolism, especially inositol phospholipid breakdown, at the presynapses and/or postsynapses during accelerated synaptic transmission. 2-AG is rapidly released from the neuronal cells to the synaptic cleft, because 2-AG is a membrane-permeable molecule. The released 2-AG then binds to the cannabinoid CB1 receptor mainly expressed in the presynapse, and suppresses neuronal excitation through inhibiting Ca^{2+} channels or activating K^+ channels, thereby diminishing subsequent neurotransmitter release (Fig. 4) [20, 21].

Such a negative feedback regulation mechanism should be effective in calming stimulated neurons after excitation. The 2-AG- and cannabinoid CB1 receptor-dependent negative feedback regulation of neurotransmission would be of great physiological significance, because sustained activation of neuronal cells is known to cause cell exhaustion and may lead to neuronal cell death. 2-AG probably plays

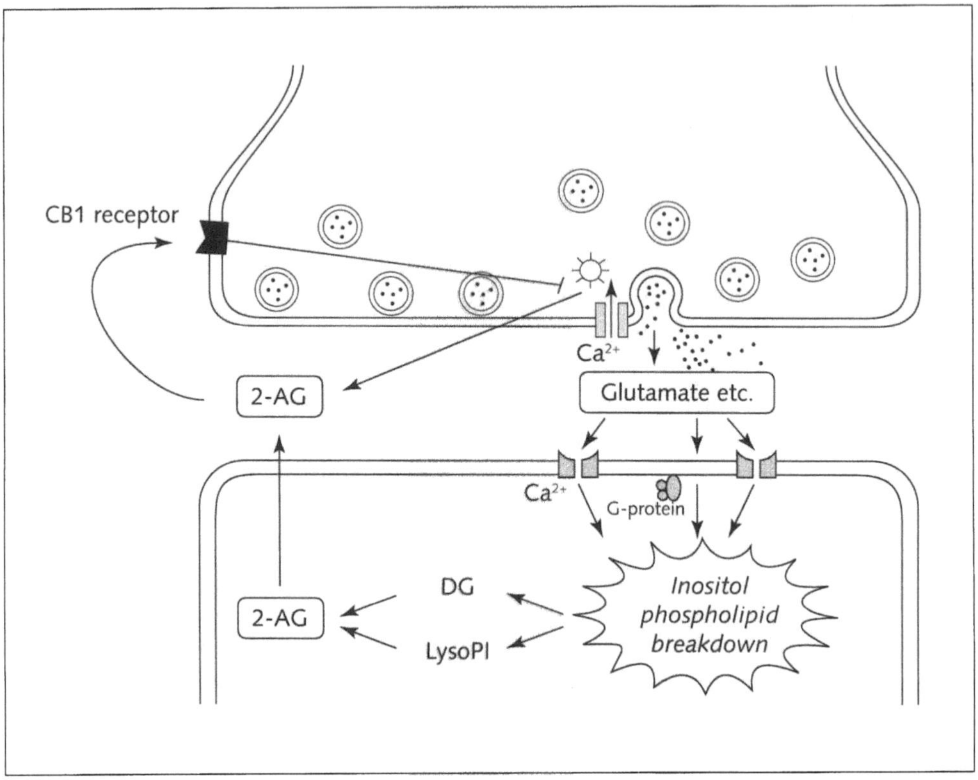

Figure 4
Physiological significance of the cannabinoid CB1 receptor and 2-AG in the synapse

such an important regulatory role in the neurotransmission in cooperation with other inhibitory neurotransmitters and neuromodulators such as GABA and adenosine *in vivo*. As for 2-AG in the brain, Stella et al. [45] demonstrated that 2-AG suppresses long-term potentiation in rat hippocampal slices, and Ameri and Simmet [89] reported that 2-AG reduces neuronal excitability in rat hippocampal slices. Panikashvili et al. [90] also recently proposed the protective role of 2-AG in the brain. On the other hand, Δ^9-THC, the major psychoactive constituent of marijuana, is a partial agonist of the cannabinoid CB1 receptor. It is possible that Δ^9-THC interferes with the actions of the physiological ligand 2-AG, thereby inducing several pharmacological effects such as altered perception and hallucination.

Recently, several investigators have reported that the endogenous cannabinoid receptor ligand derived from the postsynapse is the effector molecule of the depolarization-induced suppression (DSE) [91–93]. Based on our previous experimental results on 2-AG in the nervous system as previously mentioned, we considered that the cannabimimetic molecules involved in this response is 2-AG. Unlike anandamide, 2-AG has long been outside the limelight until recently. 2-AG is now in the spotlight. Details of the physiological functions and the regulation of the biosynthesis, degradation, release and re-uptake of 2-AG in synapses will be clarified in the near future.

Physiological roles of endogenous cannabinoid receptor ligands during inflammation and immune response

The cannabinoid CB2 receptor is abundantly expressed in various types of inflammatory cells and immune cells such as natural killer cells, B lymphocytes and macrophages [6, 8–10], suggesting that its endogenous ligand plays an important role in the regulation of inflammation and immune response. Previously, we found that the structure of 2-AG is strictly recognized by the CB2 receptor [23–25]. 2-AG acted as the most efficacious agonist among the various structural analogs; 2-AG acted as a full agonist at the CB2 receptor whereas anandamide acted as a weak partial agonist. Gonsiorek et al. [27] also demonstrated that 2-AG is a full agonist and anandamide is a partial agonist using the membrane fraction of the Sf9 cells transfected with the human CB2 receptor cDNA. These observations strongly suggest that 2-AG, and not anandamide, is the true endogenous ligand for the cannabinoid CB2 receptor as in the case of the CB1 receptor [20, 21].

Only limited information is thus far available concerning the biological activities of 2-AG toward inflammatory cells and immune cells. Previously, Lee et al. [94] reported that 2-AG affects lymphocyte proliferation. Ouyang et al. [95] also demonstrated that 2-AG suppresses the IL-2 gene expression in murine T lymphocytes through down-regulation of the nuclear factor. Very recently, Chang et al. [96] demonstrated that 2-AG inhibited the production of IL-6 in J774 macrophage-like cells. However, it remains unclear whether these effects of 2-AG are mediated through the cannabinoid receptor. On the other hand, we recently found that 2-AG induces the rapid phosphorylation and activation of the p42/44 MAP kinase in HL-60 cells [97]. 2-AG-induced activation of the p42/44 MAP kinase was abolished when the cells were pretreated with either SR144528 or PTX, indicating that the response was mediated through the CB2 receptor and Gi/Go. CP55940, a synthetic cannabinoid, also induced the activation of the p42/44 MAP kinase, whereas only a slight activation of the p42/44 MAP kinase was observed with anandamide. In addition to the p42/44 MAP kinase, we found that rapid phosphorylation of the p38 MAP kinase and c-Jun N-terminal kinase takes place in 2-AG-stimulated HL-60 cells (Fig. 5). A similar 2-AG-induced activation of the p38 MAP kinase and c-Jun N-terminal kinase has also been reported by several investigators [98, 99].

Recently, we found that 2-AG induces the migration of HL-60 cells differentiated into macrophage-like cells (Fig. 6). The 2-AG-induced migration of HL-60 cells

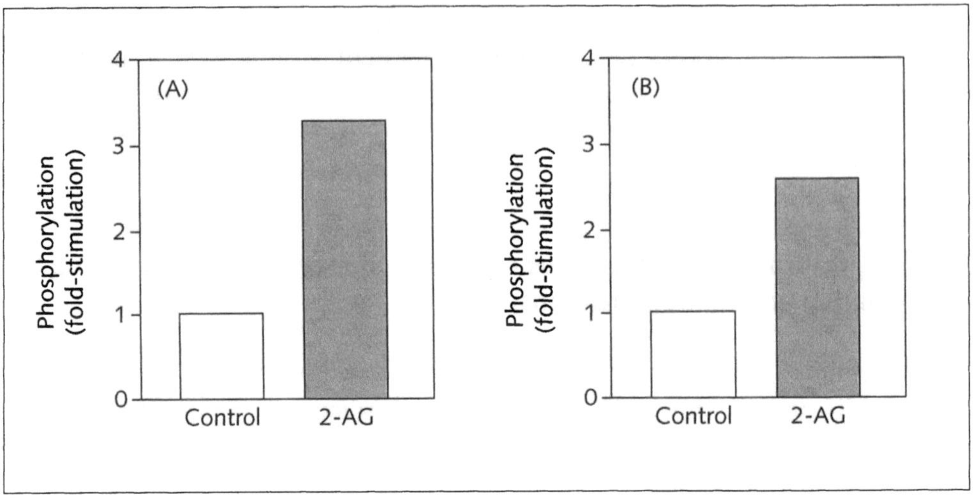

Figure 5
2-AG-induced phosphorylation of p38 MAP kinase and c-Jun N-terminal kinase in HL-60 cells
HL-60 cells were incubated in the presence or absence of 1 μM 2-AG for 30 min. The phosphorylation of p38 MAP kinase (A) and c-Jun N-terminal kinase (B) was determined by Western blotting. The data are the means of two determinations.

was markedly reduced when the cells were pretreated with either SR144528 or PTX (S. Kishimoto, M. Gokoh and T. Sugiura, unpublished results), suggesting that the migration was mediated through the CB2 receptor and Gi/Go as in the cases of Ca^{2+} transients [25] and the activation of the p42/44 MAP kinase [97]. Arachidonic acid and its metabolites are not involved in the 2-AG-induced migration, because free arachidonic acid was not capable of inducing the migration. This was also confirmed by the fact that 2-AG ether, a metabolically stable analog of 2-AG, was able to induce the migration (S. Kishimoto, M. Gokoh and T. Sugiura, unpublished results), although its activity was significantly weak compared with that of 2-AG. The migration of HL-60 cells induced by 2-AG was assumed to mainly involve chemotaxis rather than chemokinesis. In contrast to 2-AG, anandamide did not exhibit appreciable activity to induce the migration. Recently, Jorda et al. [100] also demonstrated that 2-AG induces the migration of mouse splenocytes and several myeloid cells.

Previously, Gallily et al. [101] reported that 2-AG suppresses the production of the tumor-necrosis factor α (TNF-α) in LPS-stimulated mouse macrophages *in vitro* and in LPS-administered mice *in vivo*, although whether these effects of 2-AG are mediated through the CB2 receptor is uncertain. On the other hand, we found that

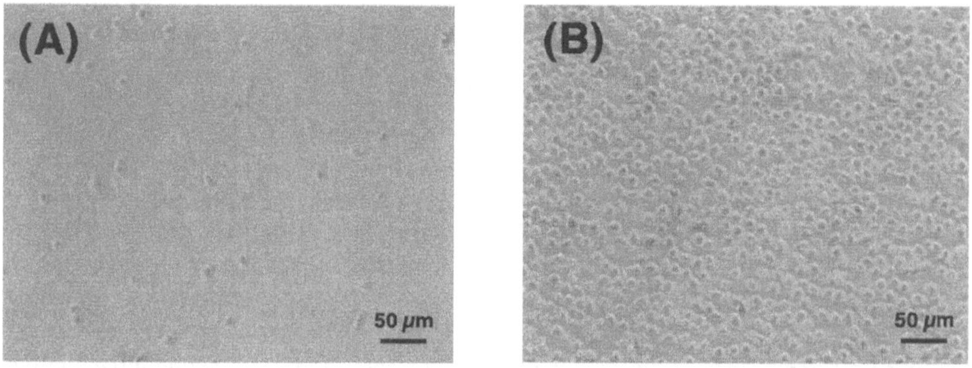

Figure 6
2-AG-induced migration of HL-60 cells differentiated into macrophage-like cells
The migration of HL-60 cells differentiated into macrophage-like cells (in the upper compartment) in response to 1 μM 2-AG (in the lower compartment) was examined using Costar Transwells (pore size, 5 μm) and 24-well culture plates. Cells were incubated in an atmosphere of 5% CO_2 and 95% air for 4 h. (A), cells migrated from the upper compartment to the lower compartment in the presence of vehicle alone; (B), cells migrated from the upper compartment to the lower compartment in the presence of 1 μM 2-AG. The magnification was × 200.

the addition of 2-AG to HL-60 cells enhanced the production of chemokines such as IL-8 and MCP-1 through a CB2 receptor- and Gi/Go-dependent mechanism (S. Kishimoto and T. Sugiura, unpublished results). CP55940 also caused the accelerated production of IL-8 and MCP-1. Meanwhile, anandamide as well as the free arachidonic acid failed to induce the augmented production of chemokines. Based on the experimental results concerning cell migration and chemokine production, we propose that 2-AG has a positive rather than negative role during inflammatory reactions and immune responses (Fig. 7).

It is well known that Δ^9-THC affects inflammatory reactions and immune response *in vivo* [8–10, 102]. For example, orally administered Δ^9-THC exhibited anti-inflammatory activity in rat paws injected with several stimulants such as carrageenan [103], although contradictory results have been reported as well [104]. The administration of Δ^9-THC to experimental animals has also been shown to result in a decreased resistance to viral and bacterial infection [8–10, 102]. The impairment of cell-mediated immunity has been assumed to be involved in the decreased resistance to microorganisms. A number of investigators have already demonstrated that Δ^9-THC suppresses several cellular functions of inflammatory cells and immune cells such as phagocytosis, antigen presentation, cytotoxicity, pro-

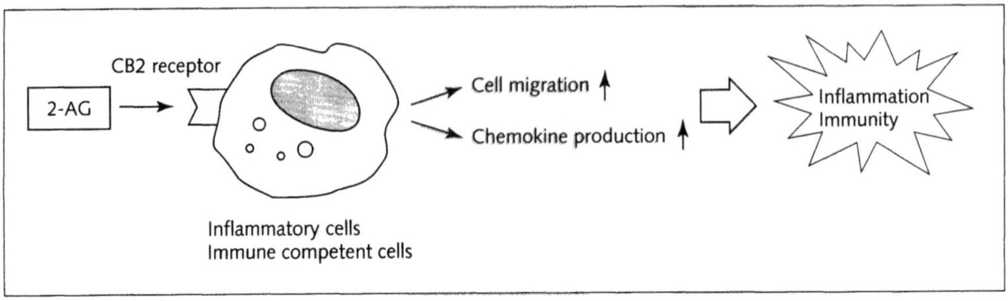

Figure 7
Physiological significance of the cannabinoid CB2 receptor and 2-AG during inflammation and immune response

duction of nitric oxide and cytokines such as TNF-α and mitogen-induced blastogenesis *in vitro* [8–10, 102], although the required concentrations are relatively high. The mechanism by which Δ^9-THC suppresses inflammatory cell and immune cell functions has not been fully elucidated. It has been assumed that the effects of Δ^9-THC are partly mediated by the cannabinoid receptors (CB2 and CB1) and partly by cannabinoid receptor-independent mechanisms. Previously, we demonstrated that Δ^9-THC is a weak partial agonist of the cannabinoid CB2 receptor [25]. Bayewitch et al. [105] also reported that Δ^9-THC acted as an antagonist toward the CB2 receptor. Therefore, it is possible that Δ^9-THC interferes with the actions of the physiological ligand 2-AG thereby causing suppression or inhibition of several functions of inflammatory cells and immune competent cells at least in some cases.

Previously, it has been demonstrated that low concentrations of cannabinoids such as CP55940 induce the augmented production of chemokines such as IL-8 and MCP-1 in HL-60 cells [106]. Δ^9-THC was also found to elicit the accelerated production of several cytokines such as IL-4 and IL-10 [107]. On the other hand, Berdyshev et al. [108] reported that micromolar concentrations of Δ^9-THC induce the increased production of TNF-α, IL-6 and IL-8 in LPS-stimulated human peripheral blood mononuclear cells. These results are generally consistent with our experimental results on the stimulative effects of 2-AG on chemokine production as previously mentioned. Noticeably, low concentrations of CP55940 were shown to induce the proliferation [109] and differentiation [110] of B lymphocytes. This strongly suggests that the cannabinoid receptors are involved in the regulation of the functions of B lymphocytes and humoral immunity. Recently, Iwamura et al. [111] reported that JTE-907, a CB2 receptor antagonist/inverse agonist, inhibited inflammation in the paw evoked by the injection of carrageenan in mice. A similar effect was observed with SR144528 [111], suggesting close involvement and mandatory roles of the cannabinoid receptors in inflammation reaction. It is thus clear that the endogenous ligand of the cannabinoid receptors (we are assuming that it is 2-AG) plays important positive roles during the course of inflammation and immune response, although details still remain to be determined. Further studies are necessary for a thorough elucidation of the physiological and pathophysiological significances of the cannabinoid receptor and its endogenous ligand 2-AG in the immune system. Such studies would be quite helpful to better understand the precise regulatory mechanisms of inflammation and immune homeostasis.

Concluding remarks

The cannabinoid receptors (CB1 and CB2) and their endogenous ligands, especially 2-AG, are assumed to play essential roles in the nervous system and the immune system as addressed in this review, although subsequent intensive studies are indispensable for a comprehensive understanding. In addition to this, the cannabinoid receptor system is also suggested to play important roles in other biological systems such as the cardiovascular system [112, 113] and the reproductive system [114, 115]. However, sufficient information has not yet been accumulated as to this issue and details still remain rather obscure. Apparently, the cannabinoid receptors and their endogenous ligands are attractive targets for new drug development such as anti-inflammation drugs, yet the research is just beginning. Thus, future studies on the cannabinoid receptors and their endogenous ligands will open promising new areas in the biological and biomedical sciences.

References

1 Gaoni Y, Mechoulam R (1964) Isolation, structure and partial synthesis of an active constituent of hashish. *J Am Chem Soc* 86: 1646–1647
2 Dewey WL (1986) Cannabinoid pharmacology. *Pharmacol Rev* 38: 151–178
3 Devane WA, Dysarz FA III, Johnson MR, Melvin LS, Howlett AC (1998) Determination and characterization of a cannabinoid receptor in rat brain. *Mol Pharmacol* 34: 605–613
4 Matsuda LA, Lolait SJ, Brownstein MJ, Young AC, Bonner TI (1990) Structure of a cannabinoid receptor and functional expression of the cloned cDNA. *Nature* 346: 561–564
5 Pertwee RG, Ross RA (2002) Cannabinoid receptors and their ligands. *Prostaglandins Leukot Essent Fatty Acids* 66: 101–121
6 Herkenham M (1995) Localization of cannabinoid receptors in brain and periphery. In: RG Pertwee (ed): *Cannabinoid Receptors*. Academic Press, London, 145–166
7 Munro S, Thomas KL, Abu-Shaar M (1993) Molecular characterization of a peripheral receptor for cannabinoids. *Nature* 365: 61–65

8 Berdyshev EV (2000) Cannabinoid receptors and the regulation of immune response. *Chem Phys Lipids* 108: 169–190

9 Parolaro D, Massi P, Rubino T, Monti E (2002) Endocannabinoids in the immune system and cancer. *Prostaglandins Leukot Essent Fatty Acids* 66: 319–332

10 Cabral GA (2002) Marijuana and cannabinoid effects on immunity and AIDS. In: ES Onaivi (ed): *Biology of Marijuana*. Taylor & Francis, London, 282–307

11 Devane WA, Hanus L, Breuer A, Pertwee RG, Stevenson LA, Griffin G, Gibson D, Mandelbaum A, Etinger A, Mechoulam R (1992) Isolation and structure of a brain constituent that binds to the cannabinoid receptor. *Science* 258: 1946–1949

12 Mechoulam R, Fride E, Di Marzo V (1998) Endocannabinoids. *Eur J Pharmacol* 359: 1–18

13 Piomelli D, Beltramo M, Giuffrida A, Stella N (1998) Endogenous cannabinoid signaling. *Neurobiol Dis* 5: 462–473

14 Howlett AC, Mukhopadhyay S (2000) Cellular signal transduction by anandamide and 2-arachidonoylglycerol. *Chem Phys Lipids* 108: 53–70

15 Di Marzo V, De Petrocellis L, Bisogno T, Berger A, Mechoulam R (2002) Biology of endocannabinoids. In: ES Onaivi (ed.): Biology of Marijuana. Taylor & Francis, London, 125–173

16 Sugiura T, Kobayashi Y, Oka S, Waku K (2002) Biosynthesis and degradation of anandamide and 2-arachidonoylglycerol and their possible physiological significance. *Prostaglandins Leukot Essent Fatty Acids* 66: 173–192

17 Sugiura T, Itoh K, Waku K, Hanahan DJ (1994) Biological activities of N-acylethanolamine and N-acylethanolamine phosphate analogues (in Japanese). *Proc Jpn Conf Biochem Lipids* 36: 71–74

18 Sugiura T, Kondo S, Sukagawa A, Nakane S, Shinoda A, Itoh K, Yamashita A, Waku K (1995) 2-Arachidonoylglycerol: a possible endogenous cannabinoid receptor ligand in brain. *Biochem Biophys Res Commun* 215: 89–97

19 Mechoulam R, Ben-Shabat S, Hanus L, Ligumsky M, Kaminski NE, Schatz AR, Gopher A, Almog S, Martin BR, Compton DR et al (1995) Identification of an endogenous 2-monoglyceride, present in canine gut, that binds to cannabinoid receptors. *Biochem Pharmacol* 50: 83–90

20 Sugiura T, Waku K (2000) 2-Arachidonoylglycerol and the cannabinoid receptors. *Chem Phys Lipids* 108: 89–106

21 Sugiura T, Kobayashi Y, Oka S, Waku K (2002) Biosynthesis and degradation of anandamide and 2-arachidonoylglycerol and their possible physiological significance. *Prostaglandins Leukot Essent Fatty Acids* 66: 173–192

22 Sugiura T, Kodaka T, Kondo S, Tonegawa T, Nakane S, Kishimoto S, Yamashita A, Waku K (1996) 2-Arachidonoylglycerol, a putative endogenous cannabinoid receptor ligand, induces rapid, transient elevation of intracellular free Ca^{2+} in neuroblastoma x glioma hybrid NG108-15 cells. *Biochem Biophys Res Commun* 229: 58–64

23 Sugiura T, Kodaka T, Kondo S, Nakane S, Kondo H, Waku K, Ishima Y, Watanabe K, Yamamoto I (1997) Is the cannabinoid CB1 receptor a 2-arachidonoylglycerol receptor?

Structural requirements for triggering a Ca^{2+} transient in NG108-15 cells. *J Biochem* 122: 890–895
24 Sugiura T, Kodaka T, Nakane S, Miyashita T, Kondo S, Suhara Y, Takayama H, Waku K, Seki C, Baba N et al (1999) Evidence that the cannabinoid CB1 receptor is a 2-arachidonoylglycerol receptor. Structure-activity relationship of 2-arachidonoylglycerol, ether-linked analogues, and related compounds. *J Biol Chem* 274: 2794–2801
25 Sugiura T, Kondo S, Kishimoto S, Miyashita T, Nakane S, Kodaka T, Suhara Y, Takayama H, Waku K (2000) Evidence that 2-arachidonoylglycerol but not N-palmitoylethanolamine or anandamide is the physiological ligand for the cannabinoid CB2 receptor. Comparison of the agonistic activities of various cannabinoid receptor ligands in HL-60 cells. *J Biol Chem* 275: 605–612
26 Hillard CJ (2000) Biochemistry and pharmacology of the endocannabinoids arachidonylethanolamide and 2-arachidonylglycerol. *Prostaglandins Other Lipid Mediat* 61: 3–18
27 Gonsiorek W, Lunn C, Fan X, Narula S, Lundell D, Hipkin RW (2000) Endocannabinoid 2-arachidonyl glycerol is a full agonist through human type 2 cannabinoid receptor: Antagonism by anandamide. *Mol Pharmacol* 57: 1045–1050
28 Savinainen JR, Jarvinen T, Laine K Laitinen JT (2001) Despite substantial degradation, 2-arachidonoylglycerol is a potent full efficacy agonist mediating CB(1) receptor-dependent G-protein activation in rat cerebellar membranes. *Br J Pharmacol* 134: 664–672
29 Hanus S, Abu-Lafi S, Fride E, Breuer A, Vogel Z, Shalev DE, Kustanovich I, Mechoulam R (2001) 2-Arachidonyl glyceryl ether, an endogenous agonist of the cannabinoid CB1 receptor. *Proc Natl Acad Sci USA* 98: 3662–3665
30 Fezza F, Bisogno T, Minassi A, Appendino G, Mechoulam R Di Marzo V (2002) Noladin ether, a putative novel endocannabinoid: Inactivation mechanisms and a sensitive method for its quantification in rat tissues. *FEBS Lett* 513: 294–298
31 Horrocks LA (1970) Content, composition, and metabolism of mammalian and avian lipids that contain ether groups. In: F Snyder (ed): *Ether Lipids*. Academic Press, New York, 177–272
32 Horrocks LA, Sharma M (1982) Plasmalogens and O-alkyl glycerophospholipids. In: JN Hawthorne, GB Ansell (eds): *Phospholipids*. Elsevier, Amsterdam, 51–93
33 Sugiura T, Waku K (1987) Composition of alkyl ether-linked phospholipids in mammalian tissues. In: F Snyder (ed): *Platelet-Activating Factor and Related Lipid Mediators*. Plenum, New York, 55–85
34 Schmid PC, Krebsbach RJ, Perry SR, Dettmer TM, Maasson JL, Schmid HHO (1995) Occurrence and postmortem generation of anandamide and other long-chain N-acylethanolamines in mammalian brain. *FEBS Lett* 375: 117–120, Corrigendum, (1996) *FEBS Lett* 385: 124–130
35 Kempe K, Hsu FF, Bohrer A, Turk J (1996) Isotope dilution mass spectrometric measurements indicate that arachidonylethanolamide, the proposed endogenous ligand of the cannabinoid receptor, accumulates in rat brain tissue post mortem but is contained at low levels in or is absent from fresh tissue. *J Biol Chem* 271: 17287–17295

36 Felder CC, Nielsen A, Briley EM, Palkovits M, Priller J, Axelrod J, Nguyen DN, Richardson JM, Riggin RM, Koppel GA et al (1996) Isolation and measurement of the endogenous cannabinoid receptor agonist, anandamide, in brain and peripheral tissues of human and rat. *FEBS Lett* 393: 231–235
37 Sugiura T, Kondo S, Sukagawa A, Tonegawa T, Nakane S, Yamashita A, Ishima Y, Waku K (1996) Transacylase-mediated and phosphodiesterase-mediated synthesis of N-arachidonoylethanolamine, an endogenous cannabinoid-receptor ligand, in rat brain microsomes. Comparison with synthesis from free arachidonic acid and ethanolamine. *Eur J Biochem* 240: 53–62
38 Di Marzo V, Goparaju SK, Wang L, Liu J, Batkai S, Jarai Z, Fezza F, Miura GI, Palmiter RD, Sugiura T et al (2001) Leptin-regulated endocannabinoids are involved in maintaining food intake. *Nature* 410: 822–825
39 Giuffrida A, Parsons LH, Kerr TM, Rodriguez de Fonseca F, Navarro M, Piomelli D (1999) Dopamine activation of endogenous cannabinoid signaling in dorsal striatum. *Nature Neurosci* 2: 358–363
40 Baker D, Pryce G, Croxford JL, Brown P, Pertwee RG, Makriyannis A, Khanolkar A, Layward L, Fezza F, Bisogno T et al (2001) Endocannabinoids control spasticity in a multiple sclerosis model. *FASEB J* 15: 300–302
41 Schmid PC, Kuwae T, Krebsbach RJ, Schmid HHO (1997) Anandamide and other N-acylethanolamines in mouse peritoneal macrophages. *Chem Phys Lipids* 87: 103–110
42 Yang HY, Karoum F, Felder C, Badger H, Wang TC, Markey SP (1999) GC/MS analysis of anandamide and quantification of N-arachidonoylphosphatidylethanolamides in various brain regions, spinal cord, testis, and spleen of the rat. *J Neurochem* 72: 1959–1968
43 Kondo S, Kondo H, Nakane S, Kodaka T, Tokumura A, Waku K, Sugiura T (1998) 2-Arachidonoylglycerol, an endogenous cannabinoid receptor agonist: identification as one of the major species of monoacylglycerols in various rat tissues, and evidence for its generation through Ca^{2+}-dependent and -independent mechanisms. *FEBS Lett* 429: 152–156
44 Sugiura T, Yoshinaga N, Kondo S, Waku K, Ishima Y (2000) Generation of 2-arachidonoylglycerol, an endogenous cannabinoid receptor ligand, in picrotoxinin-administered rat brain. *Biochem Biophys Res Commun* 271: 654–658
45 Stella N, Schweitzer P, Piomelli D (1997) A second endogenous cannabinoid that modulates long-term potentiation. *Nature* 388: 773–778
46 Bisogno T, Berrendero F, Ambrosino G, Cebeira M, Ramos JA, Fernandez-Ruiz JJ, Di Marzo V (1999) Brain regional distribution of endocannabinoids: Implications for their biosynthesis and biological function. *Biochem Biophys Res Commun* 256: 377–380
47 Huang SM, Strangman NM, Walker JM (1999) Liquid chromatographic-mass spectrometric measurement of the endogenous cannabinoid 2-arachidonylglycerol in the spinal cord and peripheral nervous system. *Acta Pharmacol Sinica* 20: 1098–1102
48 Schmid PC, Schwartz KD, Smith CN, Krebsbach RJ, Berdyshev EV, Schmid HHO

(2000) A sensitive endocannabinoid assay. The simultaneous analysis of N-acylethanolamines and 2-monoacylglycerols. *Chem Phys Lipids* 104: 185–191
49 Sugiura T, Yoshinaga N, Waku K (2001) Rapid generation of 2-arachidonoylglycerol, an endogenous cannabinoid receptor ligand, in rat brain after decapitation. *Neurosci Lett* 297: 175–178
50 Di Marzo V, Hill MP, Bisogno T, Crossman AR, Brotchie JM (2000) Enhanced levels of endogenous cannabinoids in the globus pallidus are associated with a reduction in movement in an animal model of Parkinson's disease. *FASEB J* 14: 1432–1438
51 Di Marzo V, Fontana A, Cadas H, Schinelli S, Cimino G, Schwartz JC, Piomelli D (1994) Formation and inactivation of endogenous cannabinoid anandamide in central neurons. *Nature* 372: 686–691
52 Di Marzo V, De Petrocellis L, Sepe N, Buono A (1996) Biosynthesis of anandamide and related acylethanolamides in mouse J774 macrophages and N18 neuroblastoma cells. *Biochem J* 316: 977–984
53 Bisogno T, Maurelli S, Melck D, De Petrocellis L, Di Marzo V (1997) Biosynthesis, uptake, and degradation of anandamide and palmitoylethanolamide in leukocytes. *J Biol Chem* 272: 3315–3323
54 Hansen HS, Lauritzen L, Strand AM, Moesgaard B, Frandsen A (1995) Glutamate stimulates the formation of N-acylphosphatidylethanolamine and N-acylethanolamine in cortical neurons in culture. *Biochim Biophys Acta* 1258: 303–308
55 Hansen HS, Lauritzen L, Strand AM, Vinggaard AM, Frandsen A, Schousboe A (1997) Characterization of glutamate-induced formation of N-acylphosphatidylethanolamine and N-acylethanolamine in cultured neocortical neurons. *J Neurochem* 69: 753–761
56 Wagner JA, Varga K, Ellis EF, Rzigalinski BA, Martin BR, Kunos G (1997) Activation of peripheral CB1 cannabinoid receptors in haemorrhagic shock. Nature 390: 518–521
57 Pestonjamasp VK, Burstein SH (1998) Anandamide synthesis is induced by arachidonate mobilizing agonists in cells of the immune system. *Biochim Biophys Acta* 1394: 249–260
58 Basavarajappa BS, Hungund BL (1999) Chronic ethanol increases the cannabinoid receptor agonist anandamide and its precursor N-arachidonoylphosphatidylethanolamine in SK-N-SH cells. *J Neurochem* 72: 522–528
59 Walker JM, Huang SM, Strangman NM, Tsou K, Sanudo-Pena MC (1999) Pain modulation by release of the endogenous cannabinoid anandamide. *Proc Natl Acad Sci USA* 96: 12198–12203
60 Kondo S, Sugiura T, Kodaka T, Kudo N, Waku K, Tokumura A (1998) Accumulation of various N-acylethanolamines including N-arachidonoylethanolamine (anandamide) in cadmium chloride-administered rat testis. *Arch Biochem Biophys* 354: 303–310
61 Kuwae T, Shiota Y, Schmid PC, Krebsbach R, Schmid HHO (1999) Biosynthesis and turnover of anandamide and other N-acylethanolamines in peritoneal macrophages. *FEBS Lett* 459: 123–127
62 Deutsch DG, Chin SA (1993) Enzymatic synthesis and degradation of anandamide, a cannabinoid receptor agonist. *Biochem Pharmacol* 46: 791–796

63 Devane WA, Axelrod J (1994) Enzymatic synthesis of anandamide, an endogenous ligand for the cannabinoid receptor, by brain membranes. *Proc Natl Acad Sci USA* 91: 6698–6701

64 Kruszka KK, Gross RW (1994) The ATP- and CoA-independent synthesis of arachidonoylethanolamide. A novel mechanism underlying the synthesis of the endogenous ligand of the cannabinoid receptor. *J Biol Chem* 269: 14345–14348

65 Schmid HHO, Schmid PC Natarajan V (1990) N-acylated glycerophospholipids and their derivatives. *Prog Lipid Res* 29: 1–43

66 Cadas H, di Tomaso E, Piomelli D (1997) Occurrence and biosynthesis of endogenous cannabinoid precursor, N-arachidonoyl phosphatidylethanolamine, in rat brain. *J Neurosci* 17: 1226–1242

67 Ueda N, Kurahashi Y, Yamamoto S, Tokunaga T (1995) Partial purification and characterization of the porcine brain enzyme hydrolyzing and synthesizing anandamide. *J Biol Chem* 270: 23823–23827

68 Ueda N, Yamanaka K, Yamamoto S (2001) Purification and characterization of an acid amidase selective for N-palmitoylethanolamine, a putative endogenous anti-inflammatory substance. *J Biol Chem* 276: 35552–35557

69 Cravatt BF, Giang DK, Mayfield SP, Boger DL, Lerner RA, Gilula NB (1996) Molecular characterization of an enzyme that degrades neuro-modulatory fatty-acid amides. *Nature* 384: 83–87

70 Prescott SM, Majerus PW (1983) Characterization of 1,2-diacylglycerol hydrolysis in human platelets. Demonstration of an arachidonoyl-monoacylglycerol intermediate. *J Biol Chem* 258: 764–769

71 Hasegawa-Sasaki H (1985) Early changes in inositol lipids and their metabolites induced by platelet-derived growth factor in quiescent Swiss mouse 3T3 cells. *Biochem J* 232: 99–109

72 Gammon CM, Allen AC, Morell P (1989) Bradykinin stimulates phosphoinositide hydrolysis and mobilization of arachidonic acid in dorsal root ganglion neurons. *J Neurochem* 53: 95–101

73 Bisogno T, Sepe N, Melck D, Maurelli S, De Petrocellis L, Di Marzo V (1997) Biosynthesis, release and degradation of the novel endogenous cannabimimetic metabolite 2-arachidonoylglycerol in mouse neuroblastoma cells. *Biochem J* 322: 671–677

74 Sugiura T, Kodaka T, Nakane S, Kishimoto S, Kondo S, Waku K (1998) Detectionof an endogenous cannabimimetic molecule, 2-arachidonoylglycerol, and cannabinoid CB1 receptor mRNA in human vascular cells: is 2-arachidonoylglycerol a possible vasomodulator? *Biochem Biophys Res Commun* 243: 838–843

75 Mechoulam R, Fride E, Ben-Shabat S, Meiri U, Horowitz M (1998) Carbachol, an acetylcholine receptor agonist, enhances production in rat aorta of 2-arachidonoyl glycerol, a hypotensive endocannabinoid. *Eur J Pharmacol* 362: R1–3

76 Basavarajappa BS, Saito M, Cooper TB, Hungund BL (2000) Stimulation of cannabinoid receptor agonist 2-arachidonylglycerol by chronic ethanol and its modulation by

specific neuromodulators in cerebellar granule neurons. *Biochim Biophys Acta* 1535: 78–86
77 Varga K, Wagner JA, Bridgen DT, Kunos G (1998) Platelet- and macrophage-derived endogenous cannabinoids are involved in endotoxin-induced hypotension. *FASEB J* 12: 1035–1044
78 Di Marzo V, Bisogno T, De Petrocellis L, Melck D, Orlando P, Wagner JA, Kunos G (1999) Biosynthesis and inactivation of the endocannabinoid 2-arachidonoylglycerol in circulating and tumoral macrophages. *Eur J Biochem* 264: 258–267
79 Berdyshev EV, Schmid PC, Krebsbach RJ, Schmid HH (2001) Activation of PAF receptors results in enhanced synthesis of 2-arachidonoylglycerol (2-AG) in immune cells. *FASEB J* 15: 2171–2178
80 Tsutsumi T, Kobayashi T, Miyashita M, Watanabe S, Homma Y, Okuyama H (1995) A lysophosphoinositide-specific phospholipase C distinct from other phospholipase C families in rat brain. *Arch Biochem Biophys* 317: 331–336
81 Nakane S, Oka S, Arai S, Waku K, Ishima Y, Tokumura A, Sugiura T (2002) 2-Arachidonoyl-sn-glycero-3-phosphate, an arachidonic acid-containing lysophosphatidic acid: Occurrence and rapid enzymatic conversion to 2-arachidonoyl-sn-glycerol, a cannabinoid receptor ligand, in rat brain. *Arch Biochem Biophys* 402: 51–58
82 Bisogno T, Melck D, De Petrocellis L, Di Marzo V (1999) Phosphatidic acid as the biosynthetic precursor of the endocannabinoid 2-arachidonoylglycerol in intact mouse neuroblastoma cells stimulated with ionomycin. *J Neurochem* 72: 2113–2119
83 Goparaju SK, Ueda N, Taniguchi K, Yamamoto S (1999) Enzymes of porcine brain hydrolyzing 2-arachidonoylglycerol, an endogenous ligand of cannabinoid receptors. *Biochem Pharmacol* 57: 417–23
84 Goparaju SK, Ueda N, Yamaguchi H, Yamamoto S (1998) Anandamide amidohydrolase reacting with 2-arachidonoylglycerol, another cannabinoid receptor ligand. *FEBS Lett* 422: 69–73
85 Di Marzo V, Bisogno T, Sugiura T, Melck D, De Petrocellis L (1998) The novel endogenous cannabinoid 2-arachidonoylglycerol is inactivated by neuronal- and basophil-like cells: Connections with anandamide. *Biochem J* 331: 15–19
86 Zygmunt PM, Petersson J, Andersson DA, Chuang H, Sorgard M, Di Marzo V, Julius D, Hogestatt ED (1999) Vanilloid receptors on sensory nerves mediate the vasodilator action of anandamide. *Nature* 400: 452–457
87 Maingret F, Patel AJ, Lazdunski M, Honore E (2001) The endocannabinoid anandamide is a direct and selective blocker of the background K+ channel TASK-1. *EMBO J* 20: 47–54
88 Jarrahian A, Hillard CJ (1997) Arachidonylethanolamide (anandamide) binds with low affinity to dihydropyridine binding sites in brain membranes. *Prostaglandins Leukot Essent Fatty Acids* 57: 551–554
89 Ameri A, Simmet T (2000) Effects of 2-arachidonylglycerol, an endogenous cannabinoid, on neuronal activity in rat hippocampal slices. *Naunyn Schmiedebergs Arch Pharmacol* 361: 265–272

90 Panikashvili D, Simeonidou C, Ben-Shabat S, Hanus L, Breuer A, Mechoulam R, Shohami E (2001) An endogenous cannabinoid (2-AG) is neuroprotective after brain injury. *Nature* 413: 527–531
91 Wilson RI, Nicoll RA (2001) Endogenous cannabinoids mediate retrograde signalling at hippocampal synapses. *Nature* 410: 588–592
92 Kreitzer AC, Regehr WG (2001) Retrograde inhibition of presynaptic calcium influx by endogenous cannabinoids at excitatory synapses onto Purkinje cells. *Neuron* 29: 717–727
93 Ohno-Shosaku T, Maejima T, Kano M (2001) Endogenous cannabinoids mediate retrograde signals from depolarized postsynaptic neurons to presynaptic terminals. *Neuron* 29: 729–738
94 Lee M, Yang KH, Kaminski NE (1995) Effects of putative cannabinoid receptor ligands, anandamide and 2-arachidonyl-glycerol, on immune function in B6C3F1 mouse splenocytes. *J Pharmacol Exp Ther* 275: 529–536
95 Ouyang Y, Hwang SG, Han SH, Kaminski NE (1998) Suppression of interleukin-2 by the putative endogenous cannabinoid 2-arachidonyl-glycerol is mediated through down-regulation of the nuclear factor of activated T cells. *Mol Pharmacol* 53: 676–683
96 Chang YH, Lee ST, Lin WW (2001) Effects of cannabinoids on LPS-stimulated inflammatory mediator release from macrophages: Involvement of eicosanoids. *J Cell Biochem* 81: 715–723
97 Kobayashi Y, Arai S, Waku K, Sugiura T (2001) Activation by 2-arachidonoylglycerol, an endogenous cannabinoid receptor ligand, of p42/44 mitogen-activated protein kinase in HL-60 cells. *J Biochem* 129: 665–669
98 Derkinderen P, Ledent C, Parmentier M, Girault JA (2001) Cannabinoids activate p38 mitogen-activated protein kinases through CB1 receptors in hippocampus. *J Neurochem* 77: 957–960.
99 Rueda D, Galve-Roperh I, Haro A, Guzman M (2000) The CB1 cannabinoid receptor is coupled to the activation of c-Jun N-terminal kinase. *Mol Pharmacol* 58: 814–820
100 Jorda MA, Verbakel SE, Valk PJ, Vankan-Berkhoudt YV, Maccarrone M, Finazzi-Agro A, Lowenberg B, Delwel R (2002) Hematopoietic cells expressing the peripheral cannabinoid receptor migrate in response to the endocannabinoid 2-arachidonoylglycerol. *Blood* 99: 2786–2793
101 Gallily R, Breuer A, Mechoulam R (2000) 2-Arachidonylglycerol, an endogenous cannabinoid, inhibits tumor necrosis factor-α production in murine macrophages, and in mice. *Eur J Pharmacol* 406: R5–7
102 Klein TW, Newton C, Friedman H. (1998) Cannabinoid receptors and immunity. *Immunol Today* 19: 373–381
103 Sofia RD, Nalepa SD, Vassar HB, Knobloch LC (1974) Comparative anti-phlogistic activity of delta 9-tetrahydrocannabinol, hydrocortisone and aspirin in various rat paw edema models. *Life Sci* 15: 251–260
104 Koserky DS, Dewey WL, Harris LS (1973) Antipyretic, analgesic and anti-inflammatory effects of Δ9-tetrahydrocannabinol in the rat. *Eur J Pharmacol* 24: 1–7

105 Bayewitch M, Rhee MH, Avidor-Reiss T, Breuer A, Mechoulam R (1996) (-)-Δ9-Tetrahydrocannabinol antagonizes the peripheral cannabinoid receptor-mediated inhibition of adenylyl cyclase. *J Biol Chem* 271: 9902–9905
106 Derocq JM, Jbilo O, Bouaboula M, Segui M, Clere C, Casellas P (2000) Genomic and functional changes induced by the activation of the peripheral cannabinoid receptor CB2 in the promyelocytic cells HL-60. Possible involvement of the CB2 receptor in cell differentiation. *J Biol Chem* 275: 15621–15628
107 Klein TW, Newton C, Friedman H (1998) Cannabinoid receptors and the cytokine network. *Adv Exp Med Biol* 437: 215–222
108 Berdyshev EV, Boichot E, Germain N, Allain N, Anger JP, Lagente V (1997) Influence of fatty acid ethanolamides and Δ9-tetrahydrocannabinol on cytokine and arachidonate release by mononuclear cells. *Eur J Pharmacol* 330: 231–240
109 Derocq JM, Segui M, Marchand J, Le Fur G, Casellas P (1995) Cannabinoids enhance human B-cell growth at low nanomolar concentrations. *FEBS Lett* 369: 177–182
110 Carayon P, Marchand J, Dussossoy D, Derocq JM, Jbilo O, Bord A, Bouaboula M, Galiegue S, Mondiere P, Penarier G et al (1998) Modulation and functional involvement of CB2 peripheral cannabinoid receptors during B-cell differentiation. *Blood* 92: 3605–3615
111 Iwamura H, Suzuki H, Ueda Y, Kaya T, Inaba T (2001) *In vitro* and *in vivo* pharmacological characterization of JTE-907, a novel selective ligand for cannabinoid CB2 receptor. *J Pharmacol Exp Ther* 296: 420–425
112 Kunos G, Jarai Z, Batkai S, Goparaju SK, Ishac EJN, Liu J, Wang L, Wagner JA (2000) Endocannabinoids as cardiovascular modulators. *Chem Phys Lipids* 108: 159–168
113 Randall MD, Harris D, Kendall DA (2001) The vascular pharmacology of endocannabinoids. In: ES Onaivi (ed.): *Biology of Marijuana*. Taylor & Francis, London, 542–553
114 Paria BC, Dey SK (2000) Ligand-receptor signaling with endocannabinoids in preimplantation embryo development and implantation. *Chem Phys Lipids* 108: 211–220
115 Schuel H, Burkman LJ, Lippes J, Crickard K, Mahony MC, Giuffrida A, Picone RP, Makriyannis A (2002) Evidence that anandamide-signaling regulates human sperm functions required for fertilization. *Mol Reprod Dev* 63: 376–387

Index

MIX
Papier aus verantwortungsvollen Quellen
Paper from responsible sources
FSC® C105338

If you have any concerns about our products,
you can contact us on
ProductSafety@springernature.com

In case Publisher is established outside the EU,
the EU authorized representative is:
Springer Nature Customer Service Center GmbH
Europaplatz 3, 69115 Heidelberg, Germany

Printed by Libri Plureos GmbH
in Hamburg, Germany